MW00718719

Data Analysis and Approximate Models

Model Choice, Location-Scale, Analysis of Variance, Nonparametric Regression and Image Analysis

MONOGRAPHS ON STATISTICS AND APPLIED PROBABILITY

General Editors

F. Bunea, V. Isham, N. Keiding, T. Louis, R. L. Smith, and H. Tong

1. Stochastic Population Models in Ecology and Epidemiology *M.S. Barlett* (1960)
2. Queues *D.R. Cox and W.L. Smith* (1961)
3. Monte Carlo Methods *J.M. Hammersley and D.C. Handscomb* (1964)
4. The Statistical Analysis of Series of Events *D.R. Cox and P.A.W. Lewis* (1966)
5. Population Genetics *W.J. Ewens* (1969)
6. Probability, Statistics and Time *M.S. Barlett* (1975)
7. Statistical Inference *S.D. Silvey* (1975)
8. The Analysis of Contingency Tables *B.S. Everitt* (1977)
9. Multivariate Analysis in Behavioural Research *A.E. Maxwell* (1977)
10. Stochastic Abundance Models *S. Engen* (1978)
11. Some Basic Theory for Statistical Inference *E.J.G. Pitman* (1979)
12. Point Processes *D.R. Cox and V. Isham* (1980)
13. Identification of Outliers *D.M. Hawkins* (1980)
14. Optimal Design *S.D. Silvey* (1980)
15. Finite Mixture Distributions *B.S. Everitt and D.J. Hand* (1981)
16. Classification *A.D. Gordon* (1981)
17. Distribution-Free Statistical Methods, 2nd edition *J.S. Maritz* (1995)
18. Residuals and Influence in Regression *R.D. Cook and S. Weisberg* (1982)
19. Applications of Queueing Theory, 2nd edition *G.F. Newell* (1982)
20. Risk Theory, 3rd edition *R.E. Beard, T. Pentikäinen and E. Pesonen* (1984)
21. Analysis of Survival Data *D.R. Cox and D. Oakes* (1984)
22. An Introduction to Latent Variable Models *B.S. Everitt* (1984)
23. Bandit Problems *D.A. Berry and B. Fristedt* (1985)
24. Stochastic Modelling and Control *M.H.A. Davis and R. Vinter* (1985)
25. The Statistical Analysis of Composition Data *J. Aitchison* (1986)
26. Density Estimation for Statistics and Data Analysis *B.W. Silverman* (1986)
27. Regression Analysis with Applications *G.B. Wetherill* (1986)
28. Sequential Methods in Statistics, 3rd edition *G.B. Wetherill and K.D. Glazebrook* (1986)
29. Tensor Methods in Statistics *P. McCullagh* (1987)
30. Transformation and Weighting in Regression *R.J. Carroll and D. Ruppert* (1988)
31. Asymptotic Techniques for Use in Statistics *O.E. Bandorff-Nielsen and D.R. Cox* (1989)
32. Analysis of Binary Data, 2nd edition *D.R. Cox and E.J. Snell* (1989)
33. Analysis of Infectious Disease Data *N.G. Becker* (1989)
34. Design and Analysis of Cross-Over Trials *B. Jones and M.G. Kenward* (1989)
35. Empirical Bayes Methods, 2nd edition *J.S. Maritz and T. Lwin* (1989)
36. Symmetric Multivariate and Related Distributions *K.T. Fang, S. Kotz and K.W. Ng* (1990)
37. Generalized Linear Models, 2nd edition *P. McCullagh and J.A. Nelder* (1989)
38. Cyclic and Computer Generated Designs, 2nd edition *J.A. John and E.R. Williams* (1995)
39. Analog Estimation Methods in Econometrics *C.F. Manski* (1988)
40. Subset Selection in Regression *A.J. Miller* (1990)
41. Analysis of Repeated Measures *M.J. Crowder and D.J. Hand* (1990)
42. Statistical Reasoning with Imprecise Probabilities *P. Walley* (1991)
43. Generalized Additive Models *T.J. Hastie and R.J. Tibshirani* (1990)
44. Inspection Errors for Attributes in Quality Control *N.L. Johnson, S. Kotz and X. Wu* (1991)
45. The Analysis of Contingency Tables, 2nd edition *B.S. Everitt* (1992)

46. The Analysis of Quantal Response Data *B.J.T. Morgan* (1992)
47. Longitudinal Data with Serial Correlation—A State-Space Approach *R.H. Jones* (1993)
48. Differential Geometry and Statistics *M.K. Murray and J.W. Rice* (1993)
49. Markov Models and Optimization *M.H.A. Davis* (1993)
50. Networks and Chaos—Statistical and Probabilistic Aspects
 O.E. Barndorff-Nielsen, J.L. Jensen and W.S. Kendall (1993)
51. Number-Theoretic Methods in Statistics *K.-T. Fang and Y. Wang* (1994)
52. Inference and Asymptotics *O.E. Barndorff-Nielsen and D.R. Cox* (1994)
53. Practical Risk Theory for Actuaries *C.D. Daykin, T. Pentikäinen and M. Pesonen* (1994)
54. Biplots *J.C. Gower and D.J. Hand* (1996)
55. Predictive Inference—An Introduction *S. Geisser* (1993)
56. Model-Free Curve Estimation *M.E. Tarter and M.D. Lock* (1993)
57. An Introduction to the Bootstrap *B. Efron and R.J. Tibshirani* (1993)
58. Nonparametric Regression and Generalized Linear Models *P.J. Green and B.W. Silverman* (1994)
59. Multidimensional Scaling *T.F. Cox and M.A.A. Cox* (1994)
60. Kernel Smoothing *M.P. Wand and M.C. Jones* (1995)
61. Statistics for Long Memory Processes *J. Beran* (1995)
62. Nonlinear Models for Repeated Measurement Data *M. Davidian and D.M. Giltinan* (1995)
63. Measurement Error in Nonlinear Models *R.J. Carroll, D. Rupert and L.A. Stefanski* (1995)
64. Analyzing and Modeling Rank Data *J.J. Marden* (1995)
65. Time Series Models—In Econometrics, Finance and Other Fields
 D.R. Cox, D.V. Hinkley and O.E. Barndorff-Nielsen (1996)
66. Local Polynomial Modeling and its Applications *J. Fan and I. Gijbels* (1996)
67. Multivariate Dependencies—Models, Analysis and Interpretation *D.R. Cox and N. Wermuth* (1996)
68. Statistical Inference—Based on the Likelihood *A. Azzalini* (1996)
69. Bayes and Empirical Bayes Methods for Data Analysis *B.P. Carlin and T.A Louis* (1996)
70. Hidden Markov and Other Models for Discrete-Valued Time Series *I.L. MacDonald and W. Zucchini* (1997)
71. Statistical Evidence—A Likelihood Paradigm *R. Royall* (1997)
72. Analysis of Incomplete Multivariate Data *J.L. Schafer* (1997)
73. Multivariate Models and Dependence Concepts *H. Joe* (1997)
74. Theory of Sample Surveys *M.E. Thompson* (1997)
75. Retrial Queues *G. Falin and J.G.C. Templeton* (1997)
76. Theory of Dispersion Models *B. Jørgensen* (1997)
77. Mixed Poisson Processes *J. Grandell* (1997)
78. Variance Components Estimation—Mixed Models, Methodologies and Applications *P.S.R.S. Rao* (1997)
79. Bayesian Methods for Finite Population Sampling *G. Meeden and M. Ghosh* (1997)
80. Stochastic Geometry—Likelihood and computation
 O.E. Barndorff-Nielsen, W.S. Kendall and M.N.M. van Lieshout (1998)
81. Computer-Assisted Analysis of Mixtures and Applications—Meta-Analysis, Disease Mapping and Others
 D. Böhning (1999)
82. Classification, 2nd edition *A.D. Gordon* (1999)
83. Semimartingales and their Statistical Inference *B.L.S. Prakasa Rao* (1999)
84. Statistical Aspects of BSE and vCJD—Models for Epidemics *C.A. Donnelly and N.M. Ferguson* (1999)
85. Set-Indexed Martingales *G. Ivanoff and E. Merzbach* (2000)
86. The Theory of the Design of Experiments *D.R. Cox and N. Reid* (2000)
87. Complex Stochastic Systems *O.E. Barndorff-Nielsen, D.R. Cox and C. Klüppelberg* (2001)
88. Multidimensional Scaling, 2nd edition *T.F. Cox and M.A.A. Cox* (2001)
89. Algebraic Statistics—Computational Commutative Algebra in Statistics
 G. Pistone, E. Riccomagno and H.P. Wynn (2001)
90. Analysis of Time Series Structure—SSA and Related Techniques
 N. Golyandina, V. Nekrutkin and A.A. Zhigljavsky (2001)
91. Subjective Probability Models for Lifetimes *Fabio Spizzichino* (2001)
92. Empirical Likelihood *Art B. Owen* (2001)

93. Statistics in the 21st Century *Adrian E. Raftery, Martin A. Tanner, and Martin T. Wells* (2001)
94. Accelerated Life Models: Modeling and Statistical Analysis
Vilijandas Bagdonavicius and Mikhail Nikulin (2001)
95. Subset Selection in Regression, Second Edition *Alan Miller* (2002)
96. Topics in Modelling of Clustered Data *Marc Aerts, Helena Geys, Geert Molenberghs, and Louise M. Ryan* (2002)
97. Components of Variance *D.R. Cox and P.J. Solomon* (2002)
98. Design and Analysis of Cross-Over Trials, 2nd Edition *Byron Jones and Michael G. Kenward* (2003)
99. Extreme Values in Finance, Telecommunications, and the Environment
Bärbel Finkenstädt and Holger Rootzén (2003)
100. Statistical Inference and Simulation for Spatial Point Processes
Jesper Møller and Rasmus Plenge Waagepetersen (2004)
101. Hierarchical Modeling and Analysis for Spatial Data
Sudipto Banerjee, Bradley P. Carlin, and Alan E. Gelfand (2004)
102. Diagnostic Checks in Time Series *Wai Keung Li* (2004)
103. Stereology for Statisticians *Adrian Baddeley and Eva B. Vedel Jensen* (2004)
104. Gaussian Markov Random Fields: Theory and Applications *Håvard Rue and Leonhard Held* (2005)
105. Measurement Error in Nonlinear Models: A Modern Perspective, Second Edition
Raymond J. Carroll, David Ruppert, Leonard A. Stefanski, and Ciprian M. Crainiceanu (2006)
106. Generalized Linear Models with Random Effects: Unified Analysis via H-likelihood
Youngjo Lee, John A. Nelder, and Yudi Pawitan (2006)
107. Statistical Methods for Spatio-Temporal Systems
Bärbel Finkenstädt, Leonhard Held, and Valerie Isham (2007)
108. Nonlinear Time Series: Semiparametric and Nonparametric Methods *Jiti Gao* (2007)
109. Missing Data in Longitudinal Studies: Strategies for Bayesian Modeling and Sensitivity Analysis
Michael J. Daniels and Joseph W. Hogan (2008)
110. Hidden Markov Models for Time Series: An Introduction Using R
Walter Zucchini and Iain L. MacDonald (2009)
111. ROC Curves for Continuous Data *Wojtek J. Krzanowski and David J. Hand* (2009)
112. Antedependence Models for Longitudinal Data *Dale L. Zimmerman and Vicente A. Núñez-Antón* (2009)
113. Mixed Effects Models for Complex Data *Lang Wu* (2010)
114. Intoduction to Time Series Modeling *Genshiro Kitagawa* (2010)
115. Expansions and Asymptotics for Statistics *Christopher G. Small* (2010)
116. Statistical Inference: An Integrated Bayesian/Likelihood Approach *Murray Aitkin* (2010)
117. Circular and Linear Regression: Fitting Circles and Lines by Least Squares *Nikolai Chernov* (2010)
118. Simultaneous Inference in Regression *Wei Liu* (2010)
119. Robust Nonparametric Statistical Methods, Second Edition
Thomas P. Hettmansperger and Joseph W. McKean (2011)
120. Statistical Inference: The Minimum Distance Approach
Ayanendranath Basu, Hiroyuki Shioya, and Chanseok Park (2011)
121. Smoothing Splines: Methods and Applications *Yuedong Wang* (2011)
122. Extreme Value Methods with Applications to Finance *Serguei Y. Novak* (2012)
123. Dynamic Prediction in Clinical Survival Analysis *Hans C. van Houwelingen and Hein Putter* (2012)
124. Statistical Methods for Stochastic Differential Equations
Mathieu Kessler, Alexander Lindner, and Michael Sørensen (2012)
125. Maximum Likelihood Estimation for Sample Surveys
R. L. Chambers, D. G. Steel, Suojin Wang, and A. H. Welsh (2012)
126. Mean Field Simulation for Monte Carlo Integration *Pierre Del Moral* (2013)
127. Analysis of Variance for Functional Data *Jin-Ting Zhang* (2013)
128. Statistical Analysis of Spatial and Spatio-Temporal Point Patterns, Third Edition *Peter J. Diggle* (2013)
129. Constrained Principal Component Analysis and Related Techniques *Yoshio Takane* (2014)
130. Randomised Response-Adaptive Designs in Clinical Trials *Anthony C. Atkinson and Atanu Biswas* (2014)
131. Theory of Factorial Design: Single- and Multi-Stratum Experiments *Ching-Shui Cheng* (2014)
132. Quasi-Least Squares Regression *Justine Shults and Joseph M. Hilbe* (2014)
133. Data Analysis and Approximate Models: Model Choice, Location-Scale, Analysis of Variance, Nonparametric Regression and Image Analysis *Laurie Davies* (2014)

Monographs on Statistics and Applied Probability 133

Data Analysis and Approximate Models

Model Choice, Location-Scale, Analysis of Variance, Nonparametric Regression and Image Analysis

Laurie Davies

University of Duisburg-Essen, Germany

CRC Press is an imprint of the
Taylor & Francis Group, an **informa** business
A CHAPMAN & HALL BOOK

CRC Press
Taylor & Francis Group
6000 Broken Sound Parkway NW, Suite 300
Boca Raton, FL 33487-2742

© 2014 by Taylor & Francis Group, LLC
CRC Press is an imprint of Taylor & Francis Group, an Informa business

No claim to original U.S. Government works

Printed on acid-free paper
Version Date: 20140505

International Standard Book Number-13: 978-1-4822-1586-1 (Hardback)

This book contains information obtained from authentic and highly regarded sources. Reasonable efforts have been made to publish reliable data and information, but the author and publisher cannot assume responsibility for the validity of all materials or the consequences of their use. The authors and publishers have attempted to trace the copyright holders of all material reproduced in this publication and apologize to copyright holders if permission to publish in this form has not been obtained. If any copyright material has not been acknowledged please write and let us know so we may rectify in any future reprint.

Except as permitted under U.S. Copyright Law, no part of this book may be reprinted, reproduced, transmitted, or utilized in any form by any electronic, mechanical, or other means, now known or hereafter invented, including photocopying, microfilming, and recording, or in any information storage or retrieval system, without written permission from the publishers.

For permission to photocopy or use material electronically from this work, please access www.copyright.com (http://www.copyright.com/) or contact the Copyright Clearance Center, Inc. (CCC), 222 Rosewood Drive, Danvers, MA 01923, 978-750-8400. CCC is a not-for-profit organization that provides licenses and registration for a variety of users. For organizations that have been granted a photocopy license by the CCC, a separate system of payment has been arranged.

Trademark Notice: Product or corporate names may be trademarks or registered trademarks, and are used only for identification and explanation without intent to infringe.

Library of Congress Cataloging-in-Publication Data

Davies, Patrick Laurie, author.
Data analysis and approximate models : model choice, location-scale, analysis of variance, nonparametric regression and image analysis / Patrick Laurie Davies.
pages cm. -- (Monographs on statistics and applied probability)
Summary: "This book presents a philosophical study of statistics via the concept of data approximation. Developed by the well-regarded author, this approach discusses how analysis must take into account that models are, at best, an approximation of real data. It is, therefore, closely related to robust statistics and nonparametric statistics and can be used to study nearly any statistical technique. The book also includes an interesting discussion of the frequentist versus Bayesian debate in statistics. "-- Provided by publisher.
Includes bibliographical references and index.
ISBN 978-1-4822-1586-1 (hardback)
1. Probabilities--Philosophy. 2. Approximation theory. I. Title.

QA273.A35D38 2014
519.5--dc23 2014015952

Visit the Taylor & Francis Web site at
http://www.taylorandfrancis.com

and the CRC Press Web site at
http://www.crcpress.com

Contents

Preface

This book stems from a dissatisfaction with what is called formal statistical inference. The dissatisfaction started with my first contact with statistics in a course of lectures given by John Kingman in Cambridge in 1963. In spite of Kingman's excellent pedagogical capabilities it was the only course in the Mathematical Tripos I did not understand. Kingman later told me that the course was based on notes by Dennis Lindley, but the approach given was not a Bayesian one.

From Cambridge I went to LSE where I did an M.Sc. course in statistics. Again, in spite of excellent teachers including David Brillinger, Jim Durbin and Alan Stuart I did not really understand what was going on. This did not prevent me from doing whatever I was doing with success and I was awarded a distinction in the final examinations. Later I found out that I was not the only person who had problems with statistics. Some years ago I asked a respected German colleague D. W. Müller of the University of Heidelberg why he had chosen statistics. He replied that it was the only subject he had not understood as a student. Frank Hampel has even written an article entitled 'Is statistics too difficult?'.

I continued at LSE and wrote my Ph. D. thesis on random entire functions under the supervision of Cyril Offord. It involved no statistics whatsoever. From London I moved to Constance in Germany, from there to Sheffield, then back to Germany to the town of Münster. All the time I continued writing papers in probability theory including some on the continuity properties of Gaussian processes. At that time Jack Cuzick now of Queen Mary, University of London, and Cancer Research UK also worked on this somewhat esoteric subject.

From 1967 until my arrival in Münster in 1975 I had managed to avoid statistics but now could no longer do so. There was a two-year cycle of courses at Münster, the first year was devoted to probability theory and the second to statistics. From then until my retirement in 2009 every second year was spent giving a course of statistics. The course was restricted to mathematical statistics and included topics such as the optimality of the t-test and minimal sufficient sigma-algebras. I made no attempt to relate any of this to real world problems. In 1977 I moved from Münster to Essen.

In 1981/2 two events caused a change of direction in research and interests. I was always looking for statistics books on which I could base my course of lectures and in 1981 Peter Huber's book 'Robust Statistics' appeared. This did not put to rest all my problems with statistics but it was a big first step in the right direction. In December 1982 a Commodore C 64 arrived on my desktop and its being there was a sufficient reason to start using it. I programmed a least squares regression which worked and encouraged by this success I went from there to the robust linear regression as described in Huber's book. I generated a data set with outliers and was disappointed when they were not found. The result of the robust regression was not very different from that of ordinary least squares. I assumed that it was a programming mistake but could not find it. In 1983 I went to a workshop in Heidelberg and heard a talk by Peter Rousseeuw on robust regression. It was my first contact with what became known as Least Median of Squares, LMS. The idea is due to Frank Hampel but based on earlier work by John Tukey. I returned to Essen, programmed the LMS procedure and was somewhat surprised when the outliers were found. This started my research interests in robust statistics which I still have.

Although my research interests had changed there still remained, in spite of Huber's book, the problem of teaching statistics. In 1983 Lehmann's book 'The Theory of Point Estimation' was published and I hoped that this would prove to be the book I was waiting for. It was at the right academic level and written by one of the most respected statisticians. It was a disappointment. It answered none of the difficulties I had with statistics. Something was clearly wrong but I was unable to state precisely what.

At about the same time I was having problems with robust statistics. Some estimators including the LMS required strong assumptions to be able to prove existence and uniqueness, asymptotic normality and efficiency. These involved for example the existence of a symmetric density which is monotone decreasing on the positive real line. To have to assume the existence of a density was to me, and still is, the very antithesis of robustness. I am not alone in this opinion, Peter Huber shares it. At this point I decided the only option I had was to develop my own way of thinking about statistics. This book is the result.

The problem is how to relate a given data set and a given model. Models in statistics are, with some exceptions, ad hoc, although not arbitrary, and so should be treated consistently as approximations and not as truth. I believe there is a true but unknown amount of copper in a sample of water, but I do not believe that there is a true but unknown parameter value. Nor am I prepared to behave as if there were a 'true but unknown parameter value', here put in inverted commas as statisticians often do to demonstrate that they do not actually believe that there is such a thing. Moreover the model has to be an approximation to the data and not to some 'true but unknown generating mechanism'. There may well be some physical truth behind the data but this will be many orders of magnitude complexer than anything that can be reasonably modelled by a probability measure. Furthermore there will be a different truth behind every single observation. The simple coin toss is an example where the deterministic Newtonian dynamics behind a mechanical coin toss are complicated.

The definition of approximation used in this book is that a model is an adequate approximation to a data set if 'typical' data generated under the model 'look like' the real data. The words 'look like' mean that certain features of the data are of importance and must be exhibited by 'typical' sets generated under the model. Which features are important will depend on the very reasons for analysing the data in the first place. They cannot be dictated by any purely statistical concepts such as likelihood or efficiency. The word 'typical' is operationalized by specifying a percentage such that this percentage of samples generated under the model exhibit the features defining 'look like'.

In general there will be more than one model which is an adequate approximation to the data and consequently most statistical problems are ill-posed. They require regularization which can involve either the regularization of the model itself or the operations which are performed on the data. Regularization is a recurring theme in the book.

The word 'approximation' means 'close' in some sense which may be seen as specifying a topology in the mathematical sense. The definition of approximation is at the level of samples and in this sense the topology is the Topology of EDA, Explorative Data Analysis, of Tukey and others. Mathematically this can be associated with the Kolmogorov metric. Statistical analysis tends to be split into two separate parts. There is the EDA part where the statistician looks for an adequate model. Once this has been decided on the statistician moves on to formal inference. This is often density based: for Bayesians it is always density based. This involves a density based topology which may be characterized by the total variation metric. The total variation metric is strictly stronger than the Kolmogorov metric in that it generates more open sets. Another way of expressing it is that EDA operates at the level of distribution functions with a measure of closeness given by the Kolmogorov metric. Formal inference operates at the level of densities with a measure of closeness given by the total variation metric. The two are connected by the differential operator which in the Lebesgue case is pathologically discontinuous. This book has only one topology, the weak one associated with the Kolmogorov metric.

The approach given here is neither Bayesian nor frequentist. The concept of truth is deeply embedded in the Bayesian approach. Two different parameter values cannot both be true and from this follows the additivity of Bayesian priors. However two different parameter values can both be adequate approximations. If the Bayesian substitutes 'approximation' for 'truth' the additivity of priors is lost and in consequence coherence and in consequence Bayesian statistics. A model approximates a given data set. There is no necessity for this to be repeatable. The interpretation of a frequentist confidence interval is that if it is calculated for many data sets distributed according to the assumed model than the correct percentage of these intervals will contain the 'true but unknown' parameter values. I never cease to be surprised that this is taken seriously and even taught to students. This very definition requires repetition of the data. In contrast an approximation interval has the following interpretation: take any parameter value in the approximation interval and generate data using it. Then the correct percentage of these data sets will 'look like' the real data in the specified sense of 'look like'.

The first chapter of the book highlights problems if concepts such as likelihood and efficiency are taken seriously. The second chapter is concerned with the definition of approximation and its consequences. The third chapter on discrete data is devoted to the total variation metric and the Kullback-Leibler and chi-squared discrepancies as measures of fit. The fourth chapter is on outliers which speaks for itself. The fifth chapter on the location-scale problem is mostly concerned with approximation intervals. This is not exactly an exciting topic but it is a necessary one. It has been ignored to a large extent in the literature in spite of the fact that it is seen as being important. The results form the basis for the treatment of the one-way table in Chapter 6. The remainder of Chapter 6 gives a new definition of interaction in higher table. The standard definition is so obviously bad that the only thing to be said in its favour is that the necessary calculations involve no more than least squares regression. The next four chapters are concerned with non-parametric regression. This was a topic chosen to see if it were possible to come up with new procedures based on the idea of approximation. The first paper in this direction was published in 2001 and was a joint one with Arne Kovac. It was devoted to the problem of modality in non-parametric regression. It turned out later that Nemirovski had already published a paper with a similar approach in 1985. He attributed the idea to S. V. Shil'man. The final chapter is a critique of statistics and was the starting point of approach given here. The topics covered vary from the likelihood principle to asymptotics and model choice.

A first attempt to formulate the ideas in this book was sent to John Tukey in 1993. His response was favourable and encouraged me to continue in this direction. The first attempt to publish the ideas was less successful. It was rejected by three journals before being accepted by Richard Gill for publication in Statistica Neerlandica. In all there were about 10 referees reports only one of which was positive. Peter Hall invited me to write a paper for the *Journal of the Korean Statistical Society*. It was published in 2008. He also invited me to the University of Melbourne in 2012 where the parts of the book were written. I spent the first six months of 2010 at the University of California, Davis, at the invitation of Wolfgang Polonik and again parts of the book were written there. Finally I spent October 2011 at the University of Bristol at the invitation of Peter Green.

The book is also the work of colleagues and former students: Arne Kovac and Monika Meise made essential contributions to the work on non-parametric regression; there is joint work with Ursula Gather on outliers, breakdown and the analysis of the photon data which was provided by Dieter Mergel; the analysis of financial data is part of the Ph. D. thesis of Christian Höhenrieder and the Diplom thesis of Sarah Banasiak with contributions from Walter Krämer and myself. Other contributions are from Rahel Stichtenoth on image analysis and Evgeny Zoldin on diffusion processes.

Data Analysis and Approximate Models

Model Choice, Location-Scale, Analysis of Variance, Nonparametric Regression and Image Analysis

Chapter 1

Introduction

1.1 Approximate Models

This book is about treating probability models as approximations in a consistent manner. Although statisticians talk about models being approximations they rarely go from talking to writing. If one looks up the word 'approximation' in the index of a book on statistics it will either not be found (see [148]) or it will refer to such things as the normal approximation, a saddle point approximation or a finite sample approximation (see [103]). The only exception seems to be Peter Huber in [117] where Chapter 5 is entitled 'Approximate Models'. Whatever the reason for this it seems that this book is the first attempt to give a coherent account of statistical analysis with models being treated as approximations.

1.2 Notation

A model P is a single probability measure usually over the Borel sets of $\mathbb{R}^n$. This differs from the conventional meaning in statistics in which a model is a family $\mathscr{P}$ of probability models P in the sense of this book. In cases where P is a product measure $P = \otimes_{i=1}^n P'$ the notational burden will be eased by writing P' rather than $\otimes_{i=1}^n P'$. A parametric family of models is one of the form

$$\mathscr{P}_\Theta = \{P_\theta : \theta \in \Theta\}. \tag{1.1}$$

For the development of a concept of approximation it is important to distinguish between real data and data generated under a probability model. Real data will be denoted by a lower-case Latin letter as in x_i, a bold letter will denote a sample of size n as in $\mathbf{x}_n = (x_1,\ldots,x_n)$. Data generated under a model P will correspondingly be denoted by upper-case Latin letters as in $\mathbf{X}_n(P) = (X_1(P)\ldots,X_n(P))$ or, for a parametric model, $\mathbf{X}_n(\theta) = (X_1(\theta)\ldots,X_n(\theta))$.

The empirical distribution associated with the data $\mathbf{x}_n$ will be denoted by

$$\mathbb{P}_n = \mathbb{P}_{\mathbf{x}_n} = \frac{1}{n}\sum_{i=1}^n \delta_{x_i} \tag{1.2}$$

where δ_x denotes the point mass in x or the Dirac measure in x. The dependence of $\mathbb{P}_n$ on the data $\mathbf{x}_n$ will usually be suppressed.

1.3 Two Modes of Statistical Analysis

The starting point of any statistical analysis is a data set $\mathbf{x}_n = (x_1,\ldots,x_n)$ of finite precision in $\mathbb{R}^n$ and information as to how the data were obtained. The data could be a sequence of zeros and ones obtained from n throws of a coin, or a digital image of a brain, or the daily returns of some financial assets or the heights of tides on the Dutch coast over the last 100 years or more. Once statistics moves from simple summaries of the data such as the mean, median, standard deviation and MAD, to substantial questions such as the probability of 'heads', the location of the speech centre, the dependencies between the financial assets or an estimate of the highest tide in the next 100 years, it requires models. A model P is a candidate model for the whole data set $\mathbf{x}_n$. Many problems in statistics, for example the monitoring of patients on an intensive care unit [41], do not fit into this framework and will not be considered.

The standard model for the throws of a coin is the Bernoulli model of independent and identically random variables satisfying the $\mathfrak{b}(1,p)$ distribution where p denotes the probability of a 'head'. Within the context of the model it is possible to go beyond the summary statistic, the actual proportion of 'head' throws in the data set, to say a confidence interval for p. The modelling of the coin by $\mathfrak{b}(1, p)$ is deceptively simple and masks what may be termed deeper problems related to randomness, determinism and chaos [56], [199]. Nevertheless it is the model of choice unless there is reason for suspecting dependencies between the throws or changes in the probability for 'heads'. For the other examples given above there is no simple model of choice. In contrast to physics which has some generally accepted principles such as the conservation of energy and the conservation of momentum, statistics has none. Statistical models are often ad hoc, although not arbitrary, and yet the results of the analysis depend on the model that is chosen. The problem to be discussed in the remainder of the book is that of the relationship between the data, the model and the world.

1.3.1 Data Analysis and Formal Inference

Faced with a real data set the statistician is initially in a data analytical or explorative data analysis (EDA) mode such as that elucidated in [210], [160] and [107]. A list of possible activities in this mode is given in Table 1.1. Histograms here are meant as a

EDA 1	histograms,
EDA 2	distribution functions,
EDA 3	*q-q*-plots,
EDA 4	goodness-of-fit tests,
EDA 5	scatter plots,
EDA 6	outliers,
EDA 7	diagnostics,
EDA 8	specific model checks,
EDA 9	transformations,
EDA 10	tests of independence,
EDA 11	autocorrelation functions.

Table 1.1 *List of possible EDA activities.*

rough guide to the shape of the distribution of the data rather than a formal density estimate.

Based on this analysis a family of plausible models, often a parametric family $\mathscr{P}_\Theta$, is postulated and the statistician moves to the formal inference mode. In this mode he or she will operate within the model $\mathscr{P}_\Theta$ as if it were true, that is as if the data $\mathbf{x}_n$ were a realization $\mathbf{X}_n(\theta)(\omega_0)$ of random variables whose distribution is P_θ for some unknown $\theta \in \Theta$. This is expressed in [127] as follows:

> A *setting for statistical inference* is a collection of *possible* probability structure, exactly one of which (called the *true state of nature*) is the mechanism actually operating in the setting at hand; which one, is unknown. *Statistical inference* in the broad sense is the collection of procedures for reaching a conclusion about which possible state of nature is the true one in such a

setting, together with probabilistic properties (operating characteristics) of such procedures.

For the dangers of this attitude see [116] under the 'Scapegoat Syndrome: confuse the model with the truth'. In a sense the whole of this book is concerned with an approach to statistics where truth and model are explicitly not confused, to the extreme that truth plays no role at all.

For the frequentist the formal inference mode is very often but not always likelihood based. For the Bayesians it is always likelihood based. The frequentist will try to use estimation procedures which are optimal or at least asymptotically optimal. A main tool in this context is maximum likelihood. The Bayesian will have priors and posteriors linked by likelihood. For the Bayesian there is no option other than to operate as if the model were true for the concept of truth is too deeply embedded in the Bayesian paradigm for it to be dispensed with.

Data analysis operates at the level of data. That is, two data sets $\mathbf{x}_n$ and $\mathbf{y}_n$ which are very similar will by analysed in the same manner and the conclusions will be the same. When are two data sets $\mathbf{X}_n(P)$ and $\mathbf{Y}_n(Q)$ generated by two different models P and Q close in the data analytic sense? Given a distribution function F and a random variable U uniformly distributed on $[0,1]$ the random variable

$$X = F^{-1}(U) \tag{1.3}$$

has the distribution function F. Consider the Kolmogorov metric d_{ko} on the space of distribution functions

$$d_{\text{ko}}(F,G) := \sup_x |F(x) - G(x)| \tag{1.4}$$

and suppose that $d_{\text{ko}}(F,G)$ is very small. If now $X = F^{-1}(U)$ and $Y = G^{-1}(U)$ then the random variables X and Y will 1) be very close and 2) have distribution functions F and G respectively. If one takes finite precision into account the truncated versions of X and Y may well be equal. Thus any statistician in the first mode using procedures listed in Table 1.1 will come to the same conclusions when presented firstly with X and secondly with Y. This can be extended to data sets $\mathbf{X}_n(P)$ and $\mathbf{Y}_n(Q)$ of independent copies of X and Y generated by the models P and Q with distribution functions F and G respectively. Thus in a very wide sense the topology of the first mode of data analysis, that is the concept of closeness in the first mode, is the topology of the Kolmogorov metric. In particular if $d_{\text{ko}}(P,Q)$ is small, as is assumed here, then any model regarded as satisfactory for the data set $\mathbf{X}_n(P)$ will also be regarded as satisfactory for the data set $\mathbf{Y}_n(Q)$.

The formal inference mode is based on the densities (likelihoods) f and g of the two models P and Q. A possible metric on the space of densities is the L_1 metric

$$d_1(f,g) = \int_{-\infty}^{\infty} |f(x) - g(x)|\,dx. \tag{1.5}$$

The maximum value of d_1 is 2. The d_1 metric defines in a sense the topology of formal inference. Unfortunately $d_{\text{ko}}(F,G)$ being small does not imply that $d_1(f,g)$ is also small: for any $\varepsilon > 0$ and any given F, say the $\mathfrak{N}(0,1)$ model, there exists a distribution function G with a strictly positive, infinitely differentiable and bounded density g such that

$$d_{\text{ko}}(F,G) < \varepsilon, \quad d_1(f,g) > 2-\varepsilon. \tag{1.6}$$

Formal inference based on P and formal inference based on Q can lead to very different conclusions, in fact to conclusions which are arbitrarily far apart.

The inequalities (1.6) are an expression of the fact that the linear differential operator D

$$D : \mathscr{P}_{\text{ac}} \to L_1, D(F) = f \tag{1.7}$$

from the space of absolutely continuous distributions $\mathscr{P}_{\text{ac}}$ into the L_1 space

$$L_1 = \left\{ h : \int |h(x)|\,dx < \infty \right\}$$

is pathologically discontinuous.

Not only do the two modes of analysis operate with different topologies, once a model has been decided upon the first mode plays no further part and can be forgotten. Indeed, it is possible to write a whole book about the second mode which completely ignores the first. The following is taken from the preface of [30]:

> Most statistical work is concerned directly with the provision and implementation of methods for study design and for the analysis and interpretation of data. The theory of statistics deals in principle with the general concepts underlying all aspects of such work and from this perspective the formal theory of statistical inference is but a part of that full theory. Indeed, from the point of view of individual applications, it may seem a rather small part. Concern is likely to be more concentrated on whether models have been reasonably formulated to address the most fruitful questions, on whether the data are subject to unappreciated errors or contamination and, especially, on the subject-matter interpretation of the analysis and its relation with other knowledge of the field.
>
> Yet the formal theory is important for a number of reasons. Without some systematic structure statistical methods for the analysis of data become a collection of tricks that are hard to assimilate or relate to one another, or for that matter to teach.

A Bayesian formulation of EDA is given in [94].

1.3.2 Text Book Examples

Text book examples are often framed within the model and as such may be regarded as a part of mathematics: logical conclusions are drawn from the assumptions with the same degree of precision as within mathematics. Problems start when the results of an analysis of the model are transferred to real data which do not satisfy the assumptions imposed (see [116]).

In the frequentist framework an important problem is to estimate the 'true' θ using a statistic $T(\mathbf{X}_n(\theta))$ in some optimal manner, for example, by minimizing the expected loss. An exercise for a student within this framework is the following.

Exercise 1.3.1 *The data $X_1,\ldots,X_n$ are independently and identically distributed according to the Poisson distribution $\mathfrak{P}_o(\lambda)$. Show that $\bar{X}_n = (\sum_{i=1}^{n} X_i)/n$ is an unbiased estimate of λ, calculate the expected mean square error $\mathbf{E}((\bar{X}_n-\lambda)^2)$ and show that the mean minimizes this under all unbiased estimates of λ.*

A Bayesian variation is

Exercise 1.3.2 *The data $X_1,\ldots,X_n$ are independently and identically distributed according to the Poisson distribution $\mathfrak{P}_o(\lambda)$. The prior distribution of λ is the exponential distribution $\mathfrak{E}(\tau)$ with parameter $\tau = 1$. Calculate the posterior distribution for λ given the data.*

Exercises 1.3.1 and 1.3.2 are typical of the exercises in almost all text books for students of statistics. The student is informed of the distribution of the data and based on this assumption he or she is required to prove a result or carry out some calculation. An applied exercise in the same framework is the following:

Exercise 1.3.3 *The following data give 27 measurements of the amount of copper (milligrammes per litre) in a sample of drinking water:*

$$\begin{array}{ccccccccc}
2.16 & 2.21 & 2.15 & 2.05 & 2.06 & 2.04 & 1.90 & 2.03 & 2.06\\
2.02 & 2.06 & 1.92 & 2.08 & 2.05 & 1.88 & 1.99 & 2.01 & 1.86\\
1.70 & 1.88 & 1.99 & 1.93 & 2.20 & 2.02 & 1.92 & 2.13 & 2.13.
\end{array}$$

Calculate a 95% confidence interval for the true copper contamination based on the t-distribution.

Exercise 1.3.3 is much less explicit than Exercises 1.3.1 and 1.3.2. The student is not told that the data follow a normal distribution but is expected to model the data with the normal distribution and then to calculate the confidence interval assuming the model.

A slightly non-standard exercise in the same vein is the following:

Exercise 1.3.4 *The following data give 27 measurements of the amount of copper (milligrammes per litre) in a sample of drinking water:*

$$\begin{array}{ccccccccc}
2.16 & 2.21 & 2.15 & 2.05 & 2.06 & 2.04 & 1.90 & 2.03 & 2.06\\
2.02 & 2.06 & 1.92 & 2.08 & 2.05 & 1.88 & 1.99 & 2.01 & 1.86\\
1.70 & 1.88 & 1.99 & 1.93 & 2.20 & 2.02 & 1.92 & 2.13 & 2.13.
\end{array}$$

Determine the maximum likelihood estimators based on the Laplace model with density

$$f(x:\mu,\sigma) = \frac{1}{2\sigma}\exp\left(-\frac{|x-\mu|}{\sigma}\right).$$

Use simulations to calculate a 95% confidence interval for the amount of copper in the sample of drinking water.

A third and completely non-standard exercise in the same vein is the following:

Exercise 1.3.5 *The following data give 27 measurements of the amount of copper (milligrammes per litre) in a sample of drinking water:*

$$\begin{array}{ccccccccc}
2.16 & 2.21 & 2.15 & 2.05 & 2.06 & 2.04 & 1.90 & 2.03 & 2.06\\
2.02 & 2.06 & 1.92 & 2.08 & 2.05 & 1.88 & 1.99 & 2.01 & 1.86\\
1.70 & 1.88 & 1.99 & 1.93 & 2.20 & 2.02 & 1.92 & 2.13 & 2.13.
\end{array}$$

Use a fine grid to calculate the maximum likelihood parameters of the comb model with distribution function $F_{comb,(75,0.005,0.85)}$ given by

$$F_{comb,(k,ds,p)}(x) = p\left(\frac{1}{k}\sum_{j=1}^{k}\Phi((x-\iota_k(j))/ds)\right) + (1-p)\Phi(x) \tag{1.8}$$

where Φ is the distribution function of the standard normal distribution,

$$\iota_k(j) = \Phi^{-1}(i/(k+1)), i = 1,\ldots,k. \tag{1.9}$$

and $(k, ds, p) = (75, 0.005, 0.85)$. Use simulations to calculate a 95% confidence interval for the amount of copper in the sample of drinking water.

1.3.3 From the Model to the Data to the World: Estimation

If asked to justify the calculation of a confidence interval based on the t-distribution for the data of Exercise 1.3.3 the statistician, a frequentist, may well reply as follows. The data 'look normal' and that for normally distributed data a confidence interval based on the t-distribution has certain optimality properties. These optimality properties are mathematical theorems, their validity is not in doubt. Thus if the data really are normal and not just 'look normal' the argument is convincing, but 'look normal' is not the same as 'are normal'. The student is being instructed to act in the 'as if true' mode and in this mode to construct a 95% confidence interval. The result is $[1.970, 2.062]$ which answers Exercise 1.3.3. A Bayesian would not use optimal procedures but the procedures used would be likelihood based which, for the purpose of the present discussion, is the same thing.

A similar justification can be given for Exercise 1.3.4. The data 'look Laplace'. But again, 'look Laplace' is not the same as 'are Laplace'. The resulting confidence interval is $[1.996, 2.064]$.

Finally a similar justification can be given for Exercise 1.3.5. But again, 'look comb' is not the same as 'are comb'. The resulting confidence interval is $[2.024, 2.026]$.

When the statistician is in the 'behaving as if true' mode the word 'estimate' has a precise meaning and it is that of Exercise 1.3.1. There is a parametric family of models $\mathscr{P}_\Theta$ and a data set $\mathbf{X}_n(\theta_0)$ generated by the model P_{θ_0} with θ_0 not known. It is the 'true unknown parameter' which is to be estimated. This can be relaxed somewhat by postulating a true generating model $P \notin \mathscr{P}_\Theta$ for the data. The statistician may now choose an estimation procedure do minimize the discrepancy between P and P_θ. The Kullback-Leibler discrepancy

$$d_{\text{kl}}(P, P_\theta) = \int p(x)\log(p(x)/p_\theta(x))\,d\pi(x) \tag{1.10}$$

where π is some measure which dominates P and $\mathscr{P}_\Theta$ and p and p_θ are the respective densities. It is assumed that such a dominating measure exists but this is nothing more than making the mathematics work. The idea of representing 'truth' by some model

$P \notin \mathscr{P}_\Theta$ is a gross simplification. The truth will be different for every individual measurement and will be complicated as in the toss of a single coin [56], [199]. A second example is the attempt to locate which parts of the brain react to a visual or acoustic on-off stimulus. This is done using functional magnetic resonance imaging which measures the changes in the magnetic field caused by an increase in oxygen caused by an increase in the blood supply [220] caused by the stimulus. There is not even a well-defined mathematical theory for this. A third example is provided by the copper data and the continous models of exercises 1.3.3, 1.3.4 and 1.3.5. All current chemical and physical theories suggest that copper comes in atoms and water in molecules so that the concentration of copper in the water sample is not a continuous measure but a discrete one. In all cases the Kullback-Leibler discrepancy (1.10) is infinite . The whole exercise seem eminently pointless.

The copper data do not have a 'true unknown parameter' μ_0 but it is clear what is to be estimated, namely the quantity of copper in the water, which is something different and, in contrast to the 'true but unknown parameter θ', may be said to exist. There are therefore two different meanings of the word 'estimate' and they are typically conflated by identifying the one with the other, the parameter θ with the true state of nature, namely the quantity of copper in the water sample. The identification of a parameter θ with the state of the world can be problematic. The three different models of Exercises 1.3.3, 1.3.4 and 1.3.5 are all symmetric so that identifying the location parameter μ with the quantity of copper gives a consistent interpretation across the models. The data of the three Exercises 1.3.3, 1.3.4 and 1.3.5 comes from analytical chemistry. Tukey, who at one point in his career was a chemist, has stated that such data is often skewed [211]. Moreover outliers are common and have to be identified and either eliminated or downweighted [1]. In general identifying the quantity of interest, whether it be copper in water or cadmium in dust, with the centre of symmetry of a symmetric model when in the 'behave as if true' mode is not a good idea. To allow for skewed data the statistician may consider several models which are skewed such as the gamma, lognormal and Weibull models. Whilst in the 'behave as if true' the statisticians problem is well-defined, namely estimate the parameters. But it is now no longer clear how to move from the parameters to what the statistician really wants to estimate, the quantity of water or cadmium.

The situation is different when models are explicitly recognized as an approximation. The first meaning of the word 'estimation' no longer makes sense. An attempt will be made to use the word 'estimation' and related words only in the second sense when the aspect of the world of interest can be expressed in numerical values. This can cause problems as there is no elegant alternative to, for example, 'kernel estimator' which avoids the word 'estimator' but still conveys to the reader the meaning it has in statistics. As the idea of approximation explicitly recognizes that there are many adequate approximating models P for any data set it does not identify a parameter with the state of the world but a functional T defined on a subset of all probability models. In the case of location this will be taken to be an M-functional, or M-estimator which is the standard name. It no longer matter if the models all have a centre of symmetry, it only matters that they be close in a weak topology, for example, that they are close in the Kolmogorov metric.

1.3.4 Goodness-of-fit

In the last section the decision to use certain models was justified informally. This is very often done as in [3] where Aitkin gives some data on family incomes and writes

> The income sample above is clearly skewed with a longer right-hand tail of large values, so an approximating model with right skew would be appropriate. The gamma, lognormal and Weibull distributions are possible choices.

If it is intended that the data analysis be part of a paper to be submitted to a highly respected international journal it may be wise to replace the informal judgement by a formal goodness-of-fit test.

When goodness-of-fit tests are carried out to check whether the model is consistent with the data the word 'model' typically refers to a family $\mathscr{P}_\Theta$ rather than to a specific model P_θ. Such tests are based on some estimate $\hat{\theta}_n = \hat{\theta}(\mathbf{x}_n)$ of the parameter θ, for example the maximum likelihood estimate. A standard example is in [27] where the appropriateness of the Poisson model for the distribution of flying bomb hits in London is based on the mean number of hits per square. More generally most goodness-of-fit tests are of the form

$$D(\mathbb{P}_n, P_{\hat{\theta}_n}) \leq qt^*(D, \theta, \alpha, n) \tag{1.11}$$

where D is some discrepancy and $qt^*(D, \theta, \alpha, n)$ is given by

$$\mathbf{P}(D(\mathbb{P}_n(\mathbf{X}_n(\theta)), P_{\hat{\theta}(\mathbf{X}_n(\theta))}) \leq qt^*(D, \theta, \alpha, n)) = \alpha\,.$$

The quantile $qt^*(D, \theta, \alpha, n)$ depends on the 'true but unknown' value θ of the parameter but, as in the chi-squared goodness-of-fit test, the dependence may disappear asymptotically.

The 'look normal', 'look Laplace' and 'look comb' judgements can be replaced by a formal measure of goodness-of-fit based on the Kuiper metric [133] of order one, d_{ku} defined by

$$d_{\text{ku}}(P, Q) = \max_{a<b} |P((a,b]) - Q((a,b])|. \tag{1.12}$$

The critical values $qt^*(d_{\text{ku}}, \theta, \alpha, n)$ can be obtained by simulations. The results are as follows. The Kuiper distances are 0.204, 0.200 and 0.248 respectively and for $\alpha = 0.99$ the critical values are 0.312, 0.274 and 0.370. The p-values are 0.441, 0.304 and 0.321. In other words the formal goodness-of-fit tests based on the Kuiper distance have confirmed the informal judgements for all three models: the copper data do indeed look normal, they also look Laplace and they also look comb.

It is difficult to interpret a goodness-of-fit test of the form (1.11). Strictly speaking what has been done is to single out a particular parameter value $\hat{\theta}_n$ on the basis of the sample, and then to state that the model $P_{\hat{\theta}_n}$ is, or is not, consistent with the data. In practice the test is regarded as answering the question as to whether it is legitimate to operate within the full family $\mathscr{P}_\Theta$ of models. What it cannot mean is that if (1.11) holds then every model $P_\theta, \theta \in \Theta$, is consistent with the data. In general this is simply not the case as typically there will be values of θ for which P_θ is not consistent with the data. In particular it is possible that a subsequently calculated confidence region for θ contains values of θ not consistent with the data. If the goodness-of-fit is to be taken seriously such values ought to be excluded from further analysis. The test (1.11) however gives no information as to what these values could be. In practice this does not matter for most statisticians. A goodness-of-fit test is part of the data analytic mode of behaviour. Once a model has passed such tests the statistician moves on to the formal inference mode accompanied by the whole family $\mathscr{P}_\Theta$ and the results of the first mode are forgotten.

1.3.5 Likelihood

A student sufficiently good as to be able to correctly answer all five exercises may, but only may, be puzzled by the three different confidence intervals for the same data. He or she may have hoped to be able to make a clear decision in favour of one particular model on the basis of a formal goodness-of-fit test only to be disappointed by the result. The data have no anomalies and all models fit. An advanced student may now decide to use the AIC criterion [4] to choose the best model. As all models have two parameters AIC chooses that model with the highest likelihood. The log-likelihoods for the normal, Laplace and comb models are respectively 20.31, 20.09 and 31.37. The student must be well satisfied. There is a clear decision in favour of the comb model and not only that, the comb model yields the most precise information of the quantity of copper in the drinking water as reflected in the very short confidence interval $[2.024, 2.026]$.

1.3.6 Regularization

Alternatively the student may be very confused: he or she has done everything to be expected and based the final choice on the second most important thing in statistics, namely likelihood;

The **Likelihood Function**

$$lhd(\theta; y) = lhd(\theta; y_1, \ldots, y_n) = f(y_1|\theta) \ldots f(y_n|\theta)$$

is the second most important thing in (both Frequentist and Bayesian) Statistics, the first being common sense, of course. (see [219])

The result, the comb model, is clearly nonsense and the most important thing in statistics, namely common sense, would dictate the normal model. This is perfectly correct but a better justification is required than an unspecified appeal to common sense.

The 'correct' reading of these results is not that the comb model gives the most precise estimate of the amount of copper in the water sample. The correct reading is that efficiency depends on the model and that efficiency can be imported from the model at no cost. Tukey [211] calls this a free lunch and states that in his experience there is no such thing as a free lunch (TINSTAAFL) in statistics. To avoid free lunches one should therefore use that model which imports the least efficiency,

Copper data					
Model	Kuiper, p-value	log–lik.	95%–conf. int.	length	Fisher Inf.
Normal	0.204, 0.441	20.31	$[1.970, 2.062]$	0.092	$2.08 \cdot 10^3$
Laplace	0.200, 0.304	20.09	$[1.989, 2.071]$	0.082	$1.41 \cdot 10^4$
Comb	0.248, 0.321	31.37	$[2.0248, 2.0256]$	0.0008	$3.73 \cdot 10^7$

Table 1.2 *Kuiper distances and their associated p-values, log-likelihoods, 95% confidence intervals, their lengths and the empirical Fisher information for three different models for the copper data of Exercise 1.3.3.*

that is the model with the longest confidence interval, not the shortest. Tukey calls such models 'bland' or 'hornless' [211]. One measure of blandness is the Fisher information, the smaller the Fisher information the blander the model. The empirical Fisher informations of the three models are $2.078 \cdot 10^3$, $1.414 \cdot 10^4$ and $3.726 \cdot 10^7$ respectively. On this basis the normal model should be chosen as the blandest of the three.

The location-scale problem is ill-posed and must be regularized. Choosing a model consistent with the data with the smallest, or at least a small, Fisher information is a form of regularization. The normal model is one such model as it minimizes the Fisher information over all models with a given variance. Looked at it in this manner the mean is a minimax statistic as it minimizes the expected quadratic error under the most difficult circumstances. Other models which minimize Fisher information are given in Chapter 4.4 of [118]. See also Chapter 8.2d of [103].

The results of the analysis so far are given in Table 1.2.

1.3.7 Stability of Analysis

A further issue is that of stability in the sense that small changes in the data should not cause large changes to the analysis. There are two sorts of small changes. One is small changes in the values of the data, the other is possibly large changes in a small number of data points; a combination of the two is also to be regarded as small.

The data of Table 1.3 are a small perturbation of the original copper data of Exercise 1.3.3 in the first sense. The maximum absolute difference between corresponding observations is 0.015.

Perturbed copper data						
2.151841	2.206391	2.161802	2.065081	2.057873	2.035654	1.893120
2.029681	2.059767	2.031476	2.059147	1.920517	2.073690	2.050381
1.889667	1.982238	2.022498	1.871578	1.702510	1.879889	1.994095
1.921984	2.213070	2.019419	1.923515	2.138214	2.121672	

Table 1.3 *Slightly modified values of the copper data of Exercise 1.3.3.*

Table 1.4 shows the results for perturbed data. Whereas the normal and Laplace models exhibit stability with little change in the confidence intervals (although there is a drop in the p-values for the Kuiper distance) this not the case for the comb model.

Perturbed copper data					
Model	Kuiper, p-value	log–lik.	95%–conf. int.	length	Fisher Inf.
Normal	$0.228, 0.258$	20.32	$[1.973, 2.063]$	0.090	$2.08 \cdot 10^3$
Laplace	$0.217, 0.168$	20.27	$[1.991, 2.072]$	0.081	$1.43 \cdot 10^4$
Comb	$0.378, 0.006$	20.12	$[1.9684, 1.9694]$	0.001	$1.98 \cdot 10^7$

Table 1.4 *Kuiper distances and their associated p-values, log-likelihoods, 95% confidence intervals, their lengths and the empirical Fisher information for three different models for the perturbed copper data of Table 1.3.*

	1	2	3	4
Lab. 1	2.2	2.215	2.2	2.1
Lab. 2	2.53	2.505	2.505	2.47
Lab. 3	2.3	3.2	2.8	2.0
Lab. 4	2.33	2.35	2.3	2.325
Lab. 5	2.1	2.68	2.46	1.92
Lab. 6	2.692	2.54	2.692	2.462
Lab. 7	2.292	2.049	2.1	2.049
Lab. 8	2.120	2.2	1.79	2.12
Lab. 9	1.967	1.899	2.001	1.956
Lab. 10	3.8	3.751	3.751	3.797
Lab. 11	2.696	2.22	2.01	2.57
Lab. 12	2.191	2.213	2.21	2.228
Lab. 13	2.329	2.252	2.252	2.26
Lab. 14	2.37	2.42	2.33	2.36

Table 1.5 *The results of an interlaboratory test involving fourteen laboratories. The measurements are of the quantity of mercury (milligrammes per litre) in a water sample.*

There is a large change in the value for the standard deviation from 0.140 to 0.190 and a corresponding change in the p-values for the Kuiper distance from 0.321 to 0.006. This results from the high sensitivity of the comb model and is related to the high value of the Fisher information.

The second form of small change is possible large changes in a small number of data values. An example is to change the last two values of the copper data from 2.13 to 3.13. If this is done none of the models is stable when combined with maximum likelihood. In particular the scales react sensitively to the changes and increase by factors of 2.75 (normal), 1.85 (Laplace) and 2.00 (comb) with correspondingly large changes in the confidence intervals. Such deviant or exotic (Tukey) values are not unusual in analytic chemistry. Table 1.5 shows the results of an interlaboratory test where fourteen laboratories were required to determine the amount of mercury in milligrammes per litre in a sample of drinking water. The boxplot is shown in Figure1.1. The large variation in the measurements is apparent. Indeed the main

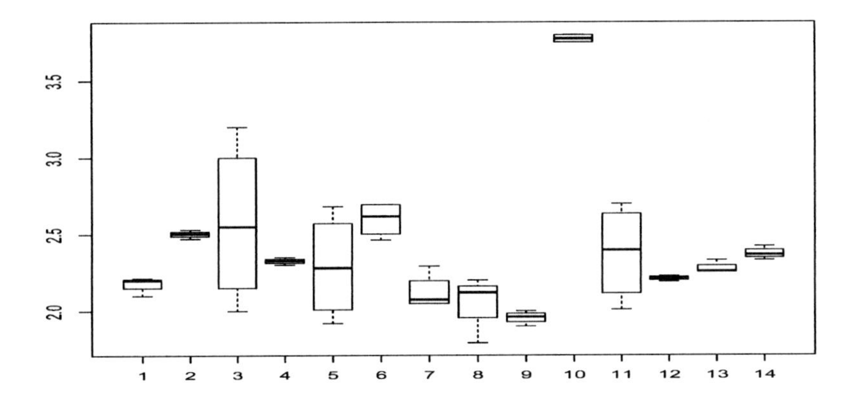

Figure 1.1 *The results of an interlaboratory test on the amount of mercury (micrograms per litre) in a sample of drinking water.*

problem when defining a procedure for the statistical analysis of interlaboratory tests is the presence of outliers (see [1]).

1.3.8 Two Topologies

1.3.8.1 Weak and Strong Metrics

Figure 1.2 shows the distribution functions of the comb model and that of the standard normal model plotted over each other. It is seen that they are very close so that at the level of distribution functions they are difficult to distinguish. The closeness can be quantified using the Kuiper metric (1.12) which is based on distribution functions.

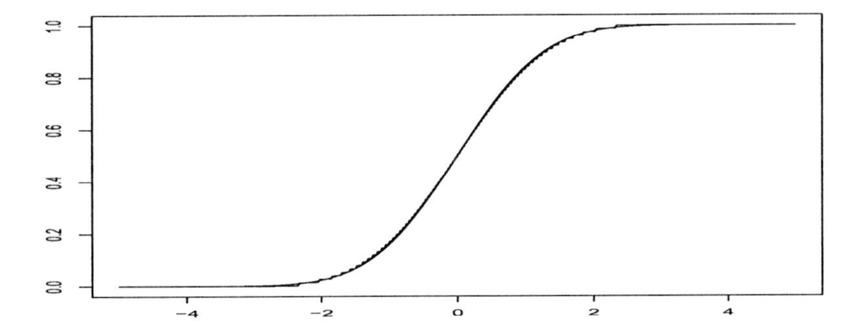

Figure 1.2 *The distribution functions of the standard normal and comb models.*

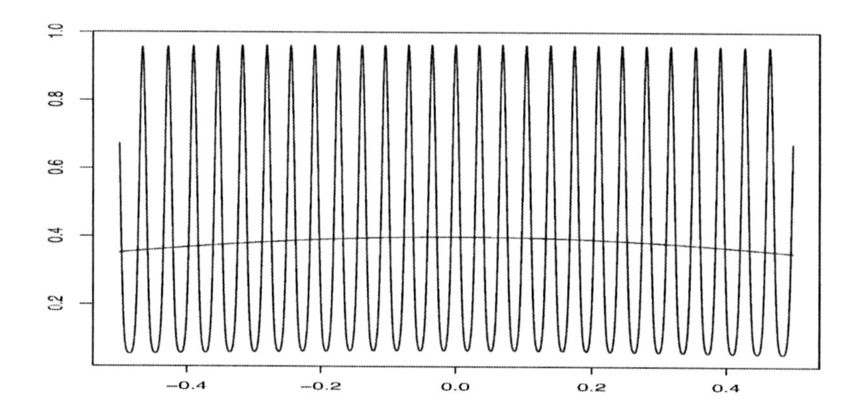

Figure 1.3 *The comb and normal densities on* $(-0.5, 0.5)$.

The result is $d_{\text{ku}}(\mathfrak{N}(0,1), \mathfrak{Comb}(0,1)) = 0.02$ so that the two distributions are indeed very close.

Although the two models have similar distribution functions their densities are very different: Figure 1.3 shows a section of the densities of the comb and normal distributions. An appropriate metric for the densities is the L_1 metric of (1.5). The value for the normal and comb densities is $d_1(\mathfrak{N}(0,1), \mathfrak{Comb}(0,1)) = 0.966$. In the space of distributions the two models are very close, in the space of densities they are far apart. In mathematical terms the Kuiper metric is weak, that is, it generates few open sets. The L_1 metric is strong, it generates strictly more open sets than the Kuiper metric.

1.3.8.2 Goodness-of-fit

It follows that an analysis of a data set based on distribution functions will give similar results for the normal and comb models. This is a partial explanation of the similarity of the results for the goodness-of-fit test for the copper data as given in Table 1.2. The strong divergence for the density based confidence intervals is consistent with the two distributions being far apart in the L_1 metric. The reason why it is only a partial explanation for the similarity of the goodness-of-fit results is that although the test is based on the Kuiper metric, the parameters involved are derived from maximum likelihood.

The problems with interpreting this form of goodness-of-fit test were stated above in Chapter 1.3.4. The simple and elegant solution is to specify all those parameter values for which the model is consistent with the data. Given a discrepancy or metric d this may be done by putting

$$\mathscr{A}_n(\mathbf{x}_n, d, \alpha, \Theta) = \{\theta : d(\mathbb{P}_n, P_\theta) \le qt(D, \theta, \alpha, n)\} \tag{1.13}$$

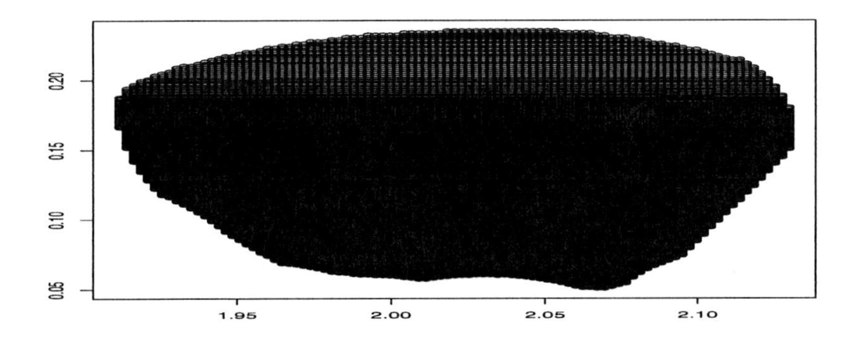

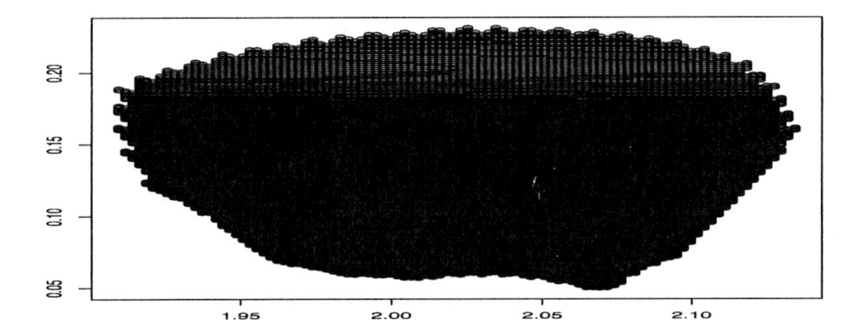

Figure 1.4 *Upper panel: The parameters (μ, σ) of the normal distribution which are consistent in the sense of the Kuiper metric d_{ku} with the copper data of Exercise 1.3.3. Lower panel: the same for the comb model.*

(compare (1.11)). For the Kuiper metric d_{ku} the expression simplifies for continuous models as the quantile $qt(d_{\text{ku}}, \theta, \alpha, n)$ does not depend on the parameter θ. The approximation $qt(d_{\text{ku}}, 0.99, n) \approx 2/\sqrt{n}$ gives

$$\mathscr{A}_n(\mathbf{x}_n, d_{\text{ku}}, 0.99, \mathbb{R} \times \mathbb{R}_+) = \{(\mu, \sigma) : d_{\text{ku}}(\mathbb{P}_n, P_{\mu,\sigma}) \leq 2/\sqrt{n}\}. \tag{1.14}$$

The results for the copper data are shown in Figure 1.4, the upper panel for the normal model, the lower for the comb model.

A further example is the following. It is a simplified version of the example in Section 3.1 of [94]. The model is

$$P_\theta = 0.5\mathfrak{N}(0,1) + 0.5\mathfrak{N}(\mu, \sigma^2) \tag{1.15}$$

with $\theta = (\mu, \sigma^2)$. Density based methods such as maximum likelihood and Bayes have problems with this model as the likelihood explodes if $\mu = x_i$ for some observation x_i and σ tends to zero. In contrast distribution based methods have no problem. An approximation region for the parameters can be defined as in (1.14) with $d = d_{\text{ku}}$. Figure 1.5 shows the results for two samples of size $n = 200$ with $(\mu, \sigma) = (1,2)$ in the upper panel and $(\mu, \sigma) = (1, 0.1)$ in the lower panel.

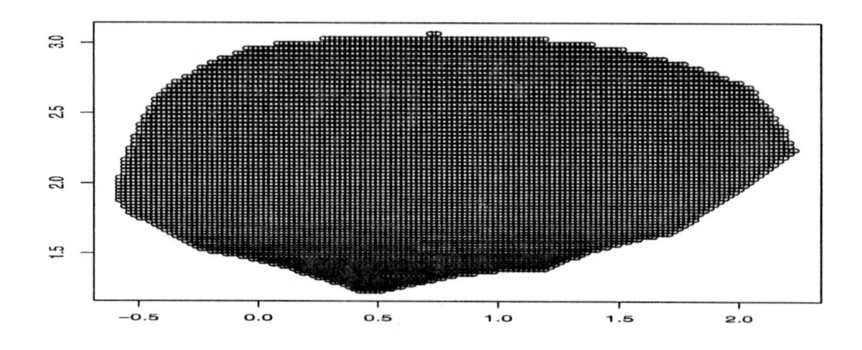

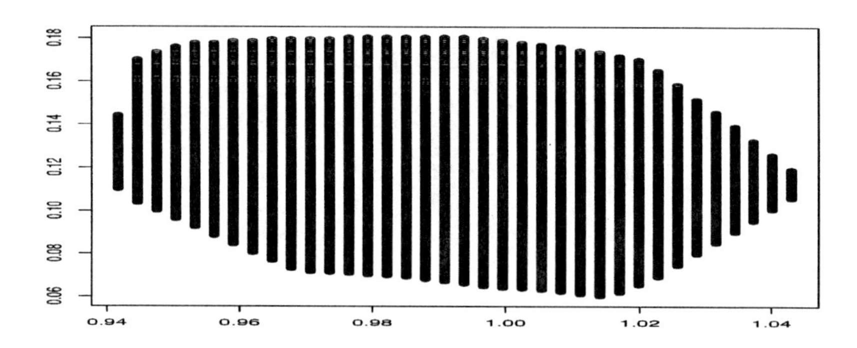

Figure 1.5 *Upper panel: The parameters* (μ, σ) *of the model (1.15) with* $(\mu, \sigma) = (1, 2)$ *based on a sample of size* $n = 200$*. Lower panel: the same for* $(\mu, \sigma) = (1, 0.1)$.

1.3.8.3 Generating Data

Data are generated at the level of distribution functions as in (1.3). Moreover if two distributions F and G are close in the Kolmogorov metric then the simulated data sets will also be close as in the discussion after (1.3) and (1.4). It follows that it is possible to generate normal random variables X and comb random variables Y which are very close to each other. One simply follows the above recipe of (1.3) using the same random variable U. Here are five normal random variables and five comb random variables generated by this method:

$$\begin{array}{rlllll} \text{normal} & 0.2754234 & -1.5703194 & 0.4376309 & -0.5924103 & 0.9045972 \\ \text{comb} & 0.2821350 & -1.5874329 & 0.4589539 & -0.6188049 & 0.9410706 \end{array} \tag{1.16}$$

The mean of the copper data is 2.016 and the standard deviation 0.116. Normal and comb data similar to the copper data can be generated by multiplying the above numbers by 0.116 and adding 2.016. This gives

$$\begin{array}{rlllll} \text{normal} & 2.047398 & 1.836984 & 2.065890 & 1.948465 & 2.119124 \\ \text{comb} & 2.048163 & 1.835033 & 2.068321 & 1.945456 & 2.123282 \end{array} \tag{1.17}$$

and finally, on rounding them to the same precision as the copper data,

$$\begin{array}{rccccc} \text{normal} & 2.05 & 1.84 & 2.07 & 1.95 & 2.12 \\ \text{comb} & 2.05 & 1.84 & 2.07 & 1.94 & 2.12 \end{array} \tag{1.18}$$

which are the same with one exception. In other words at the level of precision of the copper data it is impossible to distinguish between the normal and comb models.

An analysis of the data at the level of distribution functions, that is in the data analytic mode, respects this and leads to similar results for the two models as evidenced by Figure 1.4. Formal inference based on densities can lead to very different conclusions.

1.4 Towards One Mode of Analysis

To summarize the above, the two modes of behaviour of a statistician, the data analytic of [210] and the inferential mode of [30] are very different. The differences can be condensed to the difference between the two metrics d_{ku} of (1.12) and d_1 of (1.5). The topology of the data analytic mode is weak, that of the inference mode is strong, and there is a complete break in outlook when the statistician moves from the first to the second. This is why it is possible to write a book about the second mode without reference to the first. One aim of this book is to have only one mode of data analysis and inference whose topology is weak. There will be no density based inference without explicit regularization.

1.4.1 Looking Like the Data

The simple idea behind the whole of this book is to accept a model P if the data generated under model $\mathbf{X}_n(P)$ look like the actual data $\mathbf{x}_n$. The idea goes back at least to Donoho [58] who writes:

> No distribution which produces samples very much like those actually seen should be ruled out a priori.

A first step in this direction has already been taken. The data of (1.18) were generated under two different models to 'look like' the copper data, as indeed they would have done if 27 values had been simulated instead of just five. However if one simulates for a long time the model will eventually produce a data set which does not 'look like' the original data. Consequently 'data generated under the model' is modified to 'typical data generated under the model'. The word typical is quantified by a number $\alpha, 0 < \alpha \leq 1$, such that $100\alpha\%$ of data generated under the model are called 'typical'.

One way of thinking of the concept is the following. Given a real data set $\mathbf{x}_n$ and a model P generate 999 data sets of size n $\mathbf{X}_{1,n}(P), \ldots \mathbf{X}_{999,n}(P)$ under P. The real data set $\mathbf{x}_n$ is also included to give in all 1000 data sets. On setting $\alpha = 0.95$ the statistician is required to specify 950 typical data sets out of the 1000 or, equivalently, 50 atypical data sets. If the real data set $\mathbf{x}_n$ is one of the 950 then typical data generated under the model P look like the real data. When choosing the 950 typical data sets

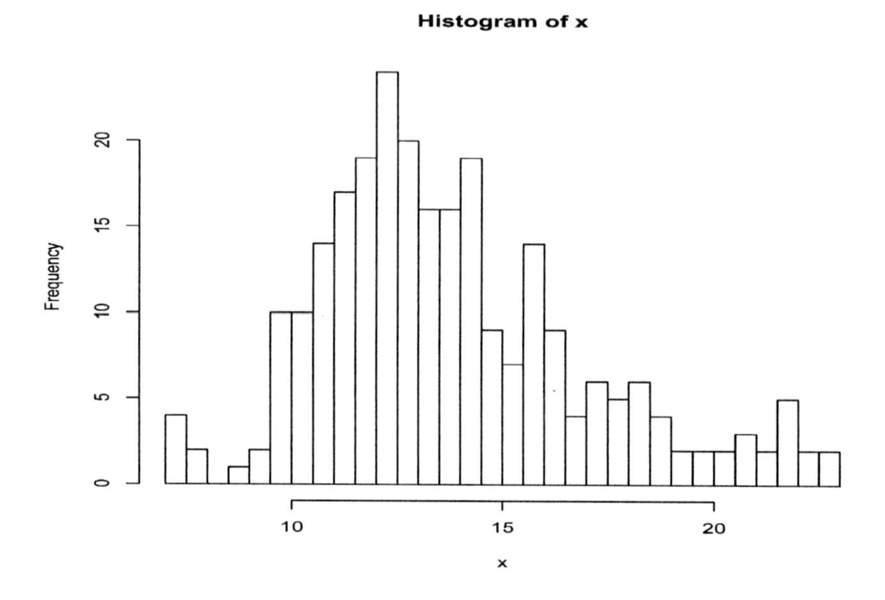

Figure 1.6 *Lengths of study in semesters of 258 German students.*

the statistician will use some specific criterion or criteria. It is this criterion which specifies what is meant by 'look like'.

The following is a reduced example of the idea. Figure 1.6 shows a histogram of the lengths of study in semesters (equivalent to six months) of 258 German students. Since this data set was obtained the degree system has been changed so that present lengths of study for a first degree are much shorter. Figure 1.7 shows the boxplots of 19 simulated $\Gamma(16,0.7)$ samples of size $n = 258$ and the data of Figure 1.6. On setting $\alpha = 0.95$ it is required to specify one atypical sample out of the 20 samples. This is not too difficult and in this sense data generated under the $\Gamma(16,0.7)$ do not look like the real data. Figure 1.8 shows the boxplots of 19 simulated samples of size 258 under the $\Gamma(18,0.75)$ model and the real data with the real data at a different position. It is now difficult to specify one atypical sample and in this sense data generated under the $\Gamma(18,0.75)$ model look like the real data.

Comparing real data with data generated under the model is not new. One example in the context of a model for the distribution of galaxies is [166] and [167]: see also [22] and, for an appraisal of their work and further examples, [23]. The idea is also to be found in [94] and Chapter 5.5 (Exhibits 5.3, 5.4 and 5.5) of [117].

1.4.1.1 An Example

The data $\mathbf{x}_n$ are given, the only feature of interest is the mean and the family of models is $\mathscr{P} = \{\mathfrak{N}(\mu,1) : \mu \in \mathbb{R}\}$. Typical will be specified by setting $\alpha = 0.95$. What is the typical behaviour of the mean $\overline{\mathbf{X}}_n\,(\mu)$ of a sample $\mathbf{X}_n(\mu)$ of size n under

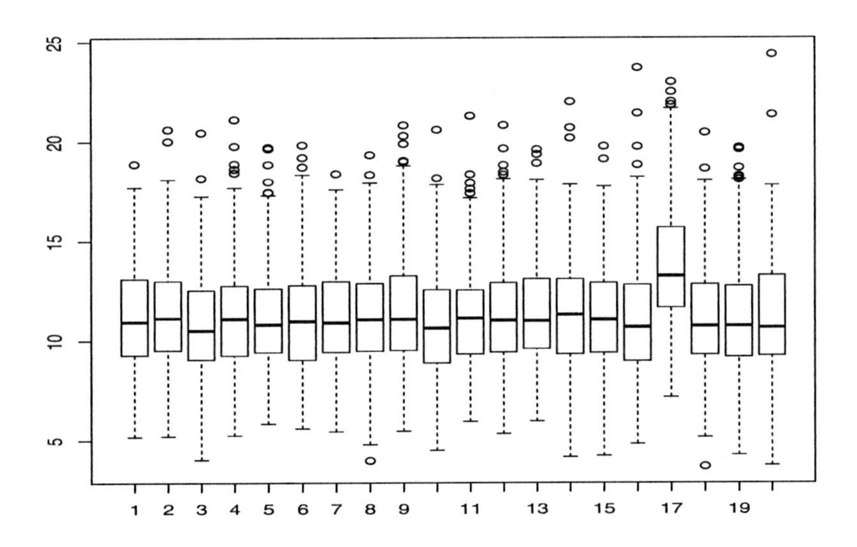

Figure 1.7 *Boxplots of nineteen simulated* $\Gamma(16, 0.7)$ *samples of size 258 and the data of Figure 1.6.*

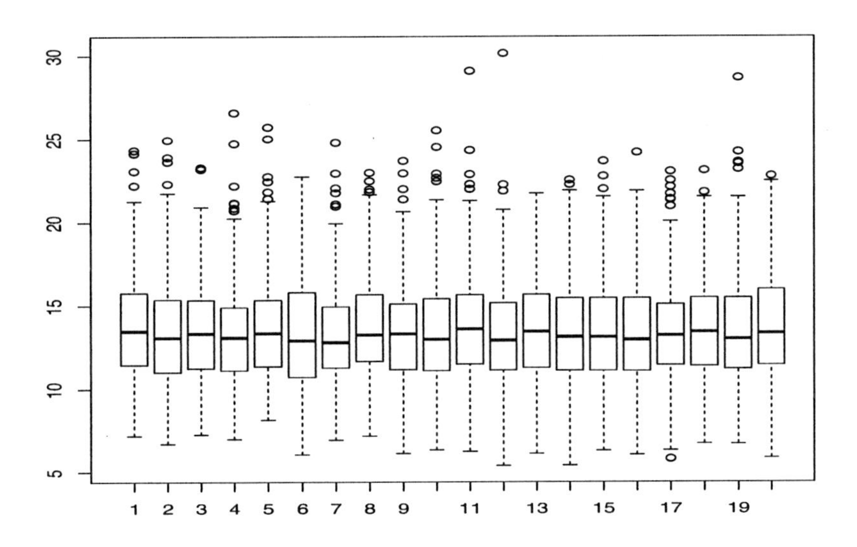

Figure 1.8 *Boxplots of nineteen simulated* $\Gamma(18, 0.75)$ *samples of size 258 and the data of Figure 1.6.*

the model $\mathfrak{N}(\mu,1)$? Typically, that is in 95% of the cases, the mean will satisfy

$$\mu - \frac{1.96}{\sqrt{n}} \leq \overline{\mathbf{X}}_n(\mu) \leq \mu + \frac{1.96}{\sqrt{n}}. \tag{1.19}$$

The data $\mathbf{x}_n$ will therefore look like a typical sample generated under the model $\mathfrak{N}(\mu,1)$ if

$$\mu - \frac{1.96}{\sqrt{n}} \leq \overline{\mathbf{x}}_n \leq \mu + \frac{1.96}{\sqrt{n}}. \tag{1.20}$$

The set of values of μ for which $\mathbf{x}_n$ looks like a typical sample generated under the model $\mathfrak{N}(\mu,1)$ is

$$\left[\overline{\mathbf{x}}_n - \frac{1.96}{\sqrt{n}}, \overline{\mathbf{x}}_n + \frac{1.96}{\sqrt{n}}\right] \tag{1.21}$$

which is the traditional 95% confidence interval for μ. The interpretation is different and to emphasize this it will be called an approximation interval.

A more complicated example has already been treated. Figure 1.4 gives the parameter values of the normal (upper panel) and comb (lower panel) models for which simulated data sets of size $n = 27$ 'look like' the copper data in the sense of the Kuiper distance. The two different definitions of 'look like' can be combined to give a criterion which depends on the values of the mean and on the Kuiper distances.

1.4.1.2 Interpreting Parameter Values

The approximation interval (1.21) of parameter values has the following interpretation. Take any μ in this interval and simulate a large number of samples under the model $\mathfrak{N}(\mu,1)$. Calculate the mean of each sample and ignore the 2.5% smallest and 2.5% largest means. The remainder define an interval which gives the typical values of the mean under the model $\mathfrak{N}(\mu,1)$. If μ lies in the interval (1.21) then $\overline{\mathbf{x}}_n$ will lie in the interval of typical values of $\overline{\mathbf{X}}_n(\mu)$. In other words, $\mathbf{x}_n$ will look like a typical sample generated under the model $\mathfrak{N}(\mu,1)$.

The following points are worthy of note:

- The data $\mathbf{x}_n$ are given. There is no assumption that they can be repeated as in a frequency approach.
- No assumptions are made about the distribution of the data $\mathbf{x}_n$. In particular it is not assumed that they are a realization of data generated under the model for $\mathfrak{N}(\mu,1)$ for some unknown μ.
- It is not even assumed that the data are random. The binary expansion of π can be successfully modelled by the $\mathfrak{b}(1,1/2)$ distribution. Central limit theorems hold in deterministic systems (see [203]).
- The statistician is not in the 'behaving as if true' mode.
- The values of the parameter μ are determined solely by the requirement that $\mathbf{x}_n$ looks like typical data sets generated under the model $\mathfrak{N}(\mu,1)$.
- There are no 'true but unknown parameter values'.
- Each value μ of the parameter space has a precise meaning with respect to the data: either the data $\mathbf{x}_n$ do or do not look like a typical sample $\mathbf{X}_n(\mu)$.

1.4.1.3 Working within the Model

Given some definition of 'look like' based on numerical values, for example the mean and the Kuiper metric, it has to be determined what the typical values are under the proposed model P. This requires calculating the quantiles of the numerical values and hence working within the model. This is not the same as 'behaving as if the model were true'. The calculation of quantiles can sometimes be done analytically within the model, or by the use of simulations or by the use of asymptotics. If an asymptotic approximation to a quantile is to be used, then the asymptotics must be sufficiently accurate for the sample size of the data. An example is the 0.99 quantile of the Kuiper metric for continuous models. The approximation is $2/\sqrt{n}$ and this is already sufficiently accurate for $n = 27$ for it to be used in (1.14).

1.4.1.4 The Weak Topology

Parameters are determined by the requirement that the actual data $\mathbf{x}_n$ 'look like' typical data sets generated and the model P_θ. As data is generated at the level of distribution functions and distribution functions which are close in say the Kuiper metric generate data sets which are also close as in (1.18). It follows that all the data analysis is to be done in a weak metric. There is no break in topology.

1.5 Approximation, Randomness, Chaos, Determinism

The concept of approximation to be proposed in this book is that of approximating a given unique data set. This is much too simplistic: the best approximation for a given data set is that model which simply reproduces the given data set. This raises the question as to why stochastic models are used at all.

Stochastic models are used in situations where experiments performed under the 'same' conditions give different results and nonpredictable results. The copper data of Example 1.3.3 are an example of this. They are repeated measurements for the same sample of drinking water and all were done under the same conditions. The fact that they are different is put down to not further specified 'measurement errors'.

Measurement is not the only source of variation: it is rare that a mistake is made when deciding whether the coin has turned up heads or tails. There are various explanations for the variability in coin tossing. It could simply be due to the the variability in the manner in which the coin is tossed. This would seem to be the case when the coin is tossed by humans but even when machines are built to toss coins the end result is not predictable . The last case is interesting as a machine tossed coin is subject to deterministic Newtonian mechanics [56]. Although the result of such a toss is in principle predictable, in practice it may not be so. Small imperceptible variations in the initial conditions can lead to different results. The abstract of [199] reads

> The dynamics of the tossed coin can be described by deterministic equations of motion, but on the other hand it is commonly taken for granted that the toss of a coin is random. A realistic mechanical model of coin tossing is constructed to examine whether the initial states leading to heads or tails are

distributed uniformly in phase space. We give arguments supporting the statement that the outcome of the coin tossing is fully determined by the initial conditions, i.e. no dynamical uncertainties due to the exponential divergence of initial conditions or fractal basin boundaries occur. We point out that although heads and tails boundaries in the initial condition space are smooth, the distance of a typical initial condition from a basin boundary is so small that practically any uncertainty in initial conditions can lead to the uncertainty of the results of tossing.

In Germany the 'Lottozahlen' are chosen by a machine which is televised during its operation. It is constructed to be chaotic in that small differences in the conditions under which the machine operates lead to different outcomes.

One possible source of 'randomness' may be chaos where the system is deterministic but there is extreme dependence on the initial conditions. On the other hand it can be argued on the basis of quantum mechanics that the world at this level is random. There are however deterministic versions of quantum mechanics, namely Bohmian mechanics [81] which are chaotic rather than random: small changes in the initial positions of a particle in the two slit experiment may cause it to go through different slits

The Solomonoff-Kolmogorov-Chaitin complexity of a sequence of binary digits is defined in limiting terms as the length of the shortest computer programme required to generate the sequence (see [140]). It was shown in [150] that any complex number in the sense of Kolmogorov is essentially random in that its binary expansion will pass all tests for randomness. The number π is not complex in the sense of Kolmogorov but all tests for randomness based on finite sections of the expansion of π have so far been passed (see [162] and [208]).

In view of the above it seems reasonable to use stochastic models and to require that the model reflect the variability of the data. If many similar data sets are available, that is, data sets obtained under similar circumstances, it should in general be possible to quantify the variability of the data. If only one data set is available, or other similar data sets are highly correlated with it, the quantification of variability can be somewhat speculative. This is the case for long range financial data such as the Standard and Poor's index.

1.6 Approximation

It is the aim of this book to provide a framework for the theory and practice of statistics that at all times explicitly recognizes that models are an approximation to the data. There is only one mode of operation and it is the mode of data analysis. There is no 'behaving as if' mode and no separate inference mode. The concept of approximation is that of the model to the data. It is not of the model to some unknown true generating mechanism which gave rise to the data. It is a concept of approximation without truth, or in the words of D. W. Müller 'approximatio sine veritate'.

A first attempt at regarding models as approximations to data can be found in [36] and a later attempt in [39] with a discussion.

Chapter 2

A Concept of Approximation

2.1 Approximation

Chapter 5 of [117] is headed 'Approximate Models'. The following is taken from pages 90-91 of the chapter:

> The header of this chapter is an intensional pleonasm: by definition, a model is not an exact counterpart of the real thing, but a judicious approximation. Mathematical statisticians, being concerned with the pure ideas, sometimes seem to forget this, despite strong admonitions to the contrary as by McCullagh and Nelder (1983, p.6): "all models are wrong". What is considered essential of course depends on the current viewpoint - the same thing may need to be modeled in various different ways.

Here McCullagh and Nelder (1983) = [152]. Later in Chapter 5.3 of [117] Huber writes

> In the opposite case, if a goodness-of-fit test does not reject a model, statisticians have become accustomed to act as if were true. Of course this is logically inadmissible, even more so if with McCullagh and Nelder one belives that all models are wrong *a priori*. ... Moreover, treating a model that has not been rejected as correct can be misleading and dangerous. Perhaps this is the main lesson we have learned from robustness.

If a model is wrong, then what is its relationship to the data? The only possible answer would seem to be that it is in some sense an approximation. If it is 'logically inadmissable' to treat a non-rejected model 'as if it were true' then there is a need to provide an account of statistical analysis which is based consistently on the concept of approximate models and not on the 'behave as it it were true' approach.

2.2 Approximating a Data Set by a Model

Given a data set $\mathbf{x}_n$ and a probability model P the model will be said to be an adequate approximation for the data if 'typical' data sets $\mathbf{X}_n(P)$ generated under P 'look like' the real data $\mathbf{x}_n$. The word 'typical' is specified by a number α whereby a proportion α of the data sets generated under the model will be regarded as typical data sets for that model. Standard values of α are 0.9, 0.95 and 0.99 but others, both smaller and larger, can be appropriate. The words 'look like' require the specification of a property or properties of a data set such that at least proportion α of data sets $\mathbf{X}_n(P)$ generated under the model exhibit the property or properties. When the real data set also exhibits this property or properties then it looks like typical data sets generated under the model. The following is taken from [211].

> **A is for Approximation**
>
> Davies's emphasis on approximation is well-chosen and surprisingly novel. While there will undoubtedly be a place for much careful work in learning how to describe the concept - - and its applications - - in detail, it is clear that Davies has taken the decisive step by asserting that there must be a formal admission that adequate approximation, of one set of observable (or simulated) values by another set, needs to be treated as practical identity.
>
> If, as is so convenient, we continue to use continuous models to describe - - or perhaps only to illuminate - - observed data, we should have to say that certain aspects of the data - - not typically, but unavoidably, including "Most

(modelled) observations have irrational values!"- - are not to be used in relating conceptual (or simulated) samples to observed samples. Thought and debate as to just which aspects are to be denied legitimacy will be both necessary and valuable.

In the quotation given above Tukey points out that data generated under for example a normal model, are irrational with probability one. In practice all numbers are of finite precision and rational, but typically the values of simulated data are given to a much higher degree of precision than that of the real data. This form of identification can be avoided by truncating the simulated samples to the same degree of precision as the real sample. This will be discussed more fully below.

Figure 2.1 is taken from thin-film physics [43] and shows 1000 values of the intensity of X-rays as measured by the number of photons as a function of the angle of diffraction as measured in degrees. The angles are equally spaced from 15.01 to 24.99. Superimposed is a kernel approximation with global bandwidth $h = 0.8$ and kernel $K(x) = (1-x^2)^3\{-1 \leq x \leq 1\}$. The model for the data is

$$X(t) = f(t) + 6.3Z(t), \quad 15.01 \leq t \leq 24.99. \tag{2.1}$$

where Z is standard Gaussian white noise. The value 6.3 is based on the data. For the purposes of the present discussion it is to be regarded as given as interest centres on the function f. Figure 2.2 shows data generated under (2.1) with f being the kernel approximation of Figure 2.1. The discrepancy is large. It is clear that out of the 1000 data sets, 999 generated under this specific model and the real data set, it would always be possible to pick out the real data set (see Chapter 1.4.1).

Figures 2.3 and 2.4 correspond respectively to Figures 2.1 and 2.2 but with bandwidth $h = 0.062$. It would now be difficult to pick out the real data sets amongst the 1000 with 999 generated under (2.1) with f being the kernel approximation of Figure 2.3. In this sense the second kernel approximation can be regarded as an adequate approximation to the data. It may be objected that the function of Figure 2.3, whilst being an adequate approximation, is nevertheless unsatisfactory because of the wriggles away from the peak. The way to overcome this objection is to regularize within the set of adequate functions. In this particular case one could look for a simplest function in terms of the number of local extreme values. The concept of regularization will play an important role in the following.

Choosing the typical data sets by inspection cannot be a basis for a concept of approximation even though such a check may be lead the statistician to accept or reject a model in particular cases. In general no matter how a statistician arrives at a model, maximum likelihood, Bayes or otherwise, it can always be checked against the real data by simulating under the model. A large, obvious and important discrepancy offers an opportunity for thought.

To make the concept precise it will have to be stated exactly which features of the data are to be included in the definition of the words 'look like'. One possibility is to imagine writing a computer programme with inputs $(\mathbf{x}_n, \alpha, P)$ and outputs 1 or 0 depending upon whether the model is an adequate approximation or not. The programme should be such that if the inputs are $(\mathbf{X}_n(P), \alpha, P)$ then the output is 1 in at least a proportion α of the cases.

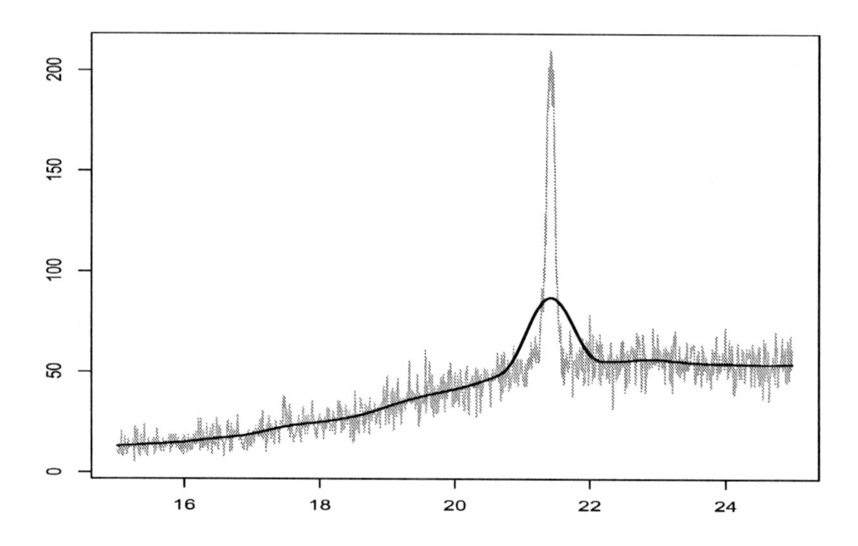

Figure 2.1 *Number of photons as a function of the angle of diffraction together with a kernel approximation with global bandwidth* $h = 0.8$.

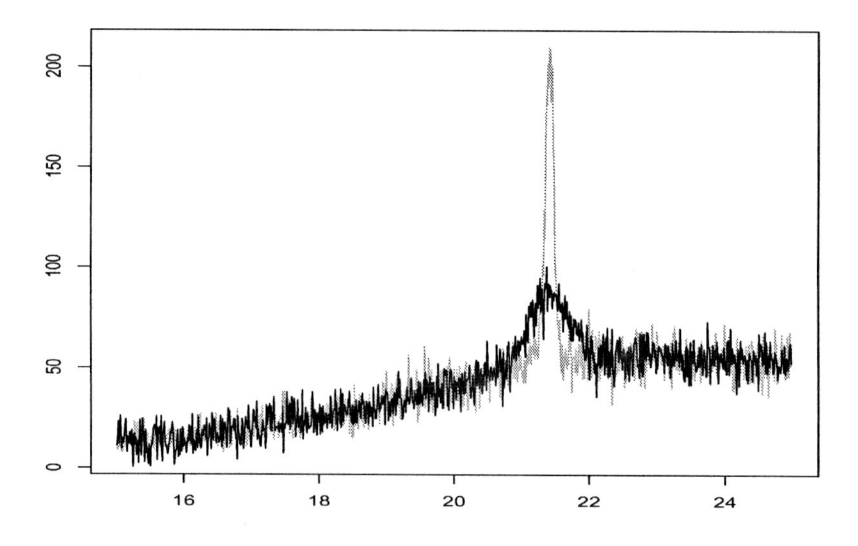

Figure 2.2 *The original data of Figure 2.1 together with data simulated under the model (2.1) with* f *being given by the kernel approximation of Figure 2.1.*

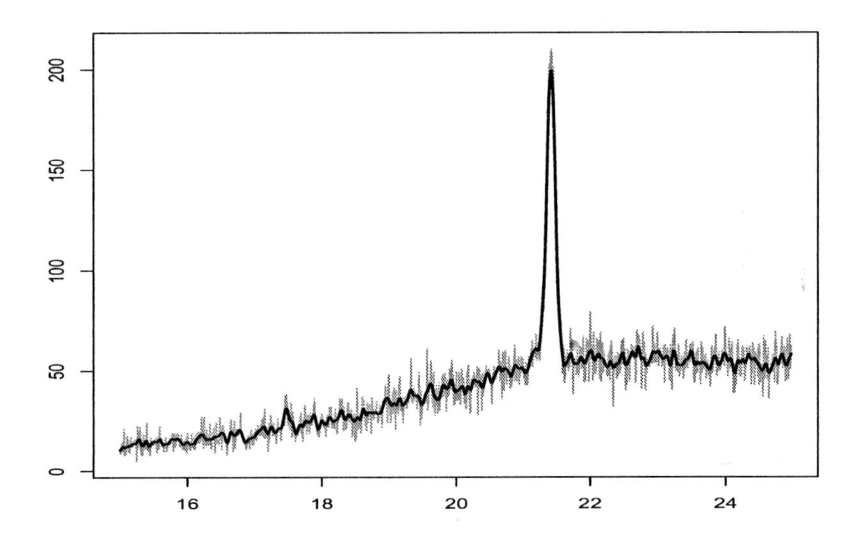

Figure 2.3 *As in Figure 2.1 but with a kernel approximation with bandwidth* $h = 0.062$.

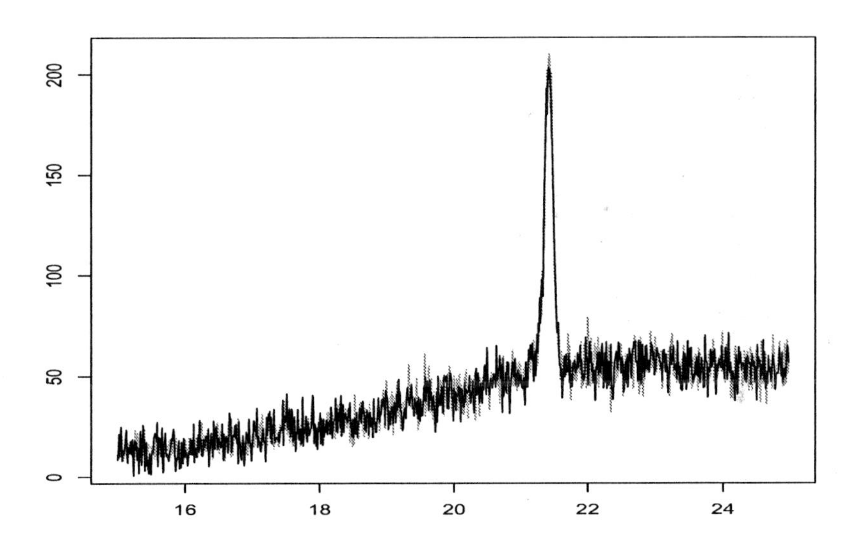

Figure 2.4 *As in Figure 2.2 but with the kernel approximation of Figure 2.3.*

A formal definition of approximation is the following. Given the data $\mathbf{x}_n$ and the model P the property or properties of interest are identified with a subset $E_n(\alpha,P) \subset \mathbb{R}^n$ such that

$$\mathbf{P}(\mathbf{X}_n(P) \in E_n(\alpha,P)) \geq \alpha\,. \tag{2.2}$$

The model is regarded as an adequate approximation for the data if $\mathbf{x}_n \in E_n(\alpha,P)$.

The data $\mathbf{x}_n$ are to be approximated as they are, the individual values and the given sample size n. In particular (2.2) is not to be replaced by asymptotic versions

$$\lim_{m\to\infty} \mathbf{P}(\mathbf{X}_m(P) \in E_m(\alpha,P)) \geq \alpha \tag{2.3}$$

in the hope that n is sufficiently large. Asymptotics can be used but only if they are sufficiently accurate for the given sample size n be it 10, 100, 1000 or 10000.

The event $\mathbf{x}_n \in E_n(\alpha,P)$ can be interpreted as a statistical test for the null hypothesis that the distribution of the data is P. This is not how it should be interpreted. Firstly such a null hypothesis is clearly wrong, no statistician really believes the data are exactly distributed as P. Given this it is misleading say that a hypothesis, known and accepted as being wrong no matter what the outcome of the test, is being tested. Secondly by accepting that the model is as at best an approximation, the statistician is free to choose the set $E_n(\alpha,P)$ based on the problem at hand. The model P cannot be allowed to dictate the form of $E_n(\alpha,P)$ for example, through optimality considerations.

2.3 Approximation Regions

The definition of approximation is given in terms of a data set $\mathbf{x}_n$ and a probability model P. In many situations interest will centre not on a single probability model P but on a family of such models $\mathscr{P}$. Given α and for each $P \in \mathscr{P}$ a region $E_n(\alpha,P)$, the approximation region $\mathscr{A}(\mathbf{x}_n,\alpha,\mathscr{P})$ is defined by

$$\mathscr{A}(\mathbf{x}_n,\alpha,\mathscr{P}) = \{P : P \in \mathscr{P},\ \mathbf{x}_n \in E_n(\alpha,P)\} \tag{2.4}$$

where (2.2) holds for all $P \in \mathscr{P}$. If $\mathscr{P}$ is a parametric family of models

$$\mathscr{P} = \mathscr{P}_\Theta = \{P_\theta : \theta \in \Theta\}$$

then

$$\mathscr{A}(\mathbf{x}_n,\alpha,\Theta) = \{\theta : \theta \in \Theta,\ \mathbf{x}_n \in E_n(\alpha,P_\theta)\} \tag{2.5}$$

is an honest [139] α-approximation region for the values of the parameter θ. It is called exact if equality holds in (2.2). If the data were generated under the model, for example in a simulation study, $\mathbf{x}_n = \mathbf{X}_n(P)$, then for all $P \in \mathscr{P}$

$$\mathbf{P}(P \in \mathscr{A}(\mathbf{X}_n(P),\alpha,\mathscr{P})) \geq \alpha. \tag{2.6}$$

If equality holds then $\mathscr{A}(\mathbf{X}_n(P),\alpha,\mathscr{P})$ is an exact α-confidence region for data generated by models P. It is always an honest [139] α-confidence region. Similarly $\mathscr{A}(\mathbf{X}_n(P_\theta),\alpha,\Theta)$ is an exact or honest α-confidence region for θ.

As an example consider the drinking water data of Exercise 1.3.3 where the family of models is taken to be

$$\mathfrak{P} = \{\mathfrak{N}(\mu, 0.0135) : \mu \in \mathbb{R}\}.$$

This has been already considered in Section 1.4.1.1. The arguments leading to (1.19) and (1.19) with the appropriate changes lead to

$$\begin{aligned} &E_{27}(0.95, \mu) \\ &= \left\{\mathbf{y}_{27} = (y_1, \ldots, y_{27}) : \mu - \frac{0.1162 \cdot 1.96}{\sqrt{27}} \le \bar{\mathbf{y}}_{27} \le \mu + \frac{0.1162 \cdot 1.96}{\sqrt{27}}\right\} \\ &= \{\mathbf{y}_{27} : \mu - 0.04383 \le \bar{\mathbf{y}}_{27} \le \mu + 0.04383\} \end{aligned}$$

and the approximation region

$$\begin{aligned} \mathscr{A}(\mathbf{x}_{27}, 0.95, \mathbb{R}) &= \{\mu : \mu \in \mathbb{R},\ \mu - 0.04383 \le 2.0159 \le \mu + 0.04383\} \\ &= [1.97207, 2.05973]. \end{aligned}$$

The second example is the length of study of German students as shown in Figure 1.6. The skewness of the data suggest a gamma, Weibull or lognormal model. Figure 2.5 shows the q-q-plots of the data for all three models, the $\Gamma(\gamma, \lambda)$ with shape parameter $\gamma = 18$ and scale parameter $\lambda = 0.75$ (see Figure1.8 for the boxplots), a translated Weibull $\mu + \mathfrak{W}(k, \lambda)$ with translation parameter $\mu = 6$, shape parameter $k = 2.60$ and scale parameter $\lambda = 8.89$ and a lognormal $\mathfrak{LN}(\mu, \sigma^2)$ with parameters $\mathbf{E}(\log(\mathfrak{LN}(\mu, \sigma^2))) = \mu = 2.61$, $\mathbf{V}(\log(\mathfrak{LN}(\mu, \sigma^2))) = \sigma^2 = 0.226$. The fits seem to be equally good for all three models and may be deemed acceptable. The problem now is to define an approximation region for the parameter values for all three models. A standard frequentist procedure would to be to estimate the parameters by maximum likelihood and derive confidence regions for the parameters perhaps using a normal approximation. The difficulty with this is that the parameters so calculated are difficult to compare. They are obtained by maximizing different functions over the parameter space and have no common interpretation.

In order to define an approximation region it has to be decided which features of the data are of interest. In the case of the student data this is length of study of the students which at the time the data were collected was judged to be much too long. The system has since been changed. The obvious and simplest measure to quantify the length of study is the mean. In practice the median is often used, possibly because it is smaller than the mean, 13.25 as against 13.89 for the data, which makes things look slightly better. Of lesser importance but also useful when summarizing the data is the standard deviation as a measure of spread. The parameters will then be chosen on the basis that data generated under that model will look like the original data in terms of the mean and standard deviation. If an α-approximation region is requires $(1-\alpha)/4$ of the total probability of $1-\alpha$ will be spent on the tails, left and right, of the mean and the standard deviation. For the gamma model this leads to the region

$$\begin{aligned} &E_n(\alpha, (\gamma, \lambda)) \\ &= \{\mathbf{y}_n : \text{qumn}((1-\alpha)/4, n, \gamma, \lambda) \le \bar{\mathbf{y}}_n \le \text{qumn}(3(1+\alpha)/4, n, \gamma, \lambda), \\ &\quad \text{qusd}((1-\alpha)/4, n, \gamma, \lambda) \le \text{sd}(\mathbf{y}_n) \le \text{qusd}(3(1+\alpha)/4, n, \gamma, \lambda)\} \qquad (2.7) \end{aligned}$$

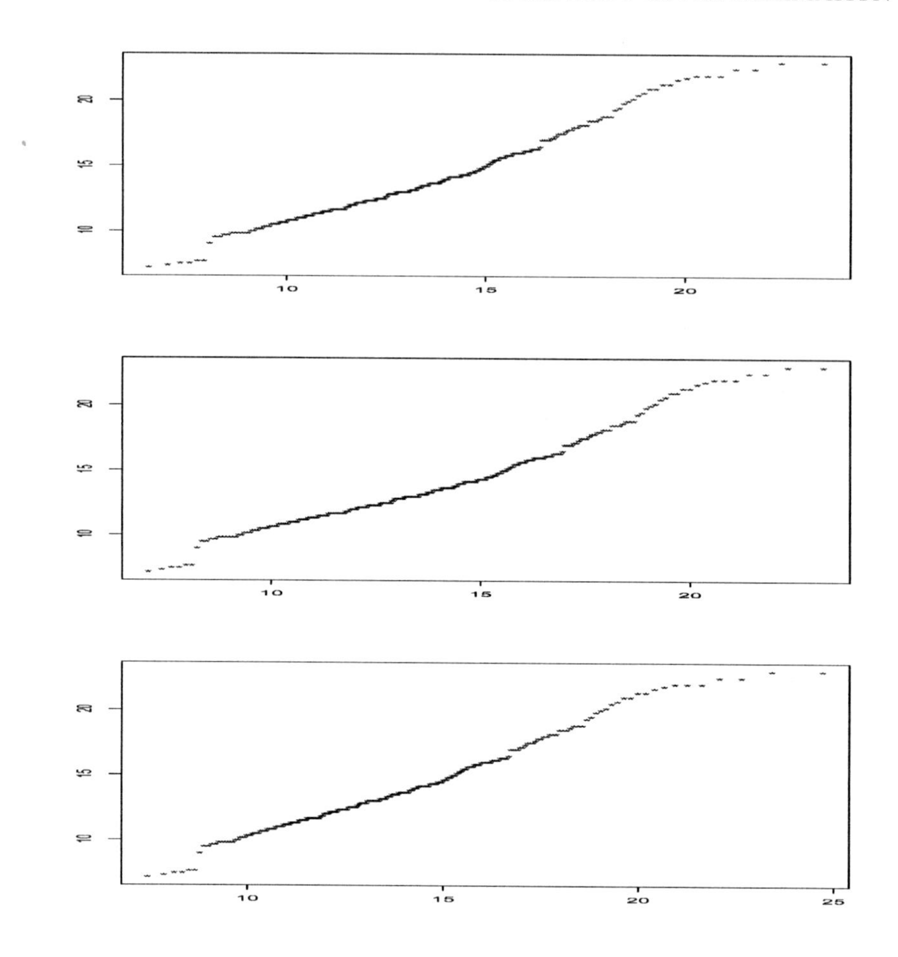

Figure 2.5 *Top: the q-q-plot of the student data Figure 1.6 and the* $\Gamma(18,0.75)$ *quantiles. Centre: The same for the* $6+\mathfrak{W}(2.60,8.89)$ *distribution. Bottom: The same for the* $\mathfrak{LN}(2.61,0.226)$ *distribution.*

where qumn$(\alpha,n,\gamma,\lambda)$ and qusd$(\alpha,n,\gamma,\lambda)$ are the α-quantiles of the mean and standard deviation respectively for gamma samples of size n with parameter values (γ,λ). For any α these can be obtained in principle by simulations. The use of simulations to obtain simple expressions for quantiles is discussed in Chapter 2.9 .

For $\alpha = 0.95$ the resulting approximation region for the values of (γ,λ) for the data of Figure 1.6 is shown in the upper panel of Figure 2.6. The lower panel shows the corresponding values of the mean and standard deviation. As mentioned above the parameter values based on maximum likelihood together with normal approximations can be used to derive a so called confidence region. In general however there is no guarantee that the parameter values so obtained will give sample means and standard deviations comparable to those of the data. One example of such a model

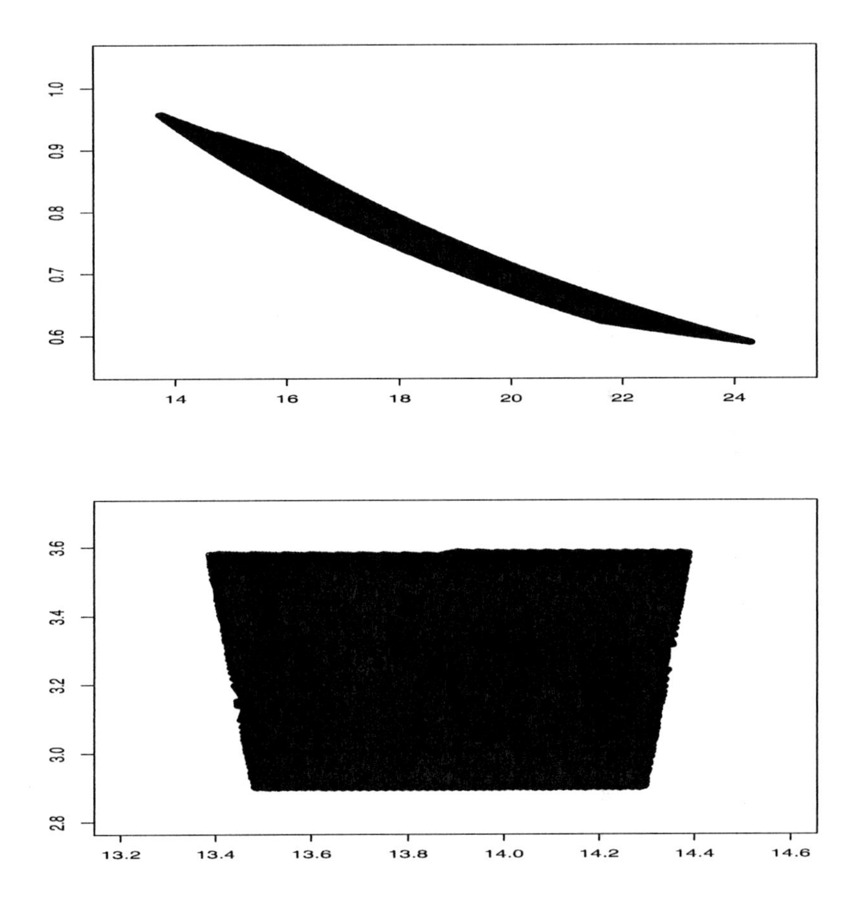

Figure 2.6 *Upper panel: the* 0.95 *approximation for the parameters* (γ, λ) *of the gamma distribution for the data of Figure 1.6 based on the mean and standard deviation of the sample. Lower panel: the corresponding* 0.95 *approximation region for the mean and standard deviation,*

is the comb model for the data of Table 1.3. The maximum likelihood value for the standard deviation is 0.192 as against a value of 0.116 for the data.

The third example is the photon data of Figure 2.1. The model is that of (2.1) where the only free parameter is the function f. Under the model the residuals are white noise. This suggests accepting a function g as being an adequate approximation for the data $(\mathbf{t}, \mathbf{x})_n$ if the residuals $x(t_j) - g(t_j), j = 1, \ldots, n$ 'look like' Gaussian white noise.The residuals for the two kernel estimators are shown in Figure 2.7. The residuals of the upper panel (bandwidth $h = 0.8$) of Figure 2.7 look like white noise for the most part, but with the clear exception of the residuals for the measurements corresponding the angles of between 21 and 22 degrees. The residuals of the lower

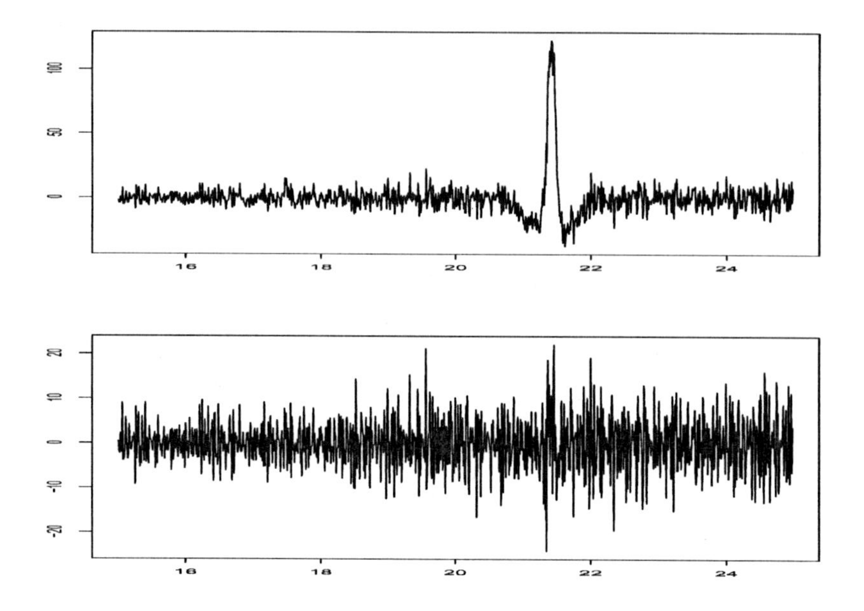

Figure 2.7 *Upper panel: Residuals for the kernel estimator with bandwidth* $h = 0.8$. *Lower panel: Residuals for the kernel estimator with bandwidth* $h = 0.062$.

panel (bandwidth $h = 0.062$) look like white noise apart from an increase in variance as the angle increases which will be ignored. This suggests a multiscale approach whereby the appropriateness of the white noise approximation is considered over all intervals $[t_j, t_k], t_j \le t_k$. Section 1.1 of [25] gives a short summary of the change in data processing through the use of multiscale methods. The authors cite amongst others [85], [24] and [97]. Multiscale techniques are closely connected with wavelets (see [62], [59], [63], [60]) and although the following makes no use of wavelets it is based on the multiscale paradigm.

Under the model

$$\frac{1}{\sqrt{|I|}} \sum_{t_i \in I} (X(t_i) - f(t_i)) = \frac{1}{\sqrt{|I|}} \sum_{t_i \in I} Z(t_i) \sim \mathfrak{N}(0, 6.3^2) \tag{2.8}$$

for any interval $I \subset [15, 25]$ where $|I|$ denotes the number of points $t_i \in I$. For any given α there exists a $\tau(\alpha, n)$ such that

$$\mathbf{P}\left(\max_{I \subset [15,25]} \frac{1}{\sqrt{|I|}} \left| \sum_{t_i \in I} Z(t_i) \right| \le \sqrt{\tau(\alpha, n) \log n} \right) = \alpha \,. \tag{2.9}$$

A typical data set $\mathbf{X}_n(P_f) = \mathbf{X}_n = (X(t_1), \ldots, X(t_n))$ therefore satisfies

$$\mathbf{P}\left(\max_{I \subset [15,25]} \frac{1}{\sqrt{|I|}} \left| \sum_{t_i \in I} (X(t_i) - f(t_i)) \right| \le 6.3\sqrt{\tau(\alpha, n) \log n} \right) = \alpha \,.$$

which leads to the approximation region

$$\mathscr{A}(\mathbf{x}_n,\alpha) = \left\{ g \in \mathscr{G} : \max_{I \subset [15,25]} \frac{1}{\sqrt{|I|}} \left| \sum_{t_i \in I} (x(t_i) - g(t_i)) \right| \le 6.3\sqrt{\tau(\alpha,n)\log n} \right\} \tag{2.10}$$

with

$$\mathscr{G} = \{g : g : [15,25] \to [0,\infty)\}.$$

Simulations give $\tau(0.9, 1000) \approx 2.72$ with a resulting upper bound of

$$6.3\sqrt{\tau(0.9,1000)\log 1000} = 27.31. \tag{2.11}$$

A computer programme can now be written with inputs the data $\mathbf{x}_n$, the design points t_i, a value for α and a function g. The programme simply checks the inequalities defining the approximation region. The function g is an adequate approximation for the data if and only if all the inequalities defining the approximation region are satisfied. For $\alpha = 0.9$ and the kernel estimator $g_{0.8}$ of Figure 2.1 with bandwidth $h = 0.8$

$$\max_{I \subset [15,25]} \frac{1}{\sqrt{|I|}} \left| \sum_{t_i \in I} (x(t_i) - g_{0.8}(t_i)) \right| = 363.5$$

with the maximum attained on the interval $I = [21.36, 21.48]$ which contains the peak. Thus $g_{0.8}$ is not an adequate approximation. For the kernel estimator $g_{0.062}$ of Figure 2.3

$$\max_{I \subset [15,25]} \frac{1}{\sqrt{|I|}} \left| \sum_{t_i \in I} (x(t_i) - g_{0.062}(t_i)) \right| = 27.05$$

which is just below the allowable upper bound of 27.31. This is of course how the bandwidth was determined.

2.4 Functionals and Equivariance

2.4.1 *Functionals*

If interest centres on a function h of the parameters of a parametric model $\mathscr{P}_\Theta$, for example $h(\mu,\sigma) = \mu/\sigma$ for the Gauss model, then the α-approximation region for the values of $h(\theta)$ is simply

$$\mathscr{A}(\mathbf{x}_n,\alpha,h(\Theta)) = \{h(\theta) : \theta \in \mathscr{A}(\mathbf{x}_n,\alpha,\Theta)\}.$$

In particular suppose the parametric family of distributions $\mathscr{P}_\Theta$ is re-parametrized to give a parametric family $\mathscr{P}_\Phi$ with a one-to-one function $\pi : \Theta \to \Phi$. Then the α-approximation region for the values of φ is simply

$$\mathscr{A}(\mathbf{x}_n,\alpha,\Phi) = \{\pi(\theta) : \theta \in \mathscr{A}(\mathbf{x}_n,\alpha,\Theta)\}.$$

More generally the approximation region can be extended to statistical functionals [215, 216], that is, functions defined on a space of probability measures. Let $\mathscr{P}_T$

be the family of distributions for which T is well-defined. For any $\mathscr{P} \subset \mathscr{P}_T$ the α-approximation region for the values of T is simply

$$\mathscr{A}(\mathbf{x}_n, \alpha, T(\mathscr{P})) = \{T(P) : P \in \mathscr{A}(\mathbf{x}_n, \alpha, \mathscr{P})\}.$$

As an example suppose $T = T_{\text{ave}}$ is the mean, that is

$$T_{\text{ave}}(P) = \int x \, dP(x) \tag{2.12}$$

and

$$\mathscr{P}_{T_{\text{ave}}} = \left\{P : \int |x| \, dP(x) < \infty \right\}.$$

If $\mathscr{P} \subset \mathscr{P}_{T_{\text{ave}}}$ the approximation region for the means of the models in $\mathscr{P}$ is

$$\mathscr{A}(\mathbf{x}_n, \alpha, T_{\text{ave}}(\mathscr{P})) = \{T_{\text{ave}}(P) : P \in \mathscr{A}(\mathbf{x}_n, \alpha, \mathscr{P})\}.$$

The standard deviation functional T_{sd} is defined for all P with

$$\int x^2 \, dP(x) < \infty \tag{2.13}$$

by

$$T_{\text{sd}}(P) = \sqrt{\int x^2 \, dP(x) - T_{\text{ave}}(P)^2}. \tag{2.14}$$

On putting $T(P) = (T_{\text{ave}}(P), T_{\text{sd}}(P))$ the approximation region is

$$\mathscr{A}(\mathbf{x}_n, \alpha, T(\mathscr{P})) = \{(T_{\text{ave}}(P), T_{\text{sd}}(P)) : P \in \mathscr{A}(\mathbf{x}_n, \alpha, \mathscr{P})\}$$

where now (2.13) holds for all $P \in \mathscr{P}$.

2.4.2 *Equivariance*

Often attention will be restricted to a class of functionals which are equivariant with respect to some group $\mathscr{G}$ operating on $\mathbb{R}^k$. The most important case for $k = 1$ is the group $\mathscr{A}$ of affine transformations $\alpha(x) = ax + b$ with $a \neq 0$. It may be thought of as a change of measurement units. A functional T is a location functional if

$$T(P^{\alpha}) = \alpha(T(P)) \text{ for all } P \text{ and for all } \alpha \in \mathscr{A} \tag{2.15}$$

and a scale functional if

$$T(P^{\alpha}) = |a| T(P) \text{ for all } P \text{ and for all } \alpha \in \mathscr{A} \tag{2.16}$$

where

$$P^{A}(B) = P(A^{-1}(B)) \text{ for all (Borel) sets } B.$$

All standard measures of location and scale are equivariant in the above sense.

In $k \geq 2$ dimensions an affine transformation α is one of the form $\alpha(x) = Ax + b$ where A is a $k \times k$ non-singular matrix and $b \in \mathbb{R}^k$. A functional T_L is a location functional

$$T_L : \mathscr{P} \to \mathbb{R}^k$$

if

$$T(P^\alpha) = \alpha(T(P)) \text{ for all } P \text{ and for all } \alpha. \tag{2.17}$$

A scale functional T_S takes values in the space of symmetric positive definite $k \times k$ matrices

$$T_S : \mathscr{P} \to \mathscr{S}_+(k)$$

and satisfies the equivariance condition

$$T(P^\alpha) = AT(P)A^t \text{ for all } P \text{ and for all } \alpha, \quad \alpha(x) = Ax + b. \tag{2.18}$$

This differs from the one-dimensional case where the value is σ rather than σ^2.

A justification for affine equivariance in $k \geq 2$ dimensions is not obvious as the concept in $\mathbb{R}^k$ goes beyond a change of units. The latter would require equivariance only for diagonal matrices A. In [118] the authors state on page 200:

> However, one ought to be aware that affine equivariance is a requirement deriving from mathematical aesthetics; it is hardly ever dictated by the scientific content of the underlying problem.

The components of k-dimensional real data may have different units length, temperature and tension for example but they are easily understood and interpreted irrespective of the units in which they are expressed. An arbitrary linear combination of the components can make no such claim although a particular linear combination may have some scientific rationale: the length of a rod of steel may depend linearly on the temperature and the tension, at least to a first approximation. Such a particular and accidental affine transformation may not be a priori known and cannot by itself be an argument for affine equivariance.

Just how strong the requirement in $k \geq 2$ dimensions is can be seen from the following. In one dimension the median as defined by (2.54) is affine equivariant and a continuous function of the measurements $\mathbf{x}_n$. If T_L is a location functional in $k \geq 2$ dimensions, that is, it satisfies (2.17), and is, in addition, a continuous function of the observations $\mathbf{x}_n$ then T_L is the mean or a constant multiple of the mean. The result is due to Rousseeuw and the proof can be found in [178].

Nevertheless the following will follow [118] and require full affine equivariance for location and scale functions in $\mathbb{R}^k$.

2.5 Regularization and Optimality

Regularization was already discussed in Chapter 1.3.6 in the context of the location-scale problem but it is included again because of its importance. There it was shown that the comb model was consistent with the copper data but when combined with optimality lead to a ridiculous conclusion. This is typical of many optimal procedures as pointed out in [116]. Nearly all statistical problems are ill posed and have to be regularized. In Chapter 1.5 it was pointed out that the 'best' approximation model is that which simply reproduces the data, but such a choice is inconsistent with regularization.

The form of regularization depends on the problem at hand and there is no one simple overriding concept which can be appealed to. In the case of non-parametric regression the function f can be regularized by the number of peaks, smoothness or the number of intervals of constancy depending on the context. The regularization need not be done with respect to the model but with respect to some functionals. Thus if the approximation region for the location and scale parameters of the comb model is based on the mean and standard deviation rather than the maximum likelihood values then it will differ little from the corresponding region for the normal model. Figure 1.4 shows the regions if the Kuiper metric is used. Regularization may then be a form of compromise (see [211] for the development of a milk bottle), in particular to guarantee stability of analysis. In general regularization coupled with optimization will not lead to an acceptable result. This is true even in the case of the normal distribution and the mean and standard deviation as demonstrated by Tukey (See Example 1.1 of [118]). Nevertheless regularization combined with optimization is useful, not as a recipe to be followed, but as a border post delimiting the possible [116]. This will be treated in more detail in Chapter 5 in the context of M-functionals where considerations of efficiency must be tempered, and to some extent overruled, by considerations of stability.

2.6 Metrics and Discrepancies

2.6.1 Some Examples

There are many metrics and discrepancies on spaces of probability distributions which are used in probability theory and statistics. Their use in statistics is largely confined to theoretical investigations, in particular to the area of robust statistics where perturbations of models and data can be quantified using metrics (see [118] and [35]). They are also of some importance in applied statistics in the context of goodness-of-fit tests.

The best known metric on the space of probability distributions on $\mathbb{R}$ is the Kolmogorov metric (1.4). It may also be written as

$$d_{\text{ko}}(P,Q) := \sup\{|P(I) - Q(I)| : I = (-\infty,x], x \in \mathbb{R}\} \tag{2.19}$$

for probability measures P and Q.

The central limit theorem can be written in terms of the Kolomogorov metric as follows. Le $X_i, i = 1,2,\ldots$ be independently and identically distributed random variables with mean μ and finite non-zero variance σ^2. The central limit theorem states

$$\lim_{n\to\infty} \mathbf{P}\left(\sqrt{n}\left(\frac{\bar{X}_n - \mu}{\sigma}\right) \le x\right) = \Phi(x) \tag{2.20}$$

where $\bar{X}_n$ denotes the mean of the first n observations and Φ is the distribution function of the standard normal distribution. It can be shown that the convergence in

(2.20) is uniform in x and hence

$$\lim_{n\to\infty} \sup_x |F_n(x) - \Phi(x)| = 0$$

where F_n is the distribution function of $\sqrt{n}\,(\bar{X}_n - \mu)/\sigma$. This is nothing more than

$$\lim_{n\to\infty} d_{\mathrm{ko}}(P_n, \mathfrak{N}(0,1)) = 0 \tag{2.21}$$

where P_n is the distribution of $\sqrt{n}\,(\bar{X}_n - \mu)/\sigma$.

Let $X_1, \ldots, X_n$ be independently distributed random variables with a common distribution P and empirical distribution $\mathbb{P}_n(P)$. For continuous P the distribution of $d_{\mathrm{ko}}(\mathbb{P}_n(P), P)$ does not depend on P: for general P the distribution of $d_{\mathrm{ko}}(\mathbb{P}_n, P)$ is smaller in distribution than for continuous P. The following inequality holds

$$\sup_P \mathbf{P}(\sqrt{n}\, d_{\mathrm{ko}}(\mathbb{P}_n(P), P) \geq x) \leq \mathbf{2}\exp(-2x^2) \tag{2.22}$$

where **2** is the best possible constant ([151]). For continuous P the asymptotic distribution is given by

$$\lim_{n\to\infty} \mathbf{P}(\sqrt{n}\, d_{\mathrm{ko}}(\mathbb{P}_n(P), P) \geq x) = \mathbf{P}(\max_{0\leq t\leq 1} |B_0(t)| \geq x) = 2\sum_{m=1}^{\infty} (-1)^{m-1} \exp(-2m^2x^2) \tag{2.23}$$

where B_0 is a Brownian bridge (see for example [65]). The convergence to the asymptotic distribution is fast.

Another useful metric is the Kuiper metric already defined by (1.12) but is given once again for ease of reference

$$d_{\mathrm{ku}}(P,Q) := \sup\{|P(I) - Q(I)| : I \text{ an interval}\}.$$

It is a simple generalization of the Kolmogorov metric but, on the basis of some limited experience, it seems to be more useful than the latter. It is seen immediately that

$$d_{\mathrm{ko}}(P,Q) \leq d_{\mathrm{ku}}(P,Q) \leq 2d_{\mathrm{ko}}(P,Q).$$

Again for continuous P the distribution of $d_{\mathrm{ku}}(\mathbb{P}_n(P), P)$ does not depend on P. The asymptotics are given by (see for example [65])

$$\begin{aligned} \lim_{n\to\infty} \mathbf{P}(\sqrt{n} d_{\mathrm{ku}}(\mathbb{P}_n(P), P) \geq x) &= \mathbf{P}(\max_{0\leq t\leq 1} B_0(t) - \min_{0\leq t\leq 1} B_0(t) \geq x) \\ &= 2\sum_{m=1}^{\infty} (4m^2x^2 - 1)\exp(-2m^2x^2). \end{aligned} \tag{2.24}$$

The Kuiper metric itself can be generalized to give Kuiper metrics d_{ki}^k of order k defined by

$$d_{\mathrm{ku}}^k(P,Q) := \sup\left\{\sum_{i=1}^{k} |P(I_i) - Q(I_i)| : I_1, \ldots, I_k \text{ disjoint intervals}\right\}. \tag{2.25}$$

An application of higher Kuiper metrics is given in [47] where they proved useful in controlling the modality of probability densities.

Another metric based directly on intervals is the Lévy metric d_{le} [138] defined by

$$d_{\mathrm{le}}(P,Q) = \inf\{\varepsilon > 0 : F_P(x-\varepsilon) - \varepsilon \le F_Q(x) \le F_P(x+\varepsilon) + \varepsilon, \text{ for all } x \in \mathbb{R}\}. \quad (2.26)$$

More generally the Prohorov metric ([174]) d_{pr} is defined by

$$d_{\mathrm{pr}}(P,Q) = \inf\{\varepsilon : P(B) \le Q(B^{\varepsilon}) + \varepsilon \text{ for all } B, B \text{ Borel}\} \quad (2.27)$$

where

$$B^{\varepsilon} = \{x : \|x-y\| < \varepsilon \text{ for some } y \in B\}.$$

Replacing the intervals $I = (-\infty, x]$ in (2.19) by the set of all Borel subsets of $\mathbb{R}$ gives rise to the total variation metric d_{tv} defined by

$$d_{\mathrm{tv}}(P,Q) := \sup_{B \in \mathfrak{B}(\mathbb{R})} |P(B) - Q(B)|. \quad (2.28)$$

In contrast to the Kolmogorov and Kuiper metrics the total variation metric does not yield any information of the closeness of the empirical distribution $\mathbb{P}_n(P)$ to P for continuous P. It is seen that

$$d_{\mathrm{tv}}(\mathbb{P}_n(P), P) = 1 \text{ for } P \text{ continuous}. \quad (2.29)$$

A general class of metrics on the space of probability measures on $\mathfrak{B}(\mathbb{R}^k)$ can be defined by

$$d_{\mathscr{C}}(P,Q) = \sup\{|P(C) - Q(C)| : C \in \mathscr{C} \subset \mathfrak{B}(\mathbb{R}^k)\}. \quad (2.30)$$

If $\mathscr{C} = \mathfrak{B}(\mathbb{R}^k)$ then $d_{\mathscr{C}}$ is simply the total variation metric and again (2.29) holds. If $\mathscr{C}$ is a Vapnik-Cervonenkis class of subsets, or equivalently, class of sets of polynomial discrimination ([172, 214]) then direct non-trivial comparisons of $\mathbb{P}_n(P)$ and P are possible. One important example is when $\mathscr{C}$ is the set of half-spaces $\mathscr{H}$:

$$\mathscr{H} = \{H : H = \{x : a^t x \le b\}, a \in \mathbb{R}^k, b \in \mathbb{R}\}, \quad (2.31)$$

$$d_{\mathscr{H}}(P,Q) = \sup\{|P(H) - Q(H)| : H \in \mathscr{H}\}. \quad (2.32)$$

For this it can be shown that

$$\mathbf{P}(d_{\mathscr{H}}(\mathbb{P}_n(P), P) \ge \lambda) \le 8 \exp(4\lambda + 4\lambda^2) n^{2k} \exp(-2n\lambda^2) \quad (2.33)$$

uniformly in P and λ (see [55]). In contrast to the one-dimensional case when $d_{\mathscr{H}} = d_{\mathrm{ko}}$ the distribution of $d_{\mathscr{H}}(\mathbb{P}_n(P), P)$ depends on P even if P is continuous. It can be shown that $\sqrt{n}\, d_{\mathscr{H}}(\mathbb{P}(P), P)$ converges to a non-degenerate limit but even this depends on P. Fortunately it turns out that a bootstrap procedure is numerically feasible and gives the correct asymptotic level whatever P may be [14].

Some metrics are defined in terms of densities. Given a dominating measure π, which in practice is either the counting measure ν or Lebesgue measure λ, the class of L_p metrics for $p > 0$ is defined by

$$d_p(Q_1, Q_2) = \left(\int |q_1 - q_2|^p \, d\pi \right)^{1/p} \tag{2.34}$$

for all Q_1 and Q_2 which are absolutely continuous with respect to π and for which the integral is finite. This is always the case for $p = 1$ which is equivalent to the total variation metric, more precisely

$$d_1 = 2d_{\mathrm{tv}}\,. \tag{2.35}$$

The L_1 metric with dominating measure the Lebesgue measure was already defined by (1.5). For the other values of p the integral may be infinite.

The Hellinger metric d_{he} is also defined in terms of densities and is given by

$$d_{\mathrm{he}}(Q_1, Q_2) = \left(\int |\sqrt{q_1} - \sqrt{q_2}|^2 \, d\pi \right)^{1/2}. \tag{2.36}$$

Two discrepancies which are commonly used in statistics are the Kullback-Leibler discrepancy

$$d_{\mathrm{kl}}(Q_1, Q_2) = \int q_1 \log(q_1/q_2) \, d\pi \tag{2.37}$$

and a grouped chi-squared discrepancy

$$d_{\chi^2}(Q_1, Q_2, \mathscr{B}) = \sum_{i=1}^{k} \frac{(Q_1(B_i) - Q_2(B_i))^2}{Q_1(B_i)}\,. \tag{2.38}$$

where $\mathscr{B} = \{B_1, \ldots, B_k\}$ is a partition of $\mathbb{R}$. If Q_1 and Q_2 are discrete distributions on say $\mathbb{N}_0$ (2.38) can be replaced by

$$d_{\chi^2}(Q_1, Q_2) = \sum_{i=0}^{\infty} \frac{(q_1(i) - q_2(i))^2}{q_2(i)}\,. \tag{2.39}$$

Both divergences upweight small probabilities $q_2(i)$. If the support of Q_2 is not compact the value of both may be dominated by one single term with a small $q_2(i)$.

2.6.2 *Invariance Considerations and Weak Convergence*

2.6.2.1 *Affine Invariance*

Affine equivariance was discussed in Chapter 2.4.2. In dimension $k = 1$ an affine transformation may be thought of as a change of units. Given an empirical distribution $\mathbb{P}_n$ and a model P they become $\mathbb{P}_n^\alpha$ and P^α after a change of units described by the affine transformation α. It seems reasonable that the distance between $\mathbb{P}_n$ and P should not be altered by this transformation. In other words the metric d used should satisfy $d(\mathbb{P}_n, P) = d(\mathbb{P}_n^\alpha, P^\alpha)$ for all affine transformations α. More generally this is required to hold for all measures P and Q and all affine transformations α

$$d(Q,P) = d(Q^\alpha, P^\alpha). \tag{2.40}$$

Metrics which satisfy (2.40) are called affinely invariant metrics.

In $k \geq 2$ dimensions affine equivariance for functionals is more difficult to justify (see Chapter 2.4.2). The same applies to affine invariance of metrics (2.40). Nevertheless it will be adopted here. Examples of affinely invariant metrics are the total variation metric and metrics based on Vapnik-Cervonenkis classes metrics (2.30) for which the set $\mathscr{C}' = \mathscr{C} \cup (\mathbb{R}^k \setminus \mathscr{C})$ is affinely invariant, that is $\alpha(\mathscr{C}') = \mathscr{C}'$ or for all affine transformations α. An example is the half-space metric $d_{\mathscr{H}}$ of (2.32).

An example of a metric which is not affinely invariant is the Prohorov metric d_{pr} of (2.27). The reason for this is that ε plays two completely different roles in (2.27). In the context of B^ε it plays the role of a measurement error which has units. In the role $+\varepsilon$ it is a dimensionless quantity, namely a probability. A physicist would never conflate the two. Unfortunately statisticians seem to have no such inhibitions, many are probably not even aware of the problem. A version of the Prohorov metric where the absolute error in B^ε is replaced by a proportional error is given in [35]. With this definition ε is always dimensionless and the resulting metric is affinely invariant.

2.6.2.2 *Weak Convergence*

A sequence P_n of probability measures converges weakly to a probability measure P if

$$\int u(x)\,dP_n(x) \rightarrow \int u(x)\,dP(x) \tag{2.41}$$

for all bounded and continuous functions u. This is written $P_n \Rightarrow P$. A metric d is said to metricize weak convergence if $P_n \Rightarrow P$ if and only if $d(P_n, P) \rightarrow 0$.

If d is an affinely invariant metric on the space $\mathscr{P}$ of probability distributions over $\mathfrak{B}(\mathbb{R})$ then it can be shown that $(\mathscr{P}, d)$ is a non-separable metric space and d does not metricize weak convergence. In contrast the Prohorov and Lévy metrics do metricize weak convergence. There is therefore a conflict between affine invariance and the wish for the metric to topologize weak convergence. On page 412 of [103] and in Chapter 2 of [118] more importance is attached to weak convergence than in this book where it will be essentially ignored. An example of a sequence of probability distributions P_n which converge weakly to a distribution P but which do not converge with respect to any affinely invariant metric is the following. Put

$$P_n = \delta_{1/n}, \quad P = \delta_0$$

where δ_x is the Dirac measure with unit mass in x. This example is constructed to give a counter example. It can be interpreted as counting $1/n$ instead of 0. However in counting there are rarely measurement errors which are not integer valued. There seem to be no relevant example where weak convergence cannot be expressed using an affine invariant metric. By far the most important result describing weak convergence, the central limit theorem, is almost always stated in terms of the affinely invariant Kolmogorov metric as in (2.21). For these reasons priority is given to affine invariance.

2.7 Strong and Weak Topologies

This topic was considered in Chapter 1.3.8.1. On a more analytical level let $\mathscr{P}$ denote the set of all probability measure over $\mathfrak{B}(\mathbb{R})$. Both $(\mathscr{P}, d_{\text{ko}})$ and $(\mathscr{P}, d_{\text{tv}})$ are metric spaces and on writing $\mathscr{O}_{\text{ko}}$ and $\mathscr{O}_{\text{tv}}$ for the open sets in $(\mathscr{P}, d_{\text{ko}})$ and $(\mathscr{P}, d_{\text{tv}})$ respectively it is seen that

$$\mathscr{O}_{\text{ko}} \subset \mathscr{O}_{\text{tv}}, \quad \mathscr{O}_{\text{ko}} \neq \mathscr{O}_{\text{tv}}. \tag{2.42}$$

In particular for every $\varepsilon > 0$ it is possible to find probability measures P and Q with $d_{\text{ko}}(P,Q) < \varepsilon$ and $d_{\text{tv}}(P,Q) = 1$. In the language of topology (2.42) states that the topology of $(\mathscr{P}, d_{\text{ko}})$ is (strictly) weaker than that of $(\mathscr{P}, d_{\text{tv}})$. Other weak metrics are the Kuiper metric d_{ku} of (1.12), the generalized Kuiper metric d_{ku}^k of (2.25), metrics based on a Vapnik-Cervonenkis class $\mathscr{C}$ as in (2.30) and the Lévy metric d_{le} of (2.26). Strong metrics are the L_p metrics of (2.34), the Hellinger metric d_{he} of (2.36), the Prohorov metric d_{pr} of (2.27) and, although not a metric, the Kullback-Leibler discrepancy $d_{\text{kl}}(P,Q))$ of (2.37).

One important property of a weak metric is that it allows a non-trivial comparison of the empirical distribution $\mathbb{P}_n$ of a data set with a possible (continuous) model P based for example on

$$\mathbf{P}\left(d_{\text{ko}}\left(\mathbb{P}_n(P), P\right) \leq 1.36/\sqrt{n}\right) \approx 0.95.$$

The total variation metric and other strong metrics yield only the useless

$$d_{\text{tv}}\left(\mathbb{P}_n(P), P\right) = 1.$$

As argued in Chapter 1.3.8.1 the formal inference mode is density (likelihood) based and its topology is that defined by the strong L_1 metric d_1 or, equivalently by the total variation metric d_{tv}. Consider the space $\mathscr{P}_{ac}$ of absolutely continuous probability measure and the space $\mathscr{F}$ of Lebesgue densities. These are the same space under different names but suppose they are metricized differently by considering $(\mathscr{P}_{ac}, d_{\text{ko}})$ and $(\mathscr{F}, d_1)$. The two spaces are connected by the invertible, unbounded linear differential operator D

$$D : (\mathscr{P}_{ac}, d_{\text{ko}}) \to (\mathscr{F}, d_1), \quad D(P) = f_P \tag{2.43}$$

where f_P is the density of P An unbounded linear operator is pathologically discontinuous. Based on this a rather pointed description of statistical practice is the following. A statistician operates in two modes. The first is an explorative data analytical mode using a weak topology followed by a formal density based inference mode using a strong topology. The two modes are connected by the pathologically discontinuous linear differential operator.

A mathematician or a numerical analyst when faced with an unbounded linear operator will regularize it, for example by using the Tikhonov regularization. A statistician when faced with an unbounded linear operator will in general also regularize it. This may be done explicitly as in deconvolution problems but very often it will be done implicitly by an appeal to common sense without a concrete acknowledgement of what is actually being done. In Chapter 1.3.6 it was done explicitly using Fisher information as a criterion for regularization.

Although the Kuiper metrics are weak and the total variation metric is strong there are distributions P and Q for which the total variation metric coincides with a Kuiper metric d_{ku}^k, that is $d_{\text{ko}}^k(P,Q) = d_{\text{tv}}(P,Q)$ for some k. Let P and Q be two probability distributions with 'simple' densities p and q respectively in the sense that there exist k_1 disjoint intervals $I_j^1, j = 1, \ldots, k_1$ such that

$$\{x : p(x) > q(x)\} = \cup_{j=1}^{k_1} I_j^1$$

and k_2 disjoint intervals $I_j^2, j = 1, \ldots, k_2$ such that

$$\{x : q(x) > p(x)\} = \cup_{j=1}^{k_2} I_j^2 .$$

Then with $k = k_1 + k_2$ it follows that $d_{\text{tv}}(P,Q) = d_{\text{ku}}^k(P,Q)$. As an example let P be the standard Cauchy distribution and Q the standard normal distribution. Then $k_1 = 2$ with $I_1^1 = (-\infty, -1.851), I_2^1 = (1.851, \infty)$ and $k_2 = 1$ with $I_1^2 = (-1.851, 1.851)$ (see Figure 2.8). Thus and $d_{\text{ku}}^3(P,Q) = 0.502 = d_{\text{tv}}(P,Q)$. Thus if $d_{\text{tv}}(P,Q)$ much larger than $d_{\text{ku}}^k(P,Q)$ for a small k then the set $\{x : |p(x) - q(x)| > 0\}$ must be complicated in that it cannot be well approximated by a small number of intervals.

2.8 On Being (almost) Honest

Figure 2.6 shows the parameters of the gamma distribution for which the gamma distribution adequately approximates the student study data shown in Figure 1.6 in the sense of the approximation region (2.7) based on the behaviour of the mean and standard deviation. The approximation region is not honest. Any data set with the same sample size, mean and standard deviation as the student data would result in the same approximation region for the parameters of the gamma distribution irrespective of the shape of the data.

The choice of a class of models, gamma, lognormal, Weibull, is initially based on the shape of the data, skewed to the right. The question arises as how this decision can be explicitly incorporated in the analysis. A Bayesian 'solution' would be to include all 'reasonable' models and then to base the decision on the relative likelihoods. A frequentist approach would be to perform a goodness-of-fit test for several possible models. Once a parametric family of models has passed the test, however narrowly, the result is treated as if the test had never been performed, or if this is taken into account, the p-value plays no further role in the analysis.

The problem of how to incorporate the result of a goodness-of-fit test into the statistical analysis was discussed in Chapter 1.3.8.2 (see Figure 1.4). There the suggestion was made to specify all those parameter values $\Theta(\mathbf{x}_n) \subset \Theta$ for which the

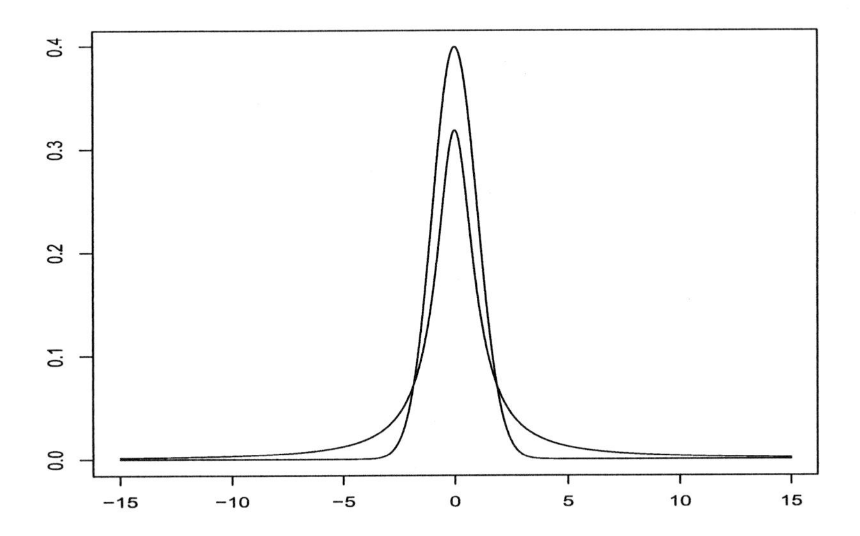

Figure 2.8 *The standard normal and Cauchy densities for which* $d_{tv}(\mathfrak{N}(0,1),\mathfrak{C}(0,1)) = d^3_{ku}(\mathfrak{N}(0,1),\mathfrak{C}(0,1)) = 0.502.$

models P_θ are consistent with the data in the sense of the Kuiper metric. In general the choice of the goodness-of-fit measure is will depend on the problem at hand, the use to which the model is to be put, on whether the data are discrete or not and whether all intervals or probabilities are to be given the same weight. The student data of Figure 1.6 will be used as an example.

It must be decided whether all students are to be treated equally or whether more weight is to be given to the right tail. Another problem is that the data are discrete whereas the model is continuous. The discreteness could be ignored, compensated for by the linearization of the empirical distribution function or explicitly recognized and a discrete measure such as total variation

$$d_{\mathrm{tv}}(P,Q) = \frac{1}{2}\sum_{j=0}^{\infty} |p(j) - q(j)| \tag{2.44}$$

used where the data generated under the model are discretized to the same precision as the real data. Once a goodness-of-fit measure has been chosen, it can be incorporated into the definition of adequacy. In the following the Kuiper metric will be used with no discretization, the effect of which is only minor for the student data.

The 0.95-approximation region of Figure 2.6 is based on spending 0.0125 of probability on the upper and lower tails of the mean and standard deviation. Figure 2.9 shows the corresponding 0.95 approximation region including a goodness-of-fit feature based on the Kuiper metric of order one. Of the available 0.05 probability

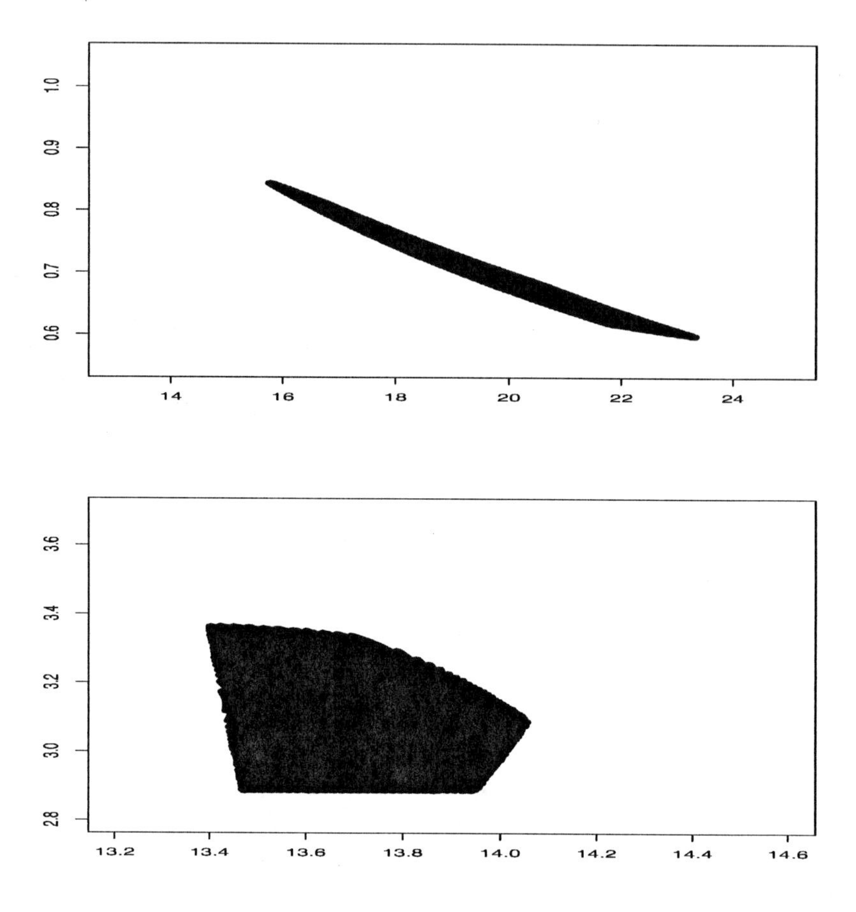

Figure 2.9 *Upper panel: the* 0.95 *approximation for the parameters* (γ, λ) *of the gamma distribution for the data of Figure 1.6 based on the mean, the standard deviation and a measure of goodness-of-fit using the Kuiper metric of order one* d^1_{ku} *defined by (2.25). Lower panel: the corresponding approximation region for the mean and standard deviation. In both cases the scale is the same as Figure 2.6.*

0.01 was spent on the goodness-of-fit and the remaining 0.04 being spent equally on the tails of the mean and standard deviation. The quantiles were calculated as described in Chapter 2.9. When compared with Figure 2.6 the approximation region is seen to be considerably smaller. This is not an indication of higher precision as would be the case for confidence regions, but an indication of lack of fit.

The standard formal method of testing the adequacy of a model is to perform a goodness-of-fit test based on the estimated parameters as described in Chapter 1.3.4. The use of an estimated parameter value instead of one given independently of the data causes a change in the distribution of the test statistic. For the gamma model

$\Gamma(\gamma,\lambda)$ the 0.99 quantile of the Kuiper metric $d_{\text{ku}}(\mathbb{P}_n(\gamma,\lambda),\Gamma(\hat{\gamma},\hat{\lambda}))$ is 0.102 for a sample of size $n = 258$ where $(\hat{\gamma},\hat{\lambda})$ are the maximum likelihood estimators bases on the sample. For the student data $d_{\text{ku}}(\mathbb{P}_n,\Gamma(\hat{\gamma},\hat{\lambda})) = 0.112$ so that the gamma model is rejected at the 0.01 level. In spite of this the approximation region shown the upper panel of Figure 2.9 is not empty. The reason is that the 0.99 quantile of the Kuiper metric $d(\mathbb{P}_n,P)$ for a continuous model P is approximately $2/\sqrt{n} = 0.125$ for a sample of size $n = 258$. Thus for some values of γ and λ, for example, the maximum likelihood values, the distance $d_{\text{ku}}(\mathbb{P}_n,\Gamma(\gamma,\lambda)) < 0.125$. For this data set and the gamma model any reasonable estimates of the parameters will give the same result. It can however happen that the maximum likelihood estimates are misleading. This is for example the case for the comb model for the data of Table1.3 as evidenced by the p-value for the Kuiper metric given in Table 1.4.

The converse so to speak is also possible. That is, even if the model is judged to be adequate on the basis a goodness-of-fit test using estimated parameter values, it may be the case that the confidence region contains values of the parameter which are not adequate in the goodness-of-fit sense.

Metrics and discrepancies can be combined with other measures for the appropriateness of the model. The metrics discussed above are of little use for checking for possible outliers. The problem of outliers is the subject of Chapter 4 and using the methods discussed there the presence or absence of outliers can be taken into consideartion. The total amount of probability available will now have to be allocated to the values of the parameter or functional of interest, to a goodness-of-fit test and to an outlier test. Alternatively the statistician can cast a surreptitious glance at a $q-q$ plot and spend the probability saved on the parameter or functional of interest.

It may not always be possible to be completely honest. Suppose a model is required for the daily returns of the Standard and Poor's index since 1928, or for other long-term financial data. It may prove difficult to specify a set of constraints which capture the appearance of the data in an adequate manner: one recognizes a cat when one sees one, without being able to give an analytical expression for 'catness'. One possibility is to test the adequacy of such a model with the help of experts who are asked to specify whether a data set is real or simulated (see [23]) but such checks can be carried out only on a small scale, not on a large scale for the determination of parameter values.

2.9 Simulations and Tables

Suppose that a data set is to be approximated by a normal distribution $\mathfrak{N}(\mu,\sigma^2)$ and that the concept of approximation is based on the Kolmogorov metric as a measure of goodness-of-fit, the mean and the standard deviation. As the distributions of all three statistics under the normal model are known the calculation of the approximation region for the parameters μ and σ does not require any simulations. This is an exceptional case. In general the distributions of the statistics of interest under the model are not known. The distribution of the mean under the gamma model is known but not the distribution of the standard deviation. The approximation regions shown in Figures 2.6 and 2.9 were based on 2000 simulations for each of those parameters' values not excluded by the values of the mean.

Although the relevant probabilities can in principle always be calculated to any degree of accuracy using simulations this can be very costly in terms of computation. If the statistic under consideration has a known asymptotic distribution for a given set of parameter values, or an asymptotic distribution which can be simulated, a more efficient manner of calculating the probabilities is sometimes available. The strategy is the following. Calculate the relevant probabilities for all small sample sizes $n \le n_0$ on a grid of parameter values and store the results. Calculate the probabilities for medium to large n, $n_0 < n \le n_1$ on a grid of n and parameter values where n_1 is such that the asymptotic approximation is judged to be sufficiently good for $n \ge n_1$. It is often the case that the relevant probabilities can be well approximated by some simple function of the samples size n for $n_0 < n \le n_1$. This part of the procedure is an art. There are no generally valid rules or strategies for finding such a simple function. The strategy is illustrated for the distribution of the standard deviation under the gamma distribution.

The parameter λ is a scale parameter so that in the case of the standard deviation it is only necessary to consider the case $\lambda = 1$. Let $X_i, i = 1,2,\ldots$ be independently and identically distributed random variables with $X_i \sim \Gamma(\gamma, 1)$. Then for all $k \in \mathbb{N}$ it holds that

$$\mathbf{E}(X^k) = \prod_{j=0}^{k-1}(\gamma + j). \tag{2.45}$$

It follows from this and the central limit theorem that

$$\frac{\sum_{i=1}^{n}((X_i - \overline{X}_n)^2 - \gamma)}{\sqrt{2n\gamma(\gamma+3)}} \Rightarrow \mathfrak{N}(0,1). \tag{2.46}$$

For large values of the shape parameter γ the X_i will be approximately $\mathfrak{N}(\gamma,\gamma)$ distributed so that it may be advantageous to use the chi-squared distribution with $n-1$ degrees of freedom rather than the normal distribution in the approximation. There is no hope of obtaining an approximation for all values of γ so a lower bound must be chosen. The one used here is $\gamma_0 = 1/2$. Similarly it is not possible to get an approximation for arbitrarily large or small values of α. The range chosen is $0.001 \le \alpha \le 0.999$ split into the three ranges $0.001 \le \alpha \le 0.2$, $0.2 \le \alpha \le 0.8$ and $0.8 \le \alpha \le 0.999$. Put

$$\mathrm{qsd}(n,\alpha) = (\mathrm{qchisq}(n-1,\alpha) - (n-1))/\sqrt{2(n-1)} \tag{2.47}$$

and denote by $\hat{q}(n,\alpha,\gamma,nsim)$ the empirical α-quantiles of

$$\frac{\sum_{i=1}^{n}((X_i - \overline{X}_n)^2 - \gamma)}{\sqrt{2n\gamma(\gamma+3)}}$$

based on $nsim$ simulations with $\lambda = 1$. The goal is to approximate the $\hat{q}(n,\alpha,\gamma,nsim)$ by a simple, in this case a quadratic, function of $\mathrm{qsd}(n,\alpha)$. For each value of (γ,n)

on the chosen grid $nsim = 100000$ simulations are performed.and the values of the constants $c_0(\gamma,n), c_1(\gamma,n)$ and $c_2(\gamma,n)$ in

$$c_0(\gamma,n) + c_1(\gamma,n)\text{qsd}(n,\alpha) + c_2(\gamma,n)\text{qsd}(n,\alpha)^2$$

were determine as the least squares approximation to $\hat{q}(n,\alpha,\gamma,nsim)$.

Figure 2.10 shows the results for $\gamma = 0.5$ and $n = 3$. The top panel gives the approximation for the range $0.001 \leq \alpha \leq 0.2$, the centre for $0.2 \leq \gamma \leq 0.8$ and the bottom panel for the range $0.8 \leq \alpha \leq 0.999$. Deviations between the empirical values obtained from the simulations and the quadratic approximations are most clearly seen in the bottom panel. As n increases the values of $c_0(\gamma,n)$ and $c_2(\gamma,n)$ both tend to zero whilst that on $c_1(\gamma,n)$ tends to one in accordance with the asymptotics. The same happens for fixed n as γ increases, again in accordance with the asymptotics. It turns out that for $\gamma \geq 100$ it is sufficient to put $c_1(\gamma,n) = 1$ and $c_0(\gamma,n) = c_2(\gamma,n) = 0$ for all n. If $0.5 \leq \gamma < 100$ and $n > 500$ the coefficients are taken to be

$$c_i(\gamma,n) = c_i(\gamma,500)\sqrt{\frac{500}{n}},\ i = 0,2 \quad c_1(\gamma,n) = 1 - (1 - c_1(\gamma,500))\sqrt{\frac{500}{n}} \tag{2.48}$$

and

$$c_1(\gamma,n) = 1 - (1 - c_1(\gamma,500))\sqrt{\frac{500}{n}}. \tag{2.49}$$

The quantile function can be inverted to give the distribution function for the same ranges of n-, γ- and α-values. The coefficients c_0, c_1 and c_2 can be stored and give a compact expression of the quantiles for all n, all $\gamma > 0.5$ and for $0.001 \leq \alpha \leq 0.999$. They are a modern version of the statistical tables.

The use of simulations to obtain estimates of probabilities is practically but not theoretically well established. It is practically well established in that it is an accepted technique. It is not theoretically established for two reasons. Firstly the random numbers generated by computers are not random and there are cases where the results of a simulation have been invalidated by a peculiarity of the number generating mechanism. Secondly apart from particular cases there are no theorems on the correctness of the interpolation or extrapolation of simulations results for certain values of the sample size to other values. There is often an assumption about the monotonicity of the approach to the asymptotic distribution which is rarely stated explicitly and which typically cannot be backed up by a theorem. The assumption is that if the results of a simulation for say $n = 100$ are very close to the asymptotic results then they will be even closer for $n = 1000$ and closer still for $n = 5000$. In Chapter 3.3 an example will be given where this is not the case.

2.10 Degree of Approximation and p-values

A confidence region is a measure of the precision with which statements can be made about the unknown parameter values based on a sample generated under the model, the more efficient the estimator the smaller the confidence region. With appropriate concern for the functionals used this carries over to approximation regions based

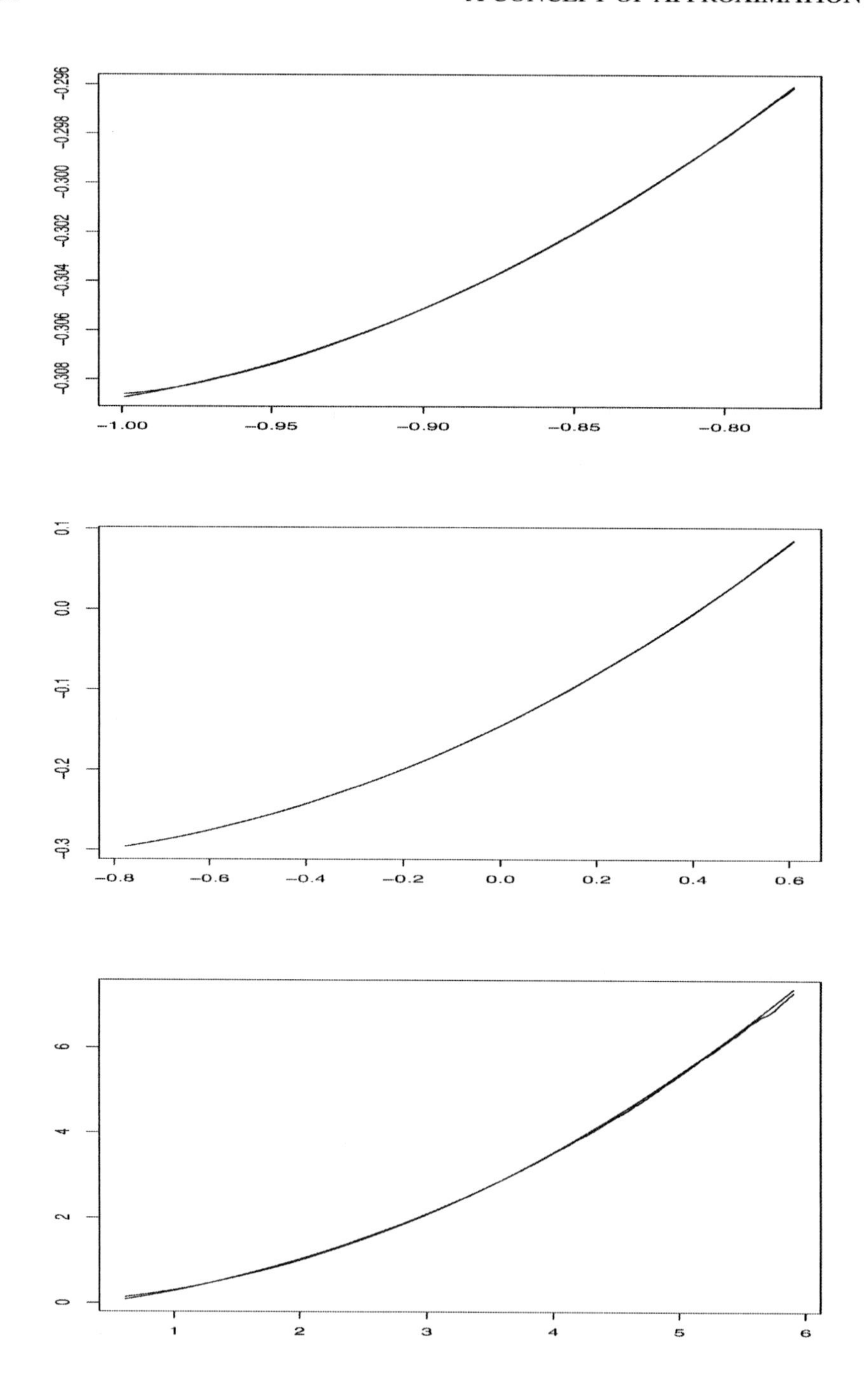

Figure 2.10 *The approximation of the empirical* α*-quantiles by the chi-squared quantiles for* $\gamma = 0.5$ *and* $n = 3$. *The top panel is the approximation for* $0.001 \leq \alpha \leq 0.2$, *the centre for* $0.2 \leq \alpha \leq 0.8$ *and the bottom panel for* $0.8 \leq \alpha \leq 0.999$.

on a family of models. As argued in the previous section the use of a particular family of models is free or at least not costed. If the adequateness of the model is taken into account by including some measure of goodness-of-fit in the definition of the approximation region the situation changes. A small approximation region as in Figure 2.9 can now be due to a lack of fit and not to the use of some efficient statistics to calculate it. In contrast to a confidence region the size of an approximation region is not only a measure of efficiency, but also a measure of the degree appropriateness of the model.

The definition of approximation given is that of a data set $\mathbf{x}_n$ by a probability measure P. The model approximates the data $\mathbf{x}_n \in E_n(\alpha,P) \subset \mathbb{R}^n$ where

$$\mathbf{P}(\mathbf{X}_n(P) \in E_n(\alpha,P)) = \alpha .$$

Suppose now that

$$E_n(\alpha_1,P) \subset E_n(\alpha_2,P), \quad \alpha_1 < \alpha_2 \tag{2.50}$$

which will be the case in general. The degree of approximation of P to $\mathbf{x}_n$ can be defined by

$$\alpha^*(\mathbf{x}_n,P) = \inf\{\alpha : \mathbf{x}_n \in E_n(\alpha,P)\} . \tag{2.51}$$

With this definition the smaller $\alpha^*(\mathbf{x}_n,P)$ the better the approximation. Alternatively one can define

$$p^*(\mathbf{x}_n,P) = 1 - \alpha^*(\mathbf{x}_n,P) \tag{2.52}$$

so that now the smaller $p^*(\mathbf{x}_n,P)$ the worse the approximation. The quantity $p^*(\mathbf{x}_n,P)$ plays the role of the attained p-value in frequentist statistics but with a different meaning, namely as a measure of the degree of approximation of the data $\mathbf{x}_n$ by the model P. Expressed in another manner it measures how weak the concept of approximation must be before the model is accepted as an adequate approximation.

The interpretation of p-values as a measure of approximation is one that is often accepted but on an informal level. In [71] it is used explicitly and formally as a measure of appropriateness in classification problems.

The relationship (2.50) holds for the $E_n(\alpha,P)$ in the definition of the approximation region for the student data. More generally it holds for all $E_n(\alpha,P)$ of the form

$$E_n(\alpha,P) = \cap_{j=1}^{k}\{\mathbf{y}_n : T_j(\mathbf{y}_n) \le \mathrm{qt}_j(\tilde{\alpha},P)\} \tag{2.53}$$

and the qt_j are the quantiles of the T_j under the model P. The $E_n(\alpha,P)$ for the student data shown in Figure 2.9 is of this form with

$$T_1(\mathbf{y}_n) = |\,\bar{\mathbf{y}}_n\,|,\ T_2(\mathbf{y}_n) = \mathrm{sd}(\mathbf{y}_n),\ T_3(\mathbf{y}_n) = d_{\mathrm{ku}}(\mathbb{P}_n(\mathbf{y}_n),P)$$

and $\tilde{\alpha} = 0.98333$.

The idea can be extended to approximation regions $\mathscr{A}(\mathbf{x}_n,\alpha,\mathscr{P})$ where the $E_n(\alpha,P)$ satisfy (2.50). To determine the best approximation P_0 to $\mathbf{x}_n$ with $P_0 \in \mathscr{P}$ the value of α in $\mathscr{A}(\mathbf{x}_n,\alpha,\mathscr{P})$ is decreased until $\mathscr{A}(\mathbf{x}_n,\alpha,\mathscr{P})$ contains just one model P_0 which is then the best approximation. This can be accomplished by decreasing $\tilde{\alpha}$ in (2.53). Doing this for the student data results in a value of $\tilde{\alpha}$ of 0.8125; the

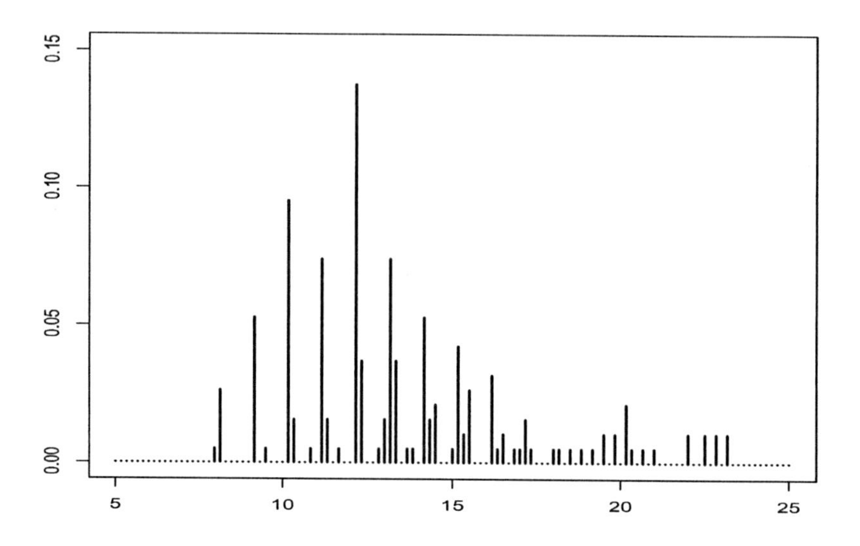

Figure 2.11 *Relative frequencies of lengths of study in semesters of 189 German students.*

best approximation has parameter values $(\gamma_0, \lambda_0) = (20.36, 0.67)$ with mean 13.89 and standard deviation 3.025. The mean and standard deviation of the data are 13.89 and 3.215 respectively. The value of α^* which corresponds to $\tilde{\alpha} = 0.8125$ can be obtained by simulation and is $\alpha^* = 0.661$ with a p-value $p^* = 0.339$.

The idea can be used in the other direction to measure how poor a fit is. Figure 2.11 shows a second data set of student study times of size $n = 189$. If again a gamma model is used together with the Kuiper metric then for $\alpha = 0.95$ the approximation regions are empty. If however $\alpha = 0.985$ then there are adequate models. In principle if α is made sufficiently large then almost any model can be made to fit the data. The smallest such value of α for which there is a fit is a measure of how poor the fit is. In practice this is not always possible as for very poor fits the value of α will be so large as to make it impossible to determine the relevant quantiles with any accuracy. A different approach due to [142] will be demonstrated for the same data set in Chapter 3.8.

2.11 Scales

The scale used in the definition of the approximation region was the probability scale. Given that the concept of approximation is to be applied to probability models this is not surprising. The data very often have their own scale and are transformed to a probability scale using the model. It may happen that differences on the data scale which are judged to be moderately large may translate into large differences on on the probability scale. In particular a difference may be highly significant on the probability scale but practically irrelevant for the purposes of the analysis. If for example the model is $\mathfrak{N}(0,1)$ then the occasional value of 5 in the data may be of no consequence although such a value translates into a tail probability of $2.87 \cdot 10^{-7}$. A physicist may regard a Kuiper distance of 0.01 between model and data as perfectly satisfactory regardless of the sample size (see Chapter 5.4 of [117]).

In general any practically relevant difference can be incorporated directly into the definition of the approximation region without reference to a probability. The situation may be mixed in that some aspects of the data are formulated on the probability scale and others directly on the data scale. The α of $\mathscr{A}(\mathbf{x}_n, \alpha, \mathscr{P})$ may then lose all or part of its role but it seems not worthwhile to adjust the notation. What is meant should be clear from the context.

2.12 Stability of Analysis

The problem of stability of analysis was already discussed in Chapter 1.3.7. There it was shown that the maximum likelihood functional for the comb distribution reacts sensitively to small changes in the individual observations x_i. The mean and standard deviation in contrast react smoothly to small changes in the values of the data, but they are sensitive to large changes in even one x_i. Consider a real-valued functional T defined on a large set of distributions $\mathscr{P}_T$, topologize $\mathscr{P}_T$ with the Kolmogorov metric d_{ko} and consider

$$T : (\mathscr{P}_T, d_{\text{ko}}) \to (\mathbb{R}, \|\cdot\|)$$

where $\|\cdot\|$ is the usual Euclidean norm. The functional T is called locally bounded if for all $P \in \mathscr{P}_T$ there exists an $\varepsilon > 0$ such that

$$\begin{aligned} -\infty < b_-(T,P,\varepsilon) = \inf\{T(Q) - T(P) : Q \in B_\varepsilon(P)\} \leq \\ \sup\{T(Q) - T(P) : Q \in B_\varepsilon(P)\} = b_+(T,P,\varepsilon) < \infty \end{aligned}$$

where

$$B_\varepsilon(P) = \{Q : Q \in \mathscr{P}_T, d_{\text{ko}}(P,Q) < \varepsilon\}.$$

The mean functional T_{ave} of (2.12) is not locally bounded as $b_-(T_{\text{ave}},P,\varepsilon) = -\infty$ and $b_+(T_{\text{ave}},P,\varepsilon) = \infty$ for every P and every $\varepsilon > 0$. It is this property of the mean functional that is responsible for the results given in [9].

The concept of stability described above is one of stability on the space of probability measures and not on the space of observations namely $\mathbb{R}^n$. The mean is a very stable and smooth function of the individual data points, in fact the smoothest such function apart from constant functions. Nevertheless this does not capture the susceptibility of the mean to outliers. There are situations however where the observations are known a priori to lie in some compact set. The student data of Figure 1.6 are one example as are dummy variables or variables with natural bounds such as age and weight. In such situations continuity in the observations will usually be required for stability of inference.

Exactly which property of a functional will be required for stability of inference depends on the situation. The median functional T_{med} is defined for all distributions P by

$$T_{\text{med}}(P) = (\ell(P) + u(P))/2 \tag{2.54}$$

where

$$\ell(P) = \inf\{x : P((-\infty, x]) \geq 1/2\},\, u(F) = \sup\{x : P((x, \infty)) \geq 1/2\}$$

and is locally bound at all P. In fact it can be seen that $b(T_{\text{med}}, P, \varepsilon) < \infty$ for all P and $\varepsilon < 1/2$ where

$$b(T, P, \varepsilon) = \sup\{|T(Q) - T(P)| : Q \in B_{\varepsilon}(P).\} \tag{2.55}$$

On the other hand each ball $B_{\varepsilon}(P)$ contains distributions at which the median is not continuous. If the detection of outliers is of importance and this is to be based on some functional T then the value of $b(T, P, \varepsilon)$ is of more relevance then any continuity properties of T. In one dimension it is known that for symmetric P the median minimizes $b(T, P, \varepsilon)$ over the set of location functionals (2.15) [114] and it is this property of the median that makes it so useful for identifying outliers. The topic will be discussed at greater length in Chapter 4.

It is more difficult to find a use for the continuity per se of a functional. Lower semi-continuity on the other hand is sometimes relevant, at least in the technical sense. Let $T_{\text{modal}}(P)$ denote the modality of the distribution P. As $1 \leq T_{\text{modal}}(P)$ for all P it is bounded from below but this bound is trivial. As $T_{\text{modal}}(P)$ is lower semi-continuous, that is

$$\liminf_{\varepsilon \to 0}\{T_{\text{modal}}(Q) : Q \in B_{\varepsilon}(P), Q \neq P\} \geq T_{\text{modal}}(P),$$

it follows that

$$\liminf_{\varepsilon \to 0}\{T_{\text{modal}}(Q) : Q \in B_{\varepsilon}(P)\} = T_{\text{modal}}(P)$$

so that the lower bound

$$\inf\{T_{\text{modal}}(Q) : Q \in B_{\varepsilon}(P)\}$$

is in general non-trivial. This is based on [58].

If more standard bounds are required for the value of a functional, that is, bounds which decrease at the rate $1/\sqrt{n}$ then more than continuity is required. The matter is somewhat technical but a sufficient condition is locally uniform Fréchet differentiability. This allows a locally uniform linearization of the functional T

$$T(Q) - T(P) = \int I(x, T, P)\, d(Q(x) - P(x)) + o(d_{\text{ku}})$$

which, on plugging in $Q = \mathbb{P}_n(P)$, gives

$$\sqrt{n}(T(\mathbb{P}_n(P)) - T(P)) = \frac{1}{\sqrt{n}} \sum_{i=1}^{n} (I(X_i, T, P) - \mathbf{E}(I(X_i, P))) + o(1).$$

From this the Central Limit Theorem for $\sqrt{n}(T(\mathbb{P}_n(P)) - T(P))$ is locally uniform over P in terms of convergence and hence gives a stable inference. The size of the approximation region is $O(1/\sqrt{n})$. Such functionals exist for the location-scale problem in $\mathbb{R}$ (see [37]) and in $\mathbb{R}^k$ (see [66]). This will be treated in Chapter 5. The function $I(x,T,P)$ is known as the influence function and is the main object of study in [103].

2.13 The Choice of $E_n(\alpha,P)$

The definition of an approximation for a data set $\mathbf{x}_n$ and a model P requires the specification of a region $E_n(\alpha,P) \subset \mathbb{R}^n$ such that P is an adequate approximation for $\mathbf{x}_n$ if $\mathbf{x}_n \in E_n(\alpha,P)$. In his discussion of [39] Hampel states

> Although somewhat downplayed, there is of course quite a bit of arbitrariness in the proposals by Davies, for example in the criteria for 'typical' samples.

In the same context Tukey [211] states

> Thought and debate as to just which aspects are to be denied legitimacy will be both necessary and valuable.

Hampel is correct in the sense that there is no single correct way of specifying $E_n(\alpha,P)$, but this does not mean that it is 'quite arbitrary'. Tukey is correct without qualification: it requires thought and debate.

South London was hit by V1 flying bombs during the Second World War and one question of interest was whether they were aimed at a particular target. If this were not the case the distribution of hits would be at random. In one sense the bombs were of course aimed at a target. After all most but not all landed in London. The hypothesis must be interpreted as meaning that they could not be accurately aimed on a scale of 250 metres. This was tested by dividing the area into 576 squares each of $1/16$ square kilometres. The assumption of randomness leads to a Poisson distribution for the number of squares with a given number of hits. This is one case where a statistical model is not ad hoc. The data are given in Table 2.1 [27].

The adequacy of the Poisson distribution was checked by Clarke using the chi-

Number of hits k	0	1	2	3	4	≥ 5
Number of squares with k hits	229	211	93	35	7	1

Table 2.1 *Data on flying bomb hits taken from [27].*

squared goodness-of-fit test (2.38)

$$d_{\chi^2}(Q_1,Q_2,\mathscr{B}) = \sum_{i=1}^{k} \frac{(Q_1(B_i) - Q_2(B_i))^2}{Q_1(B_i)}$$

with the partition

$$B_i = \{i-1\}, i = 1,\ldots,5, B_6 = \{5,6,7,\ldots\}. \tag{2.56}$$

The parameter λ was set to the mean 0.923 of the data. This is standard but it states only that the $\mathfrak{P}_o(\lambda)$ with $\lambda = 0.923$ is consistent with the data. It does not specify the range of λ values which are consistent with the data. A standard confidence interval for λ can be based on the fact that for the sample size $n = 576$ and value of $\lambda \approx 0.9$ the distribution of $\sqrt{n}(\bar{X}_n(\lambda) - \lambda)/\sqrt{\lambda}$ is approximately $\mathfrak{N}(0,1)$. This leads to the 95% confidence interval $[0.8445, 1.001]$.

The only value in the interval $[0.8445, 1.001]$ which has been tested for consistency with the data is $\lambda = 0.923$. The example of the student data and the gamma model treated above shows that including a feature on the agreement of each parameter value with the data can lead to very different approximations region as demonstrated by Figures 2.6 and 2.9. In the interests of honesty a certain amount of the 0.05 of probability to be spent must be spent on checking the agreement of the parameter values in the approximation interval with the data. If 0.025 is spent on each feature and the chi-squared discrepancy is used this leads to the approximation region

$$\begin{aligned} &\mathscr{A}(\mathbf{x}_n, 0.95, \Lambda) \\ &= \left\{\lambda : \left|\frac{\sqrt{n}(\bar{x}_n - \lambda)}{\sqrt{\lambda}}\right| \le 2.241,\ d_{\chi^2}(\mathbb{P}_n, \mathfrak{P}_o(\lambda), \mathscr{B}) \le \text{qchi}(0.975, n, \lambda)\right\} \\ &= [0.838,\ 1.017] \end{aligned} \tag{2.57}$$

with $n = 576$ and the partition $\mathscr{B}$ given by (2.56). For this particular data set the condition $d_{\chi^2}(\mathbb{P}_n, \mathfrak{P}_o(\lambda), \mathscr{B}) \le \text{qchi}(0.975, n, \lambda)$ is never active and the approximation interval is solely determined by

$$\left|\frac{\sqrt{n}(\bar{x}_n - \lambda)}{\sqrt{\lambda}}\right| \le 2.241\,.$$

Nevertheless in the interests of honesty some form of check of consistency with the data should be included. One should not simply include it or ignore it on whether the constraint becomes active or not. The allocation of the amount of probability to each feature is a matter of discussion, as is the measure of discrepancy.

Instead of the chi-squared discrepancy which upweights small probabilities the total variation metric (2.28) which treats all probabilities equally could have been used. This should be discussed. In the case of the flying bomb data the purpose of the analysis was to decide whether the bombs were being accurately aimed, at least to a certain degree. If the data were to be altered by moving some of the bombs where only one square was hit to increase the the number of squares with multiple hits this would clearly indicate some form of aiming. If the same number of bombs were moved from squares with multiple hits to squares with no hits, this would tend to counter the theory of accurate aiming. For the total variation metric the result of the two operations would be the same. There would however be large changes in the chi-squared discrepancy. In other words, the chi-squared discrepancy is more appropriate for this problem than is total variation.

As already pointed out if a gamma model is used for the student data of Figure 1.6 the inclusion of an honesty goodness-of-fit term in the definition of the approximation region reduces reduces this considerably (Figures 2.6 and 2.9). For the data of

Figure 2.11 the situation is even worse; the gamma approximation region is empty. Given the nature of both data sets it would seem to be a difficult task to find a simple parametric models that fits. An alternative is to take a non-parametric approach.

With $T=(T_{\text{ave}},T_{\text{sd}})$ as in (2.12) and (2.14) a non-parametric approximation region could be defined as

$$\begin{aligned}
&\mathscr{A}(\mathbf{x}_n,\alpha,T(\mathscr{P}))\\
&= \Big\{(T_{\text{ave}}(P),T_{\text{sd}}(P)) : \sqrt{n}|T_{\text{ave}}(\mathbb{P}_n)-T_{\text{ave}}(P)| \le \text{qnorm}((5+\alpha)/6)T_{\text{sd}}(P),\\
&\qquad \left|\sum_{i=1}^{n}(x_i-T_{\text{ave}}(P))^2/T_{\text{sd}}(P)^2-n\right| \le \text{qnorm}((5+\alpha)/6)\sqrt{2n},\\
&\qquad d_{\text{ku}}(\mathbb{P}_n,P) \le \text{qdku}((2+\alpha)/3), supp(P)\subset[5,25]\Big\}. \qquad (2.58)
\end{aligned}$$

where $\mathscr{P}=\{P: supp(P)\subset[5,25]\}$. This latter implies that under all models $P\in\mathscr{P}$ students who matriculate take between five and twenty-five semesters. This is not strictly true but there are very few exceptions. The large sample size $n=189$ together with $d_{\text{ku}}(\mathbb{P}_n,P)\le \text{qdku}((2+\alpha)/3)$ and $\mathscr{P}=\{P: supp(P)\subset[5,25]\}$ guarantee that the normal approximations to the distributions of $T_{\text{ave}}(\mathbb{P}_n)$ and $\sum_{i=1}^{n}(x_i-T_{\text{ave}}(P))^2/T_{\text{sd}}(P)^2$ hold uniformly. The resulting approximation region for $\alpha=0.95$ is shown in the upper panel of Figure 2.12. The lower panel shows the corresponding non-parametric approximation region for the student data of Figure 1.6.

The approximation region $\mathscr{A}(\mathbf{x}_n,\alpha,\mathscr{G})$ for the photon count data of Figure 2.1 is given by (2.10). The threshold is controlled by $\tau(\alpha,n)$ which for given α and n is determined by (2.9). The model (2.8) postulates a constant Gaussian white noise level with variance 6.3^2. A better model for high levels of photon counts is the Poisson model. For low levels the counts are influenced by electronic noise and the Gaussian model of constant noise is better. This leads to the approximation region

$$\begin{aligned}
&\mathscr{A}(\mathbf{x}_n,\alpha,\mathscr{G})\\
&= \Big\{g\in\mathscr{G}: \max_{I\subset[15,25]}\frac{1}{\sqrt{|I|}}\Big|\sum_{t_i\in I}\frac{x(t_i)-g(t_i)}{\max(6.3,\sqrt{g(t_i)})}\Big| \le \sqrt{\tau(\alpha,n)\log n}\Big\}. \qquad (2.59)
\end{aligned}$$

One possibility is to calculate a first approximating function $\tilde{g}$ using the approximation region of (2.10) and then to replace the $\sqrt{g(t_i)}$ in (2.59) by $\sqrt{\tilde{g}(t_i)}$. Even simpler the $\sqrt{g(t_i)}$ can be replaced by $\sqrt{x(t_i)}$. When this procedure is applied to several different data sets it becomes clear that $\tau(\alpha,n)$ is better treated as a tuning parameter without any reference to α. This is a change of scale from a probability scale to a scale at the level of noise as discussed in Chapter 2.11. The approximation region becomes

$$\begin{aligned}
&\mathscr{A}(\mathbf{x}_n,\tau_n,\mathscr{G})\\
&= \Big\{g\in\mathscr{G}: \max_{I\subset[15,25]}\frac{1}{\sqrt{|I|}}\Big|\sum_{t_i\in I}\frac{x(t_i)-g(t_i)}{\max(6.3,\sqrt{g(t_i)})}\Big| \le \sqrt{\tau_n\log n}\Big\}. \qquad (2.60)
\end{aligned}$$

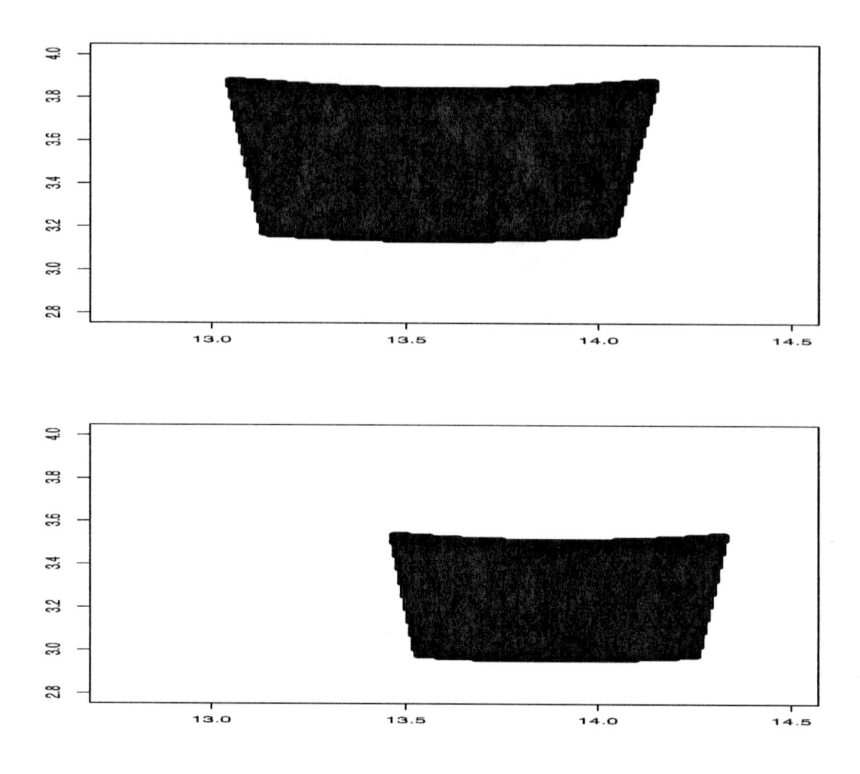

Figure 2.12 *Upper panel: the non-parametric* 0.95*-approximation region for the mean and standard deviation of the student data of Figure 2.11. Lower panel: the same for the student data of Figure 1.6.*

The value of τ_n can now be chosen to give the best results for the data sets being analysed. As there is an interest in detecting small peaks τ_n can be smaller than say $\tau(0.95, n)$. This leads to some small peaks being due to noise, but others can have a physical meaning. Which is which can be decided by the physicist who conducts the experiment. A detailed analysis of the photon data is given in [43].

The concept of approximate models developed above applies only to one data set: it is not necessary for the data to be replicated or replicable. An example of a non-replicable data set is the Standard and Poor's index for the years 1928-2012. The data are shown in the upper panel of Figure 2.13. The sample size is $n = 22378$. It has often been claimed that the data are non-stationary (see for example [96]), but what is classified as stationary or non-stationary depends on the time scale. Any sequential data set can be modelled as a section of a stationary process. The interesting but imprecise question is whether the data set can be modelled by a simple stationary process over the given time scale. The bottom two panels of Figure 2.13 show simulated data sets using a simple stationary model. The question is whether the two simulated data sets are reasonable for financial data. It may happen that experts can

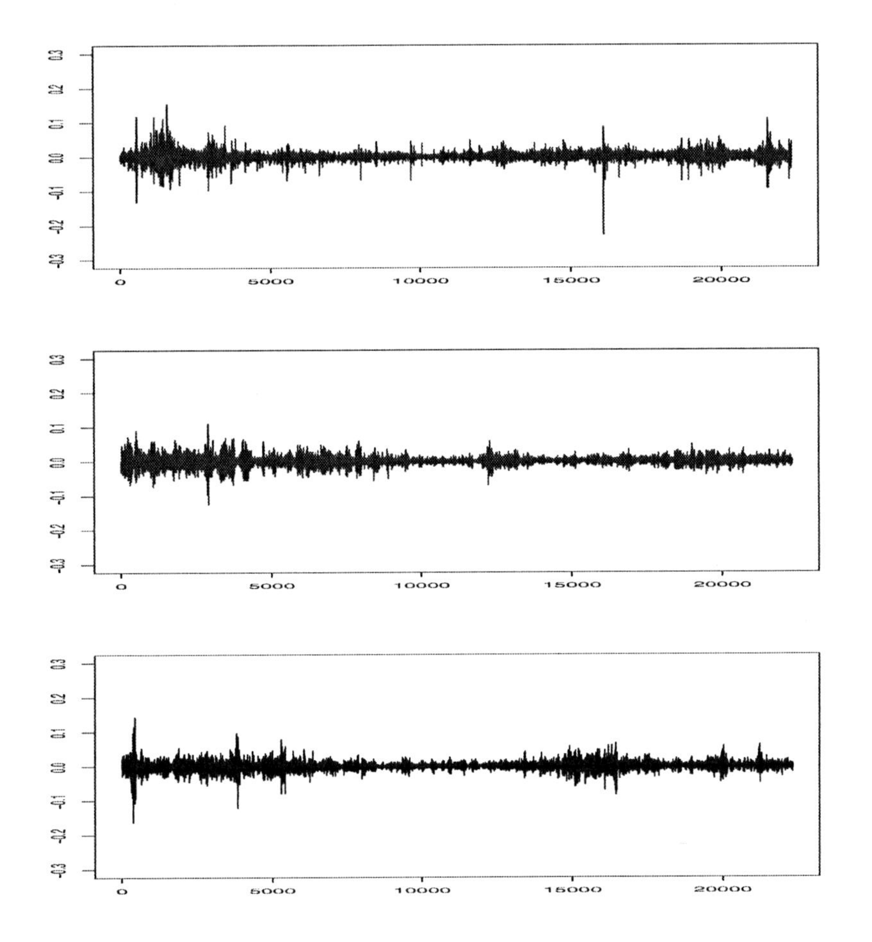

Figure 2.13 *Top panel: the daily returns of the Standard and Poor's data. Bottom two panels: two simulated sets of data.*

identify the real data set without being able to articulate their decision sufficiently precisely for this to be given a mathematical expression (see [23]). There are stylized facts about such data such as heavy tails and long term memory but they will require a mathematical expression and, in themselves, may not be sufficient. One property which can be given a mathematical expression is the number of intervals of constant, or almost constant, volatility. This will be done in Chapter 8. The problem is further confounded by the singularity of the data set. There is only one such data set and consequently no direct quantification of the variability is possible. As Tukey says, thought and debate will be necessary.

A third case to be considered is when a definition of approximation is to be applied to many data sets, present and future, in a more or less automatic manner. The standardized analysis of interlaboratory tests is one example.

2.14 Independence

Most of the models in this book assume the independence of the individual observations $X_i, i = 1, \ldots, n$. Given real data there is often no possibility of checking whether the independence assumption of the model is consistent with the data. This will be the case if there is no information about possible dependencies between individual values arising for example from the ordering of the sample or information about the spatial locations of the observations. Even if such information is available it is impossible to check the adequacy of the model assumption of independence on the basis of a single sample. Given a times series $x_1, \ldots, x_n$ and a model $X_1, \ldots, X_n$ based on independence it is possible to to test the adequacy of the independence for a given lag ℓ, that is the pair $(X_i, X_{i+\ell})$ are independent, $i = 1, \ldots, n-\ell$. The following is based on [170] for the lag $\ell = 1$ but applies for any given lag. The two real-valued random variables X and Y are independent if and only if

$$\mathbf{E}(\exp(isX + itY)) = \mathbf{E}(\exp(isX))\mathbf{E}(\exp(itY)) \tag{2.61}$$

for all s and t. This will be the case if and only if

$$\int_{-\infty}^{\infty}\int_{-\infty}^{\infty} g(s)g(t)|\mathbf{E}(\exp(isX + itY)) - \mathbf{E}(\exp(isX))\mathbf{E}(\exp(itY))|^2\, ds\, dt = 0 \tag{2.62}$$

for a strictly positive function $g : \mathbb{R} \to \mathbb{R}$. On changing the order of integration and using the empirical measures Pinkse deduces the test statistic

$$\tilde{\mathscr{C}} = \frac{1}{2}\left(\frac{\tilde{\mathscr{I}}}{(\tilde{\gamma} - \tilde{\mathscr{I}}_3)^2}\right) \tag{2.63}$$

where $\tilde{\mathscr{I}} = (\tilde{\mathscr{I}}_1 - \tilde{\mathscr{I}}_2)^2 + (\tilde{\mathscr{I}}_2 - \tilde{\mathscr{I}}_3)^2$,

$$\begin{array}{rclrcl} \tilde{\mathscr{I}}_1 &=& \frac{1}{n^2}\sum_{i,j} h_{i,j}h_{i+1,j+1}, & \tilde{\mathscr{I}}_2 &=& \frac{1}{n^3}\sum_{i,j,k} h_{i,j}h_{i+1,k}, \\ \tilde{\mathscr{I}}_3 &=& \frac{1}{n^4}\sum_{i,j,k,\ell} h_{i,j}h_{k,\ell}, & \tilde{\gamma} &=& \frac{1}{n}\sum_i \left(\frac{1}{n}\sum_j h_{i,j}\right), \end{array}$$

h is the Fourier transform of g,

$$h(x) = \int_{-\infty}^{\infty} \exp(ixu)g(u)\, du,$$

and $h_{i,j} = h(x_i - x_j)$. In the following g will be taken to be the density of the standard normal distribution so that $h(t) = \exp(-t^2/2)$. The calculation of $\tilde{\mathscr{C}}$ of order of complexity $O(n^2)$.

The empirical version of the integral in (2.62) is based on the two samples $X = (X_1, \ldots, X_{n-1})$ and $Y = (X_2, \ldots, X_n)$ which are not independent even if the $X_1, \ldots, X_n$ are independent. In [170] some finite sample corrections are given to take this into account but they will be ignored in the following. The asymptotic distribution of $\tilde{\mathscr{C}}$ is given by

$$\lim_{n\to\infty} n\tilde{\mathscr{C}} \stackrel{D}{=} \chi_1^2$$

but for a given sample critical values can be determined using random permutations.

The value of the $n\tilde{\mathscr{C}}$ is not invariant with respect to affine transformations of the data. Its value for the first 500 values of the Standard and Poor's data is 1.829274e+12 which is due to the fact that the values of the $h_{i,j}$ for this data set are of the order of $1-10^{-4}$. To function as intended the data must be on a scale comparable to the of a $\mathfrak{N}(0,1)$ random variable. For the Standard and Poor's and other similar data sets this can be achieved by standardizing the data using the median and standard deviation, $\tilde{\mathbf{x}}_n = (\mathbf{x}_n - \text{median}(\mathbf{x}_n)/\text{sd}(\mathbf{x}_n)$. If this is done for the the first 500 values of the Standard and Poor's data the value of the $n\tilde{\mathscr{C}}$ is now 30.67. The 0.9, 0.95 and 0.99 quantiles of the chi-squared distribution with one degree of freedom are 2.706, 3.841 and 6.635 respectively. The corresponding quantiles based on 5000 random permutations are respectively 2.860, 4.173 and 6.779. If the procedure is applied to the signs of the returns without any transformation the value of $n\tilde{\mathscr{C}}$ is 15.01. The 0.9, 0.95 and 0.99 quantiles based on 5000 random permutations are respectively 6.84, 10.38 and 34.67 which differ greatly from the corresponding quantiles of the chi-squared distribution with one degree of freedom. The situation is even worse. The value of $n\tilde{\mathscr{C}}$ for random ± 1 data of size $n = 500$ vary between 50 and essentially ∞. This is due to the standardization factor $(\tilde{\gamma} - \tilde{\mathscr{I}}_3)^2$ in (2.63). In general it would seem to be better to base the procedure on the non-standardized version $\tilde{\mathscr{I}}$ and to determine the quantiles using random permutations. If this is done for the first five hundred values of the normalized Standard and Poor's data the value of $\tilde{\mathscr{I}}$ is $16.9 \cdot 10^{-3}$ with quantiles $1.58 \cdot 10^{-3}$, $2.31 \cdot 10^{-3}$ and $3.75 \cdot 10^{-3}$ based on 5000 simulations. The corresponding values for the signs of the returns are $9.16 \cdot 10^{-4}$ with quantiles $4.17 \cdot 10^{-4}$, $6.613 \cdot 10^{-4}$ and $21.2 \cdot 10^{-4}$.

Pinkse's procedure is surprisingly efficacious as is witnessed by the comparisons with other procedures given in [170]. Its main drawback is the $O(n^2)$ complexity to calculate it. A procedure which reduces the computational cost is the following. Let $\psi_i, i = 1, \ldots$ be a bounded orthonormal basis for $\mathbb{R}$. If

$$\mathbf{E}(\psi_i(X)\psi_j(Y)) = \mathbf{E}(\psi_i(X))\mathbf{E}(\psi_j(Y)) \tag{2.64}$$

for all i and j then X and Y are independently distributed. For a given number ν and lag ℓ this leads to the statistic

$$\sum_{i=1}^{\nu}\sum_{j=i}^{\nu}(h_{i,j} - h_i h_j)^2 2^{-i-j} \tag{2.65}$$

where

$$h_i = \frac{1}{n-\ell}\sum_{k=1}^{n-\ell}\psi_i(x_k), \quad h_{i,j} = \frac{1}{n-\ell}\sum_{k=1}^{n-\ell}\psi_i(x_k)\psi_j(x_{k+\ell}). \tag{2.66}$$

The weights 2^{-i-j} in (2.65) can of course be replaced by some other choice. The cost of calculating the statistic (2.65) is of order $O(\nu n + \nu^2)$. One possible choice is to take the ψ_i to be the Hermite functions and to put $\nu = 6$. For the lag $\ell = 1$ the results for the first five hundred Standard and Poor's data are $4.09 \cdot 10^{-1}$ with quantiles $1.22 \cdot 10^{-1}$, $1.54 \cdot 10^{-1}$ and $2.15 \cdot 10^{-1}$ and for the signs $2.33 \cdot 10^{-1}$ with quantiles $1.66 \cdot 10^{-1}$, $2.34 \cdot 10^{-1}$ and $3.93 \cdot 10^{-1}$. The method is not as efficacious as that of Pinkse but much faster. The time required for 5000 simulations was 2.25 seconds as against 170 seconds for the Pinkse method. Given this it is possible to perform the procedure on the whole of the Standard and Poor's data set. The time required for 5000 simulations was 115 seconds with the following results. For the data set itself the value of the statistic was $227.9 \cdot 10^{-1}$ with quantiles $1.35 \cdot 10^{-1}$, $1.68 \cdot 10^{-1}$ and $2.45 \cdot 10^{-1}$. For the signs of the data the value of the statistic was $51.6 \cdot 10^{-1}$ with quantiles $1.64 \cdot 10^{-1}$, $2.35 \cdot 10^{-1}$ and $4.09 \cdot 10^{-1}$.

2.15 Procedures, Approximation and Vagueness

In statistics a procedure is a set or a sequence of actions performed on a data set which together represent the analysis or part of the analysis of the data. The procedure can include exploratory data analytic tools such as visual representations as well as more formal statistical methods such as the calculation of an approximation region. It may also contain more than one analysis of the data: it may use two or more probability models, for example the gamma and lognormal distribution; it may specify possible outliers and perform the analysis both with and without the outliers. A procedure can also have several different tasks each of which requires its own methodology. The procedure detailed in [43] for the analysis of X-ray diffractograms locates the number and positions of peaks, estimates a slowly varying base line and decomposes each peak into a sum of Gaussian or other kernels. Finally the statistical procedure may be part of a larger procedure that specifies not only the statistical analysis of the data but the manner in which the measurements are to be taken and the data collected. An example of such a larger procedure is [1] which covers all aspects of the conduct of interlaboratory tests for water, waste water and sludge.

The word 'procedure' suggests a plurality of situations in which the procedure can be applied. This does not prevent the same definition of approximation being used in the case that more than one data set has to be analysed, but any such definition will have to take the plurality of applications into account. Here is Tukey [211] on procedures:

> Since the formal consequences are consequences of the truth of the model, once we have ceased to give a model's truth a special role, we cannot allow it to "prescribe" a procedure. What we really need to do is to choose a procedure, something we can be helped in by a knowledge of the behavior of alternative procedures when various models (= various challenges) apply, most helpfully models that are (a) bland (see below) or otherwise relatively trustworthy, (b) reasonable in themselves, and (c) compatible with the data.

As an example, just because the mean is an optimal and sufficient statistic for the normal distribution this cannot be allowed to dictate the use of the mean even if the model under consideration is the normal one. The definitions of approximation used in a procedure will be guided by the available data sets and anticipated possible future data sets of a similar nature. Thought will have to be given as to the purpose of the analysis, to setting priorities regarding the properties of the data sets and to including checks, safeguards and warnings for data sets not conforming to the usual pattern.

A procedure is intended for use but not only by professional statisticians. It will often have tuning parameters whose values can be altered by the user, but to make its use as simple as possible for the non-statistician the parameters should where possible have default values which give reasonable results for most data sets. Ideally a procedure should be accompanied by a set of instructions for its use. Mathematical statistics is a branch of applied mathematics and has its own theorems. These are often used when a procedure is constructed, the central limit theorem being one example. Theorems have precise assumptions and precise conclusions. Such precise assumptions and conclusions cannot in principle be translated into precise statements for the formulation of a procedure: the instructions for the use of a procedure cannot have this precision. They must of necessity be vague and not even the vagueness can be made precise. To misquote [2] we cannot have ' rigidly defined areas of vagueness and uncertainty' or rigidly defined areas of doubt and uncertainty.

Chapter 3

Discrete Models

3.1 The Empirical Density

The chapter is restricted to counting models P with support in $\mathbb{N}_0 = \{0,1,2,\ldots\}$. The density $p(k)$ with respect to the counting measure is simply $p(k) = P(\{k\}), k \in \mathbb{N}_0$. Given a sample $\mathbf{x}_n \in \mathbb{N}_0^n$ the empirical density is given by

$$\hat{p}_n(k) = \frac{1}{n}\sum_{i=1}^{n}\{x_i = k\}\,,\, k \in \mathbb{N}_0\,. \tag{3.1}$$

For data $\mathbf{x}_n = \mathbf{X}_n = \mathbf{X}_n(P)$ generated under the model it is seen that $n\hat{p}_n(k) \sim \mathfrak{b}(n, p(k))$. This implies

$$\mathbf{E}(\hat{p}(k)) = p(k), \quad \mathbf{V}(\hat{p}(k)) = \frac{p(k)(1-p(k))}{n} \tag{3.2}$$

and

$$\mathbf{E}((\hat{p}(k)-p(k))(\hat{p}(\ell)-p(\ell))) = -\frac{p(k)p(\ell)}{n}\,, 0 \le k < \ell, \ldots \tag{3.3}$$

The asymptotic distribution of the empirical density under the model is given by

$$\begin{aligned}&\sqrt{n}\,(\hat{p}_n(0)-p(0),\hat{p}_n(1)-p(1),\ldots)\\&\quad\Rightarrow\quad (W(0)\sqrt{p(0)(1-p(0))},W(1)\sqrt{p(1)(1-p(1))},\ldots)\end{aligned}\tag{3.4}$$

where $(W(k))_{k=0}^{\infty}$ is a zero-mean Gaussian process with covariance structure

$$\mathbf{E}(W(k)^2) = 1,\tag{3.5}$$

$$\mathbf{E}(W(k)W(\ell)) = -\sqrt{\frac{p(k)p(\ell)}{(1-p(k))(1-p(\ell))}},\quad 0\le k<\ell<\infty.\tag{3.6}$$

The $(W(k))_{k=0}^{\infty}$ process can be constructed as follows. Let $(Z(k))_{k=0}^{\infty}$ be a standard white-noise Gaussian process and put

$$W = \sum_{k=0}^{\infty}\sqrt{p(k)}\,Z(k),\tag{3.7}$$

$$W(k) = (Z(k)-\sqrt{p(k)}\,W)/\sqrt{1-p(k)}\,.\tag{3.8}$$

The sum in (3.7) converges in mean square so that the random variable W is well defined.

It is seen that

$$W_k = Z_k\sqrt{1-p(k)}+R_k$$

with R_k independent of Z_k and $\mathbf{E}(R_k^2)=p(k)$. Thus if n is large and all the $p(k)$ small

$$\begin{aligned}&\sqrt{n}\,(\hat{p}_n(0)-p(0),\hat{p}_n(1)-p(1),\ldots)\\&\quad\overset{D}{\approx}\quad (Z(0)\sqrt{p(0)(1-p(0))},Z(1)\sqrt{p(1)(1-p(1))},\ldots).\end{aligned}\tag{3.9}$$

3.2 Metrics and Discrepancies

All the metrics and discrepancies of Chapter 2.6.1 can be specialized to the case that their support lies in $\mathbb{N}_0=\{0,1,\ldots\}$ as follows:

1. $d_p(P,Q)=(\Sigma_{i=0}^{\infty}|p(i)-q(i)|^p)^{1/p}$,
2. $d_{\text{tv}}(P,Q)=\sup_B|P(B)-Q(B)|=\frac{1}{2}\Sigma_{i=0}^{\infty}|p(i)-q(i)|=\frac{1}{2}d_1(P,Q)$,
3. $d_{\text{he}}(Q_1,Q_2)=\left(\Sigma_{i=0}^{\infty}\left(\sqrt{p(i)}-\sqrt{q(i)}\right)^2\right)^{1/2}$,
4. $d_{\text{kl}}(Q,P)=\Sigma_{i=0}^{\infty}q(i)\log(q(i)/p(i))$,
5. $d_{\chi^2}(P,Q,\mathscr{B}_k)=\Sigma_{i=1}^{k}(P(B_i)-Q(B_i))^2/P(B_i)$ for some finite partition $\mathscr{B}_k=\{B_1,\ldots,B_k\}$ of $\mathbb{N}_0$,
6. $d_{\chi^2}(P,Q)=\Sigma_{p(i)>0}(q(i)-p(i))^2/p(i)$.

The dominating measure π is the counting measure ν on $\mathbb{Z}$. There are differences to the case of Lebesgue densities. Firstly, relative frequencies of the data values are nothing more than the empirical density with respect to the counting measure ν. Secondly, the Prohorov metric makes little sense as there will not be an error ε when counting, 2.03 sheep instead of 2 sheep. Thirdly, the L_p metrics are finite for all $p \geq 1$. Fourthly, all L_p metrics for $p \leq 1$ and the Kolmogorov metric generate the same topology. Fifthly, all allow a direct and non-trivial comparison between the empirical measure $\mathbb{P}_n$ and a model P. In this sense they are all weak although some are weaker than others.

Following the policy of trying to be honest the definition of an approximation region should contain at least one feature of the data which is a measure of goodness-of-fit. The choice of metric or discrepancy will depend on the data and the problem at hand. The L_p metrics increase the weight of large probabilities for $p > 1$ and those of small probabilities if $p < 1$. The total variation metric $\mathrm{d_{tv}}(P,Q)$, which is equivalent to the L_1 metric, treats all density values equally and has the simplest interpretation: it is the amount of probability which has to be transferred to turn Q into P .

The Kullback-Leibler discrepancy $d_{\mathrm{ku}}(\mathbb{P}_n,P)$ and the chi-squared discrepancy $d_{\chi^2}(\mathbb{P}_n,P)$ upweight small probabilities, the chi-squared discrepancy more so than the Kullback-Leibler discrepancy: if $p(i) << 1/n$ is small but there is one observation $x_j = i$ then

$$np_n(i)\log(p_n(i)/p(i)) = -\log(np(i)) \tag{3.10}$$

and

$$n\frac{(\hat{p}_n(i)-p(i))^2}{p(i)} \approx \frac{1}{np(i)} >> -\log(np(i)). \tag{3.11}$$

The chi-squared discrepancy can be dominated by one small $p(i)$: the flying bomb data given in Table 2.1 are an example of this. Theoretically this can happen for the Kullback-Leibler discrepancy but the effect is much smaller.

3.3 The Total Variation Metric

Let $\mathbb{P}_n = \mathbb{P}_n(P)$ be the empirical distribution of n i.i.d. random variables $X_1(P),\ldots,X_n(P)$. The total variation distance is given by

$$\mathrm{d_{tv}}(\mathbb{P}_n,P) = \frac{1}{2}\sum_{k=0}^{\infty}|\hat{p}_n(k)-p(k)|$$

with expected value

$$\mathbf{E}(\mathrm{d_{tv}}(\mathbb{P}_n,P)) = \frac{1}{2}\sum_{k=0}^{\infty}\sum_{j=0}^{n}\left|\frac{j}{n}-p(k)\right|\binom{n}{j}p(k)^j(1-p(k))^{n-j} \tag{3.12}$$

which can be calculated explicitly. The second moment is

$$\mathbf{E}(\mathrm{d_{tv}}(\mathbb{P}_n,P)^2) =$$
$$\frac{1}{4}\sum_{k,\ell=0}^{\infty}\sum_{i,j=0}^{n}\left|\frac{i}{n}-p(k)\right|\left|\frac{j}{n}-p(\ell)\right|\binom{n}{i,j}p(k)^i p(\ell)^j(1-p(k)-p(\ell))^{n-i-j} \tag{3.13}$$

from which one obtains the standard deviation

$$\sigma_n(P)^2 = \mathbf{E}(\mathrm{d}_{\mathrm{tv}}(\mathbb{P}_n,P)^2) - \mathbf{E}(\mathrm{d}_{\mathrm{tv}}(\mathbb{P}_n,P))^2 \tag{3.14}$$

which can be calculated explicitly for small values of n.

The asymptotic distribution of $\mathrm{d}_{\mathrm{tv}}(\mathbb{P}_n,P)$ is given by

$$\sqrt{n}\,\mathrm{d}_{\mathrm{tv}}(\mathbb{P}_n,P) \Rightarrow \frac{1}{2}\sum_{k=0}^{\infty}|W(k)|\sqrt{p(k)(1-p(k))} \tag{3.15}$$

where $(W(k))_{k=0}^{\infty}$ is as in (3.5) and (3.6). It follows that

$$\sqrt{n}\mathbf{E}(\mathrm{d}_{\mathrm{tv}}(\mathbb{P}_n,P)) \to \frac{1}{2}\sum_{k=0}^{\infty}\mathbf{E}(|W(k)|)\sqrt{p(k)(1-p(k))} = \frac{1}{\sqrt{2\pi}}\sum_{k=0}^{\infty}\sqrt{p(k)(1-p(k))} \tag{3.16}$$

assuming the last expression to be finite. The asymptotic variance is given by

$$\begin{aligned}
\sigma_\infty^2(P) &= \lim_{n\to\infty}\mathbf{V}(\sqrt{n}\,\mathrm{d}_{\mathrm{tv}}(\mathbb{P}_n,P)) \\
&= \frac{1}{4}\sum_{k,\ell=0}\mathbf{E}\Big(\big(|W(k)|-\sqrt{2/\pi}\big)\big(|W(\ell)|-\sqrt{2/\pi}\big)\Big)\cdot \\
&\qquad\qquad \sqrt{p(k)p(\ell)(1-p(k))(1-p(\ell))} \\
&= \frac{1}{4}\sum_{k,\ell=0}(c(p(k),p(\ell))-2/\pi)\sqrt{p(k)p(\ell)(1-p(k))(1-p(\ell))}
\end{aligned} \tag{3.17}$$

with

$$c(p(k),p(\ell)) = \begin{cases} 1 &, \quad k=\ell, \\ d\left(-\sqrt{\frac{p(k)p(\ell)}{(1-p(k))(1-p(\ell))}}\right) &, \quad k\neq\ell, \end{cases}$$

and

$$d(\rho) = \mathbf{E}\left(|Z_1|\,\middle|\,\rho Z_1+\sqrt{1-\rho^2}Z_2\right|\Big) \tag{3.18}$$

where Z_1 and Z_2 are i.i.d. $\mathfrak{N}(0,1)$ random variables. The function $d(\rho)$ can be determined numerically. A plot is shown in Figure 3.1. Putting this together it is seen that

$$\sqrt{n}\,\frac{\mathrm{d}_{\mathrm{tv}}(\mathbb{P}_n,P)-\mathbf{E}(\mathrm{d}_{\mathrm{tv}}(\mathbb{P}_n,P))}{\sigma_n(P)} \Rightarrow \tag{3.19}$$

$$W_\infty(P) := \frac{\frac{1}{2}\sum_{k=0}^{\infty}\left(|W(k)|-\sqrt{\frac{2}{\pi}}\right)\sqrt{p(k)(1-p(k))}}{\sigma_\infty(P)} \tag{3.20}$$

where the latter is a zero mean and unit variance random variable. The strategy is to approximate the quantiles of the quantiles of $\sqrt{n}\,(\mathrm{d}_{\mathrm{tv}}(\mathbb{P}_n,P)-\mathbf{E}(\mathrm{d}_{\mathrm{tv}}(\mathbb{P}_n,P)))/\sigma_n(P)$ by those of $W_\infty(P)$ and to approximate the latter by a simple function of the quantiles of the standard Gaussian distribution.

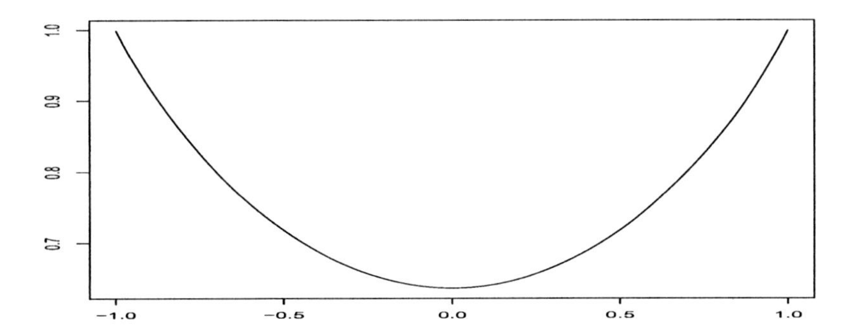

Figure 3.1 *The function $d(\rho)$ of (3.18).*

A simplification is possible if $\max_k p(k)$ is small. It follows then from (3.6) that the $W(k)$ are almost independent

$$\mathbf{E}(W(k)W(\ell)) \approx 0 \tag{3.21}$$

and (3.17) simplifies to

$$\sigma_\infty^2(P) = \frac{1}{4}\sum_{k=0}^{\infty}(1-2/\pi)p(k)(1-p(k))\,. \tag{3.22}$$

Furthermore the Lindeberg-Feller conditions imply that $W_\infty(P)$ will be approximately $\mathfrak{N}(0,1)$-distributed. An approximate α-approximation region based on d_{tv} is then given by

$$\Bigg\{P : \sqrt{n}\mathrm{d}_{\mathrm{tv}}(\mathbb{P}_n,P) \le \frac{1}{\sqrt{2\pi}}\sum_{k=0}^{\infty}\sqrt{p(k)(1-p(k))} + \mathrm{qnorm}(\alpha)\left(\frac{\pi-2}{4\pi}\left(1-\sum_{k=0}p(k)^2\right)\right)^{1/2}\Bigg\}. \tag{3.23}$$

3.4 The Kullback-Leibler and Chi-squared Discrepancies

For empirical data the normalized Kullback-Leibler and chi-squared discrepancies are

$$2nd_{\mathrm{kl}}(\mathbb{P}_n,P) = 2n\sum_{p_n(i)>0}^{\infty} p_n(i)\log(p_n(i)/p(i)) \tag{3.24}$$

and

$$nd_{\chi^2}(\mathbb{P}_n,P) = n\sum_{i=0}^{\infty}\frac{(\hat{p}_n(i)-p(i))^2}{p(i)} \tag{3.25}$$

respectively. If the support $\{i : p(i) > 0\}$ of P is finite with k-elements then the asymptotic distribution of both is χ^2_{k-1} where χ^2_{k-1} denotes the chi-squared distribution with $k-1$ degrees of freedom. If the support is not finite the discrepancies tends to infinity with increasing sample size. There seems to be no way of re-centring and re-scaling them to obtain a non-degenerate asymptotic limit. The following ad hoc approach will be taken. Put

$$k = |\{i : p(i) > 2/n\}| \tag{3.26}$$

and approximate the quantiles of the discrepancies by a quadratic function of the qunatiles of the chi-squared disribution with $k-1$ degrees of freedom.

This works well for the Kullback-Leibler discrepancy but small values of $p(i)$ as indicated by (3.11) can cause the larger quantiles of the chi-squared discrepancy to be dominated by single rare observations. The standard method of avoiding this is to group the data so that there are no small $p(i)$. But this can also have unwanted consequences. If there are several observations $x_j = i$ then the statisticians may take this to indicate that the model does not fit and would like the chi-squared discrepancy to indicate this. Grouping the observations would be counter productive. The pragmatic solution taken here is to restrict the sum to those i for which

$$np(i) \geq 2 \tag{3.27}$$

and to require

$$|\{i : np(i) \geq 2\}| \geq 2. \tag{3.28}$$

This stabilizes the quantiles and can still indicate a lack of fit for small $p(i)$. The restrictions (3.27) and (3.28) may be compared to a rule of thumb requiring an expected occupancy of 5 for each set considered in the grouped chi-squared discrepancy. This strategy works well as long as there are not too many small $p(i)$ which translates into a small support. If all of the the $p(i)$ are small then the chi-squared discrepancy becomes unstable. An example of such a model is the negative binomial model $\mathfrak{nb}(k,p)$ with p small: if $k = 5$ and $p = 0.1$ then the maximum value of $p(i)$ is 0.02. Using (3.27) and (3.28) will lead to outliers being cxcluded from the goodness-of-fit criterion. If possible outliers are important they are probably best dealt with separately by checking for and identifying them.

In the case of discrete i.i.d. models with distribution P the likelihood

$$\ell(\mathbf{x}_n, P) = \prod_{i=1}^{n} p(x(i))$$

and the Kullback-Leibler discrepancy are related by

$$D_{\mathrm{kl}}(\mathbb{P}_n, P) = \sum_i p_n(i)\log(p_n(i)) - \frac{1}{n}\log(\ell(\mathbf{x}_n, P)). \tag{3.29}$$

In particular in parametric families the maximum likelihood functional is also the minimum Kullback-Leibler discrepancy functional. This does not carry over to continuous models.

3.5 The $\mathfrak{P}_{\mathfrak{o}}(\lambda)$ Model

For models which are often used libraries can be created, referred to and extended. This is indicated here for the Poisson model.

3.5.1 The Total Variation Metric

The total variation distance is for the Poisson distribution with parameter λ is

$$\mathrm{d}_{\mathrm{tv}}(\mathbb{P}_n, \mathfrak{P}_{\mathfrak{o}}(\lambda)) = \frac{1}{2}\sum_{k=0}^{\infty}\left|\hat{p}_n(k) - \frac{\lambda^k \exp(-\lambda)}{k!}\right| . \tag{3.30}$$

For any given n and λ the quantiles of $\mathrm{d}_{\mathrm{tv}}(\mathbb{P}_n, \mathfrak{P}_{\mathfrak{o}}(\lambda))$ can be obtained by simulation. For large values of n or λ satisfactory approximations to the quantiles can be obtained by following the strategy outlined in Chapter 3.3: approximate the quantiles of

$$(\sqrt{n}\,(\mathrm{d}_{\mathrm{tv}}(\mathbb{P}_n, \mathfrak{P}_{\mathfrak{o}}(\lambda)) - \mathbf{E}(\mathrm{d}_{\mathrm{tv}}(\mathbb{P}_n, \mathfrak{P}_{\mathfrak{o}}(\lambda))))/\sigma_\infty(\mathfrak{P}_{\mathfrak{o}}(\lambda))$$

by those of $W_\infty(\mathfrak{P}_{\mathfrak{o}}(\lambda))$ of (3.19) and (3.20) with $P = \mathfrak{P}_{\mathfrak{o}}(\lambda)$. These in turn are approximated by a simple, in this case a quadratic, function of the quantiles of the standard Gaussian distribution. Estimates of the quantiles of $W_\infty(\mathfrak{P}_{\mathfrak{o}}(\lambda))$ are obtained by simulations. The range of λ must be bounded below as the convergence to $W_\infty(\mathfrak{P}_{\mathfrak{o}}(\lambda))$ is not uniform. The lower bound used here is $\lambda = 0.05$. An upper bound of $\lambda = 50$ is taken. For larger values of λ the quantiles of $W_\infty(\mathfrak{P}_{\mathfrak{o}}(\lambda))$ can be satisfactorily approximated by those of the standard normal distribution. Figure 3.2 shows the simulated quantiles of $W_\infty(\mathfrak{P}_{\mathfrak{o}}(0.05))$ (solid) and the quadratic approximation (dashed).

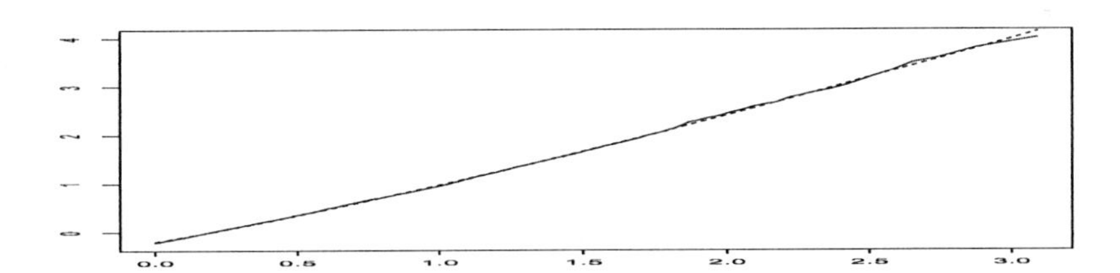

Figure 3.2 *The quantiles of the asymptotic distribution $W_\infty(\mathfrak{P}_{\mathfrak{o}}(0.05)$ of (3.19) and (3.20) (solid) and the approximation by a quadratic function of the corresponding quantiles of the standard normal distribution (dashed) plotted against the quantiles of the normal distribution.*

For small values of n and λ the distribution of $\mathrm{d}_{\mathrm{tv}}(\mathbb{P}_n, \mathfrak{P}_{\mathfrak{o}}(\lambda))$ is highly discrete and the approximation by $W_\infty(\mathfrak{P}_{\mathfrak{o}}(\lambda))$ is poor. For such values it will necessary to use the quantiles obtained from simulations and save the values for future use. However if either n or λ increases the approximation improves and becomes satisfactory. Figure 3.3 shows the simulated quantiles (solid) and the approximation (dashed) for, from top to bottom $(n, \lambda) = (500, 0.05), (100, 0.5), (40, 1), (20, 4)$.

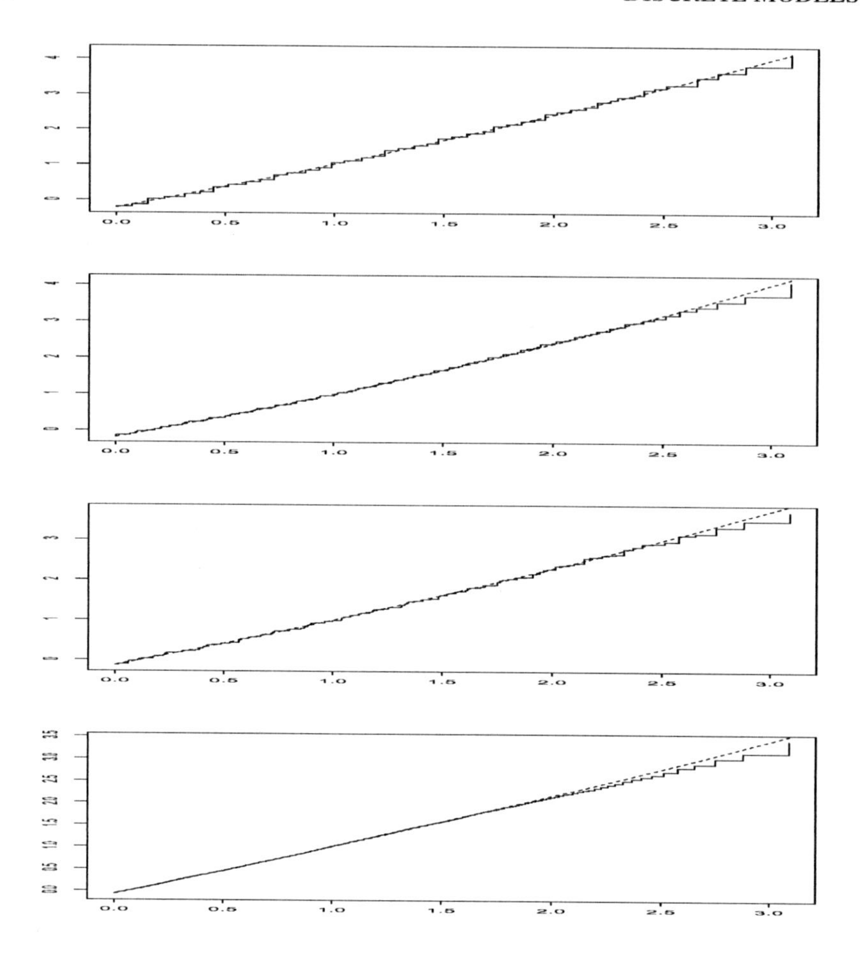

Figure 3.3 *The quantiles of (3.19) (solid) and those of the asymptotic distribution (3.20) (dashed) for, from top to bottom,* $(n,\lambda) = (500,0.05),(100,0.5),(40,1),(20,4)$, *plotted against the quantiles of the normal distribution.*

3.5.2 *The Kullback-Leibler and Chi-squared Discrepancies*

As the procedure is the same for both discrepancies only the chi-squared discrepancy will be described.

The chi-squared discrepancy for the Poisson distribution with parameter λ is

$$d_{\chi^2}(\mathbb{P}_n, \mathfrak{P}_o(\lambda)) = \sum_{k=0}^{\infty} \frac{(\hat{p}_n(k) - p(k,\lambda))^2}{p(k,\lambda)} \tag{3.31}$$

where

$$p(k,\lambda) = \frac{\lambda^k \exp(-\lambda)}{k!}.$$

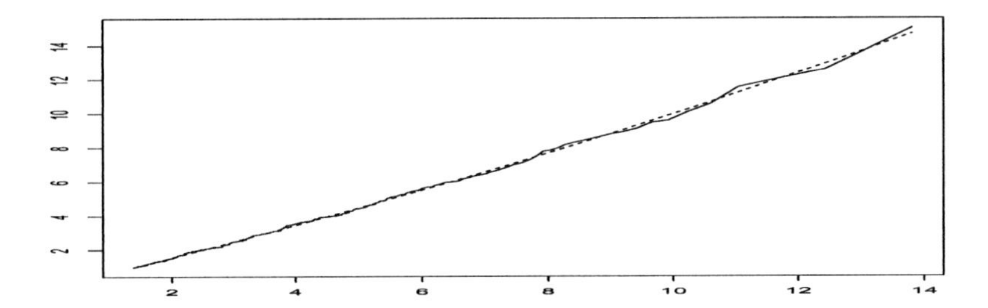

Figure 3.4 *The quantiles of the chi-squared discrepancy for the* $\mathfrak{P}_o(0.52)$ *model (solid) with* $n = 53$ *and the approximation by a quadratic function of the corresponding quantiles of the chi-squared distribution (dashed) plotted against the quantiles of the chi-squared distribution.*

As mentioned above the chi-squared discrepancy can be dominated by a single k for which $p(k,\lambda)$ is very small but $\hat{p}_n(k) \geq 1/n$. In the following the pragmatic solution of (3.27) and (3.28) will be adopted. This leads to the modified discrepancy

$$\tilde{d}_{\chi^2}(\mathbb{P}_n, \mathfrak{P}_o(\lambda)) = \sum_{np(k,\lambda)\geq 2} \frac{(\hat{p}_n(k) - p(k,\lambda))^2}{p(k,\lambda)} \tag{3.32}$$

where in addition (3.28) is required to hold. No such modification is required for the Kullback-Leibler discrepancy.

The quantiles of $\tilde{d}_{\chi^2}(\mathbb{P}_n, \mathfrak{P}_o(\lambda))$ can be approximated by a quadratic function of the quantiles of the chi-squared distribution with $k_l - 1$ degrees of freedom where k_l is defined as the number of values of k for which

$$p(k,\lambda) = \frac{\lambda^k \exp(-\lambda)}{k!} \geq \frac{2}{n} \tag{3.33}$$

holds. Figure 3.4 shows the plot of the empirical values of the quantiles and the approximation for $n = 53$ and $\lambda = 0.52$ and the 0.5 to 0.999 quantiles of the corresponding chi-squared distribution.

The chi-squared discrepancy is discrete and for a given λ there is a minimum sample size for which the constraint (3.33) holds and the approximation by a quadratic function of the quantiles is sufficiently good. For $\lambda = 0.1, 0.5$ and 2 the smallest samples size are $n = 750$, 40 and 17 respectively.

3.6 The $\mathfrak{b}(k,p)$ and $\mathfrak{nb}(k,p)$ Models

The binomial $\mathfrak{b}(k,p)$ and the negative binomial $\mathfrak{nb}(k,p)$ models are treated in the same manner as the Poisson model.

For the binomial model the range of p is restricted to $[0.05, 0.5]$, the value 0.5 is because of the asymmetry of the binomial distribution. For the negative binomial distribution the range used is $[0.05, 0.95]$.

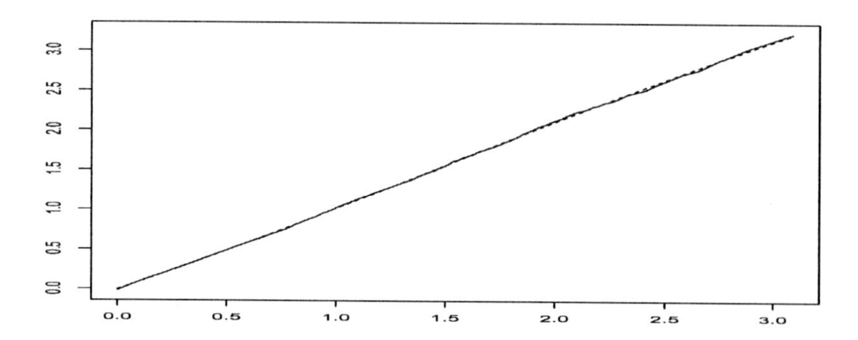

Figure 3.5 *The quantiles of the asymptotic distribution* $W_\infty(\mathfrak{g}(0.05))$ *of (3.19) and (3.20) (solid) and the approximation by a quadratic function of the corresponding quantiles of the standard normal distribution (dashed) plotted against the quantiles of the normal distribution.*

The calculation of the asymptotic distribution $W_\infty(P)$ of (3.19) and (3.20) is as for the Poisson distribution. Figure 3.5 shows approximation to the asymptotic distribution $W_\infty(\mathfrak{g}(0.05))$ of (3.19) and (3.19) by a quadratic function of the quantiles of the normal distribution.

Care must be taken when approximating the distribution of $d_{tv}(\mathbb{P}_n, \mathfrak{g}(p))$ by the asymptotic distribution $W_\infty(\mathfrak{g}(p))$ as the goodness-of-fit does not necessarily improve with increasing sample size. Figure 3.6 shows the approximation for $p = 0.75$ and for, from top to bottom, sample sizes of $n = 100$, $n = 1000$ and $n = 10000$. The lack of monotonicity in the convergence to the asymptotic distribution is apparent.

Figure 3.7 shows the results for the chi-squared distribution for the geometric model $\mathfrak{g}(0.52)$ with $n = 53$. It corresponds to Figure 3.4 for the Poisson distribution.

3.7 The Flying Bomb Data

The flying bomb data of Table 2.1 [27] were already analysed in Chapter 2.13. Using the partition of [27] and the approximation region defined by (2.57) leads to the approximation interval $[0.838, 1.017]$. In all the area was hit by 537 bombs and to make the number of bombs tally the one square with ≥ 5 hits must have had exactly seven hits. Suppose now that the raw form of the chi-squared discrepancy

$$d_{\chi^2}(P, \mathbb{P}_n) = \sum_{p(i)>0} (\hat{p}_n(i) - p(i))^2 / p(i)$$

had been used. Then as

$$\frac{576(1/576 - p(7, 0.838))^2}{p(7, 0.838)} = 67.71, \quad \frac{576(1/576 - p(7, 1.017))^2}{p(7, 0.1.017)} = 19.55.$$

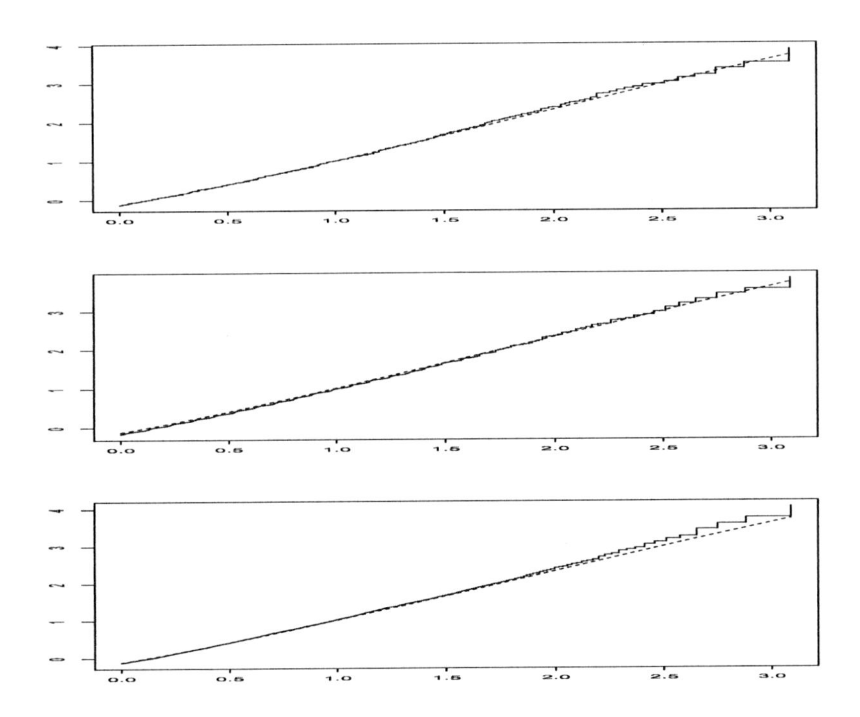

Figure 3.6 *Approximation of (3.19) for the model* $\mathfrak{g}(0.75)$ *(dashed) by the asymptotic distribution (3.20)(solid). Top panel:* $n = 100$. *Centre panel:* $n = 1000$. *Bottom panel:* $n = 10000$.

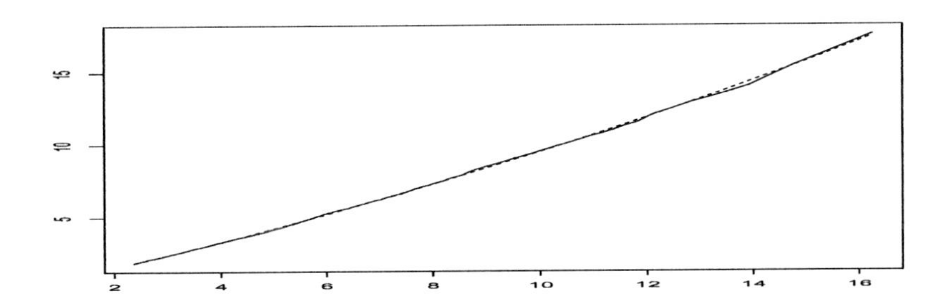

Figure 3.7 *The quantiles of the chi-squared discrepancy for the* $\mathfrak{g}(0.52)$ *model (solid) with* $n = 53$ *and the approximation by a quadratic function of the corresponding quantiles of the chi-squared distribution (dashed) plotted against the quantiles of the chi-squared distribution.*

the value of the discrepancy would have been dominated by the single square with seven hits. Note that

$$np(k,\lambda) = 576 * p(7,1) = 0.042 < 2$$

so that the value 7 is excluded in the calculation of the chi-squared statistic. The contribution of 7 to the values of the Kullback-Leibler discrepancy is much smaller, for example

$$576 \cdot \frac{1}{576} \log\left(\frac{1}{576}/p(7,1)\right) = 3.17$$

and is not excluded when calculating the Kullback-Leibler discrepancy. Although large the value 7 is not exceptional for samples of size 576 and the relevant values of λ.

The approximation region defined by (2.57) uses the normal approximation for the mean of data generated under the Poisson distribution. For this particular distribution an exact approximation based on the mean can be calculated based on the fact that

$$n\overline{\mathbf{X}}_n(\lambda) \stackrel{D}{=} \mathfrak{P}_o(n\lambda).$$

An α-approximation region using this is given by

$$\mathscr{A}_{\mathrm{ave}}(\mathbf{x}_n,\alpha,\Lambda) = \{\lambda : \mathrm{qpois}((1-\alpha)/2, n\lambda) \le n\bar{x}_n \le \mathrm{qpois}((1+\alpha)/2, n\lambda)\}. \quad (3.34)$$

For the flying bomb data the approximation interval is

$$\mathscr{A}_{\mathrm{ave}}(\mathbf{x}_n, 0.975, \Lambda) = [0.845, 1.025] \quad (3.35)$$

which is essentially the same as that using the normal approximation namely $[0.838, 1.017]$.

Repeating the analysis of Chapter 2.13 but using the version of the chi-squared discrepancy $\tilde{d}_{\chi^2}$ given by (3.32) and using the approximations to the distribution combined with the exact approximation region of (3.34) gives the approximation interval $[0.845, 1.025]$: the restriction due the chi-squared discrepancy is not active. The values of $\tilde{d}_{\chi^2}$ for $\lambda = 0.845$ and $\lambda = 1.025$ are 6.288 and 5.464 with 0.975 quantiles 11.32 and 11.16 respectively.

3.8 The Student Study Times Data

In Chapter 2.8 the data consisting of the lengths of study in semesters of 258 German students as shown in Figure 1.6 were analysed. The data were modelled as $\Gamma(\gamma,\lambda)$ and an approximations for the mean and standard deviation were derived, once with a goodness-of-fit check based on the Kuiper metric of order one $\mathrm{d}^1_{\mathrm{ku}}$ of (2.25) and once without. The approximation regions are shown in Figures 2.9 and 2.6 respectively. The difference between the two shows that the goodness-of-fit restriction was active for some parameter values in the approximation region.

The Kuiper metrics do not take the discreteness of the data into account. An alternative which does take the discreteness into account is to discretize the model to

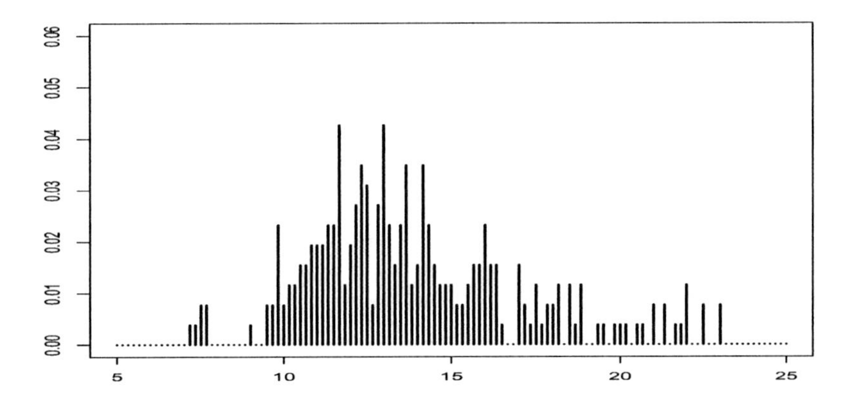

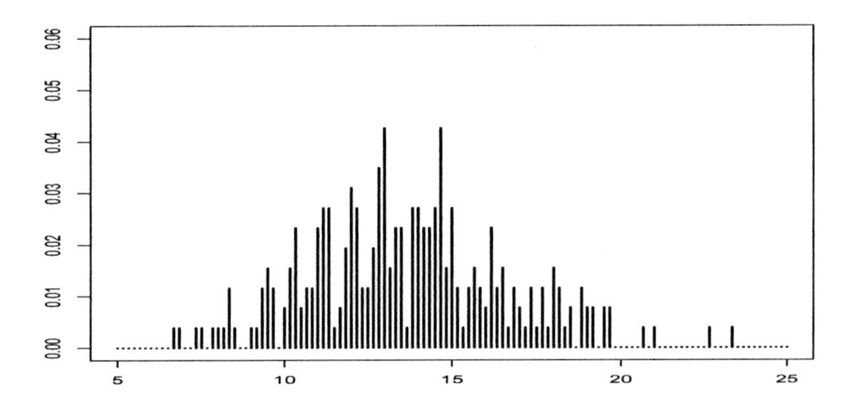

Figure 3.8 *Upper panel: the relative frequencies of the values of the student data of Figure 1.6. Lower panel: the same for a random gamma sample with shape* $\gamma = 18$ *and scale* $\lambda = 0.76$*:* $n = 258$.

the same degree of discretization as the data and then and to use a goodness-of-fit measure for discrete data. The upper panel of Figure 3.8 shows the empirical relative frequencies data, the lower panel shows the relative frequencies of a random sample from a discretized gamma distribution with shape parameter $\gamma = 18$ and scale parameter $\lambda = 0.76$. These values are consistent with the mean and standard deviation of the student data.

For the flying bomb data it was argued in Chapter 2.13 that the chi-squared discrepancy was an appropriate measure of fit. In the case of the student data there is no good reason for upweighting small probabilities and it seems reasonable to treat all probabilities equally. The total variation metric does this and is the one which will be used.

The relative frequencies of the data are all small and less than 0.05. As the sample size $n = 258$ is large it is to be expected that the distribution of the total variation metric will be well approximated by that of its asymptotic distribution W_∞ of (3.19) and (3.20). The asymptotic distribution contains the asymptotic expected values (3.16) and the asymptotic variance (3.17). For $n = 258$ these are insufficiently accurate but the deficit can be remedied by calculating the expected values exactly as in (3.12) and (3.14). If this is done the distribution of W_∞ can in turn be well approximated by the $\mathfrak{N}(0,1)$ distribution (Chapter 3.3).

On spending $0.05/3 = 0.0167$ on the mean, the standard deviation and the total deviation metric the result is that the total deviation restriction is never active. The 0.983 quantile of the total variation metric is approximately 0.28 over the range of values of the shape and scale parameters consistent with the mean and standard deviation of the data. The total variation distance from the data itself is approximately 0.26, again approximately constant over the range of values of the parameters. The approximation regions are shown in Figure 3.9. Because the total deviation restriction is not active these are larger than the approximation regions of Figure 2.6 as less has been spent on each feature.

A second set of student lengths of study, the one shown in Figure 2.11, were analysed in Chapter 2.10. It is shown again as Figure 3.10 for ease of reference. In Figure 2.11 it was shown that if the Kuiper metric is used as a measure of adequacy then there are adequate models for $\alpha = 0.985$ but not for $\alpha = 0.95$. If the total variation metric is used then the fit is so poor that even for $\alpha = 0.999999$ there are no adequate gamma models. As an example, the $\Gamma(16, 0.9)$ model adequately reflects the mean and standard deviation of the data. After discretization the total variation distance from the data is 0.584. For this model the 0.999999 quantile of the total variation distance is 0.375 (see (3.23)). The values are similar for all gamma models which adequately reflect the mean and standard deviation.

Another very different way of measuring the degree of fit of a model to the data is due to [142]. It is based on the idea that any model will be wrong if the sample size is sufficiently large. In the opposite direction, if the sample size is small then many models will be acceptable. The proposal is to determine the largest sample size such that the model is an acceptable approximation, defined as that sample size for which 50% of the bootstrapped samples of that size have at least one adequate approximation within the family of models. If this idea is applied to the student data of Figure 3.10 answer is about $n = 40$ for the total variation metric. The method has the advantage of immediate interpretability and does not require the calculation of extreme quantiles.

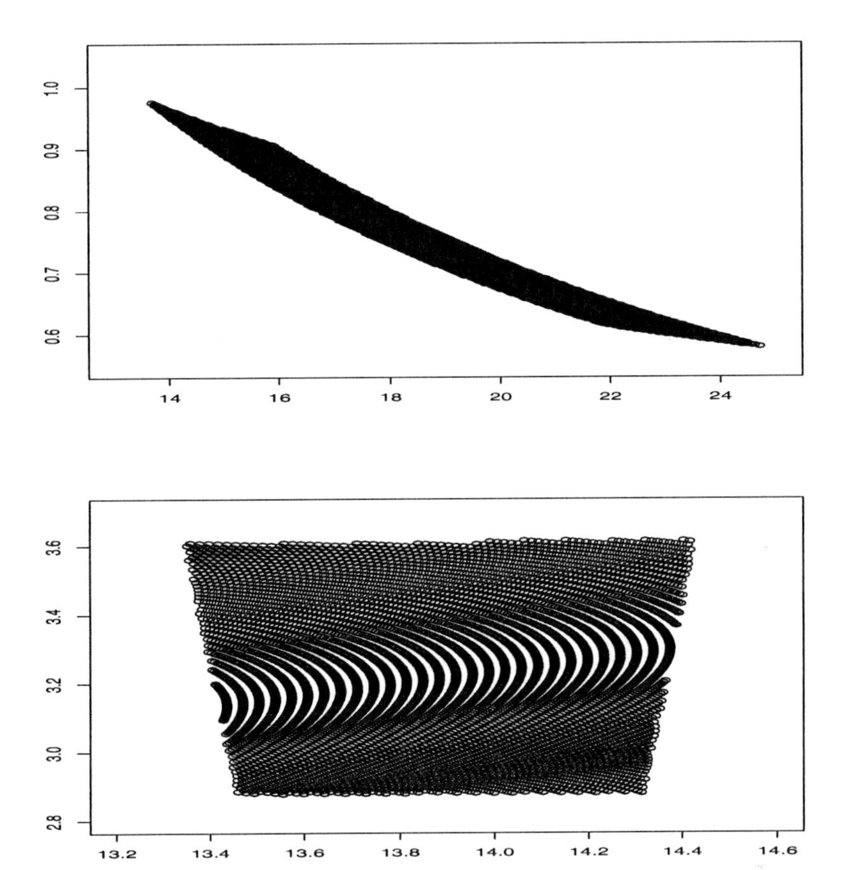

Figure 3.9 *Upper panel: the* 0.95 *approximation for the parameters* (γ, λ) *of the gamma distribution for the data of Figure 1.6 based on the mean, the standard deviation and a measure of goodness-of-fit using the total variation metric. Lower panel: the corresponding approximation region for the mean and standard deviation.*

3.9 Contingency Tables

Consideration will be restricted to tables of the form

$$
\begin{array}{c|ccccc|c}
 & 1 & 2 & \cdots & J-1 & J & \\
\hline
1 & n_{1,1} & n_{1,2} & \cdots & n_{1,J-1} & n_{1,J} & n_{1\cdot} \\
2 & n_{2,1} & n_{2,2} & \cdots & n_{2,J-1} & n_{2,J} & n_{2\cdot} \\
\cdot & \cdot & \cdot & \cdots & \cdot & \cdot & \cdot \\
\cdot & \cdot & \cdot & \cdots & \cdot & \cdot & \cdot \\
I & n_{I,1} & n_{I,2} & \cdots & n_{I,J-1} & n_{I,J} & n_{I\cdot} \\
\hline
 & n_{\cdot 1} & n_{\cdot 2} & \cdots & n_{\cdot J-1} & n_{\cdot J} & n_{\cdot\cdot}
\end{array}
\tag{3.36}
$$

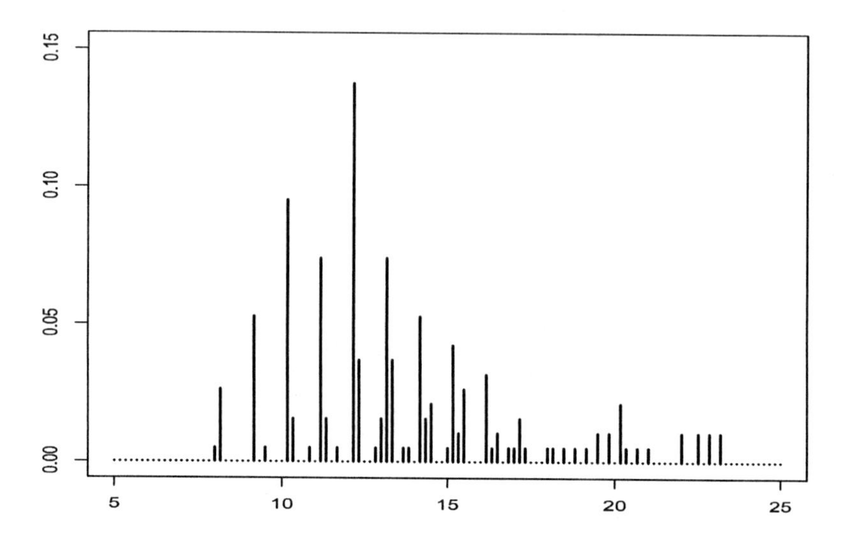

Figure 3.10 *Frequencies of lengths of study in semesters of 189 German students.*

where

$$n_{i\cdot} = \sum_{j=1}^{J} n_{j,i},\; i = 1,\ldots,I, \quad n_{\cdot j} = \sum_{i=1}^{I} n_{j,i},\; j = 1,\ldots,J$$

and

$$n_{\cdot\cdot} = \sum_{i=1}^{I} n_{i\cdot} = \sum_{j=1}^{J} n_{\cdot j}\,.$$

and the $n_{i\cdot}$ are given. The data are to be modelled as I independent samples of size J with the ith row modelled as a multinomial $\mathfrak{mult}(n_{i\cdot}, \mathbf{p}_i)$ random variable with

$$\mathbf{p}_i = (p_{i,1}, p_{i,2}, \cdots, p_{i,J}).$$

A standard question when analysing such data is whether the outcomes represented by the J columns are independent of the factors represented by the I rows. In terms of the modelling this is translated into equality of the $\mathbf{p}_i$:

$$\mathbf{p}_1 = \mathbf{p}_1 = \ldots = \mathbf{p}_I = \mathbf{p} = (p_1, \ldots, p_J)\,. \tag{3.37}$$

The independence hypothesis will be accepted, or at least not rejected, if the data can be adequately approximated by the model (3.37) with equal $\mathbf{p}_i$. This is often done by putting

$$p_j = \tilde{p}_j = \frac{n_{\cdot j}}{n_{\cdot\cdot}},\; j = 1,\ldots,J \tag{3.38}$$

	outcome				
Diet	cancers	fatal heart disease	Non-fatal heart disease	healthy	total
AHA	15	24	25	239	303
Mediterranean	7	14	8	273	302
Total	22	38	33	512	605

Table 3.1 *Health and diet.*

and then calculating the chi-squared discrepancy

$$D_{\chi^2} = \sum_j \left(\sum_i n_{i\cdot} \frac{(\hat{p}_{i,j} - \tilde{p}_j)^2}{\tilde{p}_j} \right) = \sum_j \left(\sum_i \frac{\left(n_{i,j} - \frac{n_{i\cdot} n_{\cdot j}}{n_{\cdot\cdot}} \right)^2}{\frac{n_{i\cdot} n_{\cdot j}}{n_{\cdot\cdot}}} \right) \tag{3.39}$$

between the empirical distribution

$$\hat{p}_{i,j} = n_{i,j}/n_{i\cdot} \tag{3.40}$$

and the model with

$$\mathbf{p} = \tilde{\mathbf{p}}. \tag{3.41}$$

For data generated under (3.41) the chi-squared statistics of (3.39) is approximately chi-squared distributed with $(I-1)(J-1)$ degrees of freedom.

The data in Table 3.1 are taken from

```
Online Statistics Education: A Multimedia Course of Study
(http://onlinestatbook.com/).
Project Leader: David M. Lane, Rice University.
```

Of interest is whether health is dependent on diet. The value of the chi-squared statistic is 16.55 and the number of degrees of freedom is 3. This results in a p-value of 0.0009 with the conclusion that health is dependent on diet, at least on the basis of this data. In most cases this would seem to be the end of the analysis. The null hypothesis of independence is either accepted or rejected, in this case rejected, and no more is to be said.

Before deciding on a procedure and its output thought must be given to what the experimenter wishes to know. This could include bounds on the plausible values of each p_{ij}, bounds on the differences $p_{i,j} - p_{i',j}$ and whether the responses can be adequately modelled by the same $\mathbf{p}$ for all or for some subsets of the factors. In this case, bounds for the values of the p_j may be desired. The definition of adequacy should also include an honesty term, that is some measure of the goodness-of-fit of the model to the data.

The analysis of the data of Table 3.1 fulfils none of these requirements. As stated above the independence hypothesis will accepted if there exists an adequate model of the form (3.37). The argument based on the chi-squared statistic has not however specified such a model although the model (3.37) with $\mathbf{p} = \tilde{\mathbf{p}}$ as in (3.41) would seem to be the obvious choice. For data generated under the model but with the $\mathbf{p}_i$ satisfying (3.37) with $\mathbf{p} = \mathbf{p}'$ any reasonable method of specifying those values $\mathbf{p}$ consistent with the data should include $\mathbf{p}'$ with probability α. If the chi-squared discrepancy is used with three degrees of freedom then then $\mathbf{p}'$ will accepted with a probability less than the specified α. The correct number of degrees of freedom is not three but six. If this number the p-value increases from 0.0009 to 0.0111. Thus if α were set to 0.99 there would indeed be adequate models consistent with the hypothesis of independence.

The use of the chi-squared discrepancy in the context of contingency tables is universal almost as if it were legally prescribed. As mentioned above the chi-squared discrepancy upweights small probabilities. Whether this is desirable or not depends on the real world problem behind the data. There is no a priori reason why it should be desirable. The total variation metric

$$d_{\text{tv}}(\mathbb{P}_i, \mathbf{P}_i) = 0.5 \sum_{j=1}^{J} |\hat{p}_{i,j} - p_{i,j}| \tag{3.42}$$

treats all cells equally. Here $\mathbb{P}_i$ denotes the empirical distribution of the ith row with density $\hat{p}_{i,j}$ given by (3.40) and $\mathbf{P}_i$ is the model with density $\mathbf{p}_i$. The asymptotic distribution of $d_{\text{tv}}(\mathbb{P}_i, \mathbf{P}_i)$ can be calculated but depends on $\mathbf{p}_i$. It seems plausible that the distribution of $d_{\text{tv}}(\mathbb{P}_i, \mathbf{P}_i)$ under the model is stochastically largest when the values of $p_{i,j}$ are as equal possible. This is supported by simulations. Thus if $J = 4$ and the largest $p_{i,j}$ is 0.43 the distribution is stochastically largest when $\mathbf{p}_i = (0.43, 0.19, 0.19, 0.19)$. Based on this libraries can be created parameterized by J, the sample size and the largest value of $\mathbf{p}_i$.

Given α there is $1-\alpha$ left to spend on whatever aspects of the data are deemed to be of interest. Suppose these are approximation intervals for the individual $p_{i,j}$ and for the differences $p_{i,j} - p_{i',j}$ and that an honesty term is to be included. Suppose further that amounts β_1, β_2 and β_3 respectively are to be spent on these three aspects with $\beta_1 + \beta_2 + \beta_3 = 1 - \alpha$. As there are IJ parameters $p_{i,j}$ an amount $\beta_1/(IJ)$ is available for each approximation interval being split equally between the lower and upper bounds. Similarly there are $I(I-1)K$ differences with an amount $\beta_2/(I(I-1)J)$ being available for each. Finally as there are I samples an amount β_3/I is available for a goodness-of-fit test for each sample.

The approximation region $\mathscr{A}_1$ based on the approximation intervals for the individual $p_{i,j}$ is

$$\mathscr{A}_1 = \left\{ \mathbf{P} : |\hat{p}_{i,j} - p_{i,j}| \le q_1 \sqrt{\frac{p_{i,j}(1-p_{i,j})}{n_i}} \right\} \tag{3.43}$$

where

$$q_1 = \text{qnorm}\left(1 - \frac{\beta_1}{2IJ}\right).$$

For simplicity the normal approximation has been used in (3.43) but exact bounds can be calculated if desired.

Similarly the approximation region $\mathscr{A}_2$ based on the approximation regions for the differences $p_{i,j} - p_{i',j}$ is

$$\mathscr{A}_2 = \left\{ \mathbf{P} : |\hat{p}_{i,j} - \hat{p}_{i',j} - p_{i,j} + p_{i',j}| \le q_2 \sqrt{\frac{p_{i,j}(1-p_{i,j})}{n_{i\cdot}} + \frac{p_{i',j}(1-p_{i',j})}{n_{i'\cdot}}} \right\} \tag{3.44}$$

where

$$q_2 = \text{qnorm}\left(1 - \frac{\beta_2}{2I(I-1)J}\right).$$

For simplicity the normal approximation has been used but again exact bounds can be calculated if desired.

Finally the approximation region $\mathscr{A}_3$ based on the total variation metric is given by

$$\mathscr{A}_3 = \{\mathbf{P} : d_{\text{tv}}(\hat{\mathbf{p}}_i, \mathbf{p}_i) \le q_3(\mathbf{p}_i)\} \tag{3.45}$$

where $q_3(\mathbf{p}_i)$ is the $1 - \beta_3/I$ quantile of the total variation distance $d_{\text{tv}}(\mathbb{P}_i, \mathbf{p}_i)$.

The final approximation region is

$$\mathscr{A} = \mathscr{A}_1 \cap \mathscr{A}_2 \cap \mathscr{A}_3. \tag{3.46}$$

For data generated under the model $\mathbf{P}$

$$P(\mathbf{P} \in \mathscr{A}) \ge \alpha. \tag{3.47}$$

The result of the procedure is $\mathscr{A}$ which is a subset of $\mathbb{R}^8$ and not describable analytically. It can only be turned into an output by considering a grid of values. The approximation set $\mathscr{A}_1$ provides simple bounds for the individual $p_{i,j}$ and hence the grid can be restricted to to these intervals.

For the data of Table 3.1 the choice

$$\alpha = 0.95, \beta_1 = \beta_2 = 0.02, \beta_3 = 0.01, ngrid = 30 \tag{3.48}$$

results in 42052 points in $\mathbb{R}^8$. This is not fit for intellectual digestion and hence summary values are required. These will clearly be the approximation intervals for the individual $p_{i,j}$ and the differences $p_{i,j} - p_{i',j}$. The values of the total variation distance for any particular model can be given but are not of direct interest.

For the data of Table 3.1 and the choice (3.48) the output is given in Tables 3.2 and 3.3. The approximation interval for $p_{1,4} - p_{2,4}$ does not contain 0 and hence there is no adequate model with $\mathbf{p}_1 = \mathbf{p}_2$. In other words, for this degree of approximation, $\alpha = 0.95$, and the choice of β_1, β_2 and β_3 there is no adequate model under which the outcomes are independent of the factors. In fact for $\alpha = 0.95$ there is no choice of β_1, β_2 and β_3 for which there exists an adequate independent model. This is also the case for $\alpha = 0.995$ with $\beta_1 = \beta_2 = 0.004$, $\beta_3 = 0.002$. However if

$$\alpha = 0.9975, \beta_1 = \beta_2 = 0.001, \beta_3 = 0.0005, ngrid = 30 \tag{3.49}$$

	J			
I	1	2	3	4
1	0.0219	0.0422	0.0446	0.7141
1	0.0991	0.1275	0.1283	0.8545
2	0.0068	0.0200	0.0085	0.8462
2	0.0626	0.0700	0.0674	0.9479

Table 3.2 *Lower (first line) and upper (second line) approximation bounds for the individual* $p_{i,j}$.

J			
1	2	3	4
-0.0306	-0.0293	-0.0059	-0.2008
0.0833	0.0950	0.1179	-0.0296

Table 3.3 *Lower (first line) and upper (second line) approximation bounds for the differences* $p_{1,j} - p_{2,j}$.

	J			
I	1	2	3	4
1	0.0167	0.0344	0.0366	0.6921
1	0.1149	0.1547	0.1590	0.8697
2	0.004	0.0150	0.0056	0.8278
2	0.0761	0.1107	0.0814	0.9567

Table 3.4 *Lower (first line) and upper (second line) approximation bounds for the individual* $p_{i,j}$ *for the choice (3.49).*

J			
1	2	3	4
-0.0567	-0.0629	-0.0353	-0.2345
0.1093	0.1287	0.1474	0.0041

Table 3.5 *Lower (first line) and upper (second line) approximation bounds for the differences* $p_{1,j} - p_{2,j}$ *for the choice (3.49).*

J			
1	2	3	4
0.0167	0.0344	0.0366	0.8278
0.0761	0.1107	0.0814	0.8596

Table 3.6 *Lower (first line) and upper (second line) approximation bounds for the* p_j *for the independent models for the choice (3.49).*

	J			
I	1	2	3	4
		chi-squared		
1	1.439	1.335	4.344	1.075
2	1.465	1.301	4.400	1.188
		total variation $\cdot 10^3$		
1	6.571	8.303	13.98	27.35
2	6.653	8.226	14.12	28.85

Table 3.7 *The individual cell values for the chi-squared discrepancy and the total variation metric.*

then there do exist independent models which are adequate approximations for the data. The results are shown in Tables 3.4 and 3.5.

All the intervals in Table 3.5 contain 0 and hence it seem plausible that there exist adequate independent models. This can be confirmed by checking the goodness-of-fit over a grid of values of the intervals of Table 3.5. There are 6316 adequate models on the grid from which it is possible to obtain lower and upper bounds for the p_j. These are given in Table 3.6.

It may be noted that it is the difference $p_{1,4} - p_{2,4}$ for the fourth outcome which requires such a high value of α. The approximation intervals for the other three differences contain 0 even for $\alpha = 0.95$ as may be seen in Table 3.3. This contrasts with the chi-squared analysis where it is the third outcome which has by far the largest of the individual chi-squared values. These are given in Table 3.7 together with the individual values of the total variation metric. The former is dominated by the values for the third outcome and the latter by the values for the fourth outcome.

Chapter 4

Outliers

4.1 Outliers, Data Analysis and Models

Data often contain outliers, possibly more often than is realized as they are not always detected. Outliers may be due to transmission mistakes such as an incorrect decimal place, or they may be due to faulty measurement without any deeper significance or they may be the most important observations in the data (see [175] on the Antarctic ozone hole[1]). Undetected outliers can lead to a misleading analysis of the data or to a failure to detect the most important observations.

In some situations there is no need for statistical methodology to detect outliers as a simple check of each observation is sufficient. This is the case for the student data exhibited in Figure 1.6. At the time the data were collected the range of plausible values for the length of study was known and any non-plausible value could be simply discarded. In analytical chemistry the situation is different. The range of plausible values is not known a priori, or is too large to be of practical use, and outliers are common. Indeed one of the main problems when evaluating interlaboratory tests is the identification and treatment of outliers (see [1]). An example of such data is given in Table 1.5: the boxplots are shown in Figure 1.1. The 'abbey' data of the R-package MASS give thirty-one determinations of the nickel content (ppm) in a Canadian syenite rock. There is one clear outlier. The problem of outliers is also one of the main difficulties when monitoring patients on-line on an intensive care unit (see [41]).

[1] The author was present at one such Oberwolfach meeting and can confirm Pukelsheim's account.

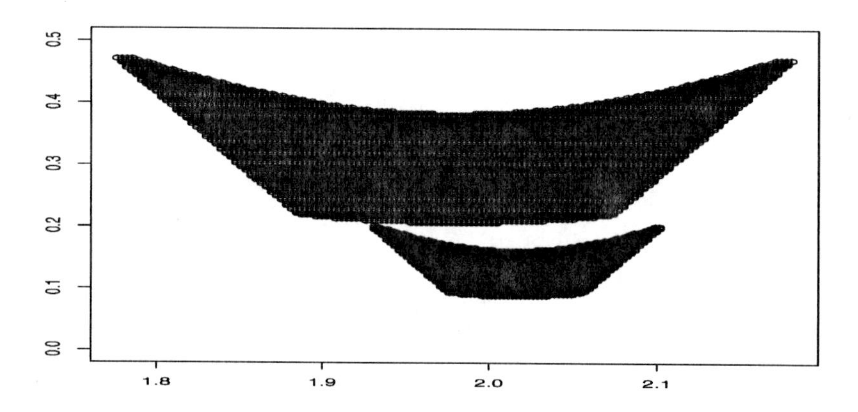

Figure 4.1 *Approximation regions for the mean and standard deviation for the copper data (small) and for the copper data with 1.7 replaced by 0.7 (large).*

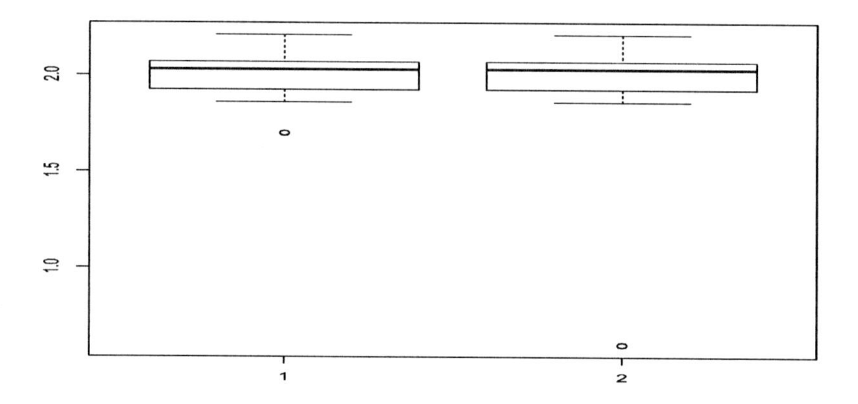

Figure 4.2 *Left: the boxplot for the original copper data. Right: boxplot with 1.7 replaced by 0.7.*

The effect of a single outlier can be seen as follows. The smallest value in the copper data of Exercise 1.3.3 is 1.7 and although considerably smaller than all the other measurements it is not sufficiently small to be classified as an outlier. If the 1.7 value is replaced by 0.7 then this is clearly an outlier. The 0.95-approximation region for the mean and standard deviation are shown in Figure 4.1. The smaller region is for the original data, the larger region if 1.7 is replaced by 0.7. The regions are not honest in that no shape constraint has been imposed. The boxplots are shown in Figure 4.2.

The identification of outliers assumes some idea of how large an observation must be to qualify as an outlier. In what follows 'large' will mean 'large compared to the other observations' and this requires some underlying coarse model. The identification of outliers for data which can be roughly approximated in their majority by a normal model differs from that required for an exponential model, although the basic ideas are the same (see [32] for the normal model and [187] for the exponential case). There are often limitations placed on methods for detecting outliers. One important one is that of invariance with respect to the units of measurements. Thus the same outliers should be identified when the temperature is measured in Fahrenheit and when it is measured in Centigrade.

Outliers is a topic which is treated in text books in a superficial manner, if it is treated at all, and the advice given is often simply wrong, even in a book devoted to the subject [12]. The reason for this state of affairs is probably the fact that outliers do not fit easily into either the frequentist or the Bayesian approaches to statistics, as may be seen by referring to [169], [12] and the reaction of some discussants to [32]. A reasonable treatment of outliers requires concepts such as breakdown point and bias (2.55) which have no place in either of those approaches. This is all the more surprising since such concepts were introduced some thirty years ago [61, 101]. An excellent discussion of outliers is given in Chapter 1.4 of [103].

4.2 Breakdown Points and Equivariance

The concept of breakdown point was introduced in [99]. The following simplified and finite sample version is to be found in [61]. Given a sample $\mathbf{x}_n$ and an integer $k, 0 \le k \le n$, a generic replacement sample to be denoted by $\mathbf{x}_n^k$ is one for which

$$\sum_{i=1}^{n} \{x_i^k \neq x_i\} \le k.$$

In other words $\mathbf{x}_n^k$ can be obtained from $\mathbf{x}_n$ by replacing at most k of the x_i by other values. Its empirical distribution will be denoted by $\mathbb{P}_n^k$. The finite sample breakdown point at $\mathbf{x}_n$ of a real-valued statistical functional T is defined by

$$\varepsilon^*(T, \mathbb{P}_n) = \min\{k/n : \sup_{\mathbb{P}_n^k} |T(\mathbb{P}_n) - T(\mathbb{P}_n^k)| = \infty\}. \tag{4.1}$$

The mean functional T_{ave} of (2.12) has a breakdown point $\varepsilon^*(T_{\text{ave}}, \mathbb{P}_n) = 1/n$ for any data $\mathbf{x}_n$. A little thought shows that the median functional T_{med} as defined by (2.54) has a breakdown point

$$\varepsilon^*(T_{\text{med}}, \mathbb{P}_n) = \lfloor (n+1)/2 \rfloor / n \tag{4.2}$$

for any data set $\mathbf{x}_n$.

It is a simple matter to define a functional T with a finite sample breakdown point of 1 namely T_0 defined by $T_0(P) = 0$ for all P. Such trivial functionals can be excluded by requiring that the functional satisfies certain conditions. In the location-scale problem is seem reasonable to require affine equivariance defined as defined by (2.15) and (2.16) for location and scale functionals respectively. The following theorem holds:

Theorem 4.2.1 *Let T_L be a measure of location, that is T_L satisfies (2.15). Then for any sample $\mathbf{x}_n$ it holds that*

$$\varepsilon^*(T_L, \mathbb{P}_n) \leq \lfloor (n+1)/2 \rfloor / n.$$

To prove the theorem it is not necessary to make full use of the affine equivariance. The theorem continues to hold for functionals which are translation equivariant, that is $T(P^b) = T(P) + b$ where $P^b(B) = P(\{x : x + b \in B\})$.

The mean functional T_{ave} is a location functional as is the median functional T_{med} defined by (2.54). From (4.2) and Theorem 4.2.1 it follows that T_{med} has the highest possible finite sample breakdown point.

The same ideas can be applied to scale functionals T_S defined by (2.16). As in the location case, breakdown will be said to occur if arbitrarily large values are possible. The question is whether values tending to zero should also be regarded as breakdown. One reason for doing this is that normalization of residuals requires dividing by some measure of scale. The following definition of breakdown takes an implosion into account

$$\varepsilon^*(T_S, \mathbb{P}_n) = \min\{k/n : \sup_{\mathbb{P}_n^k} |\log(T_S(\mathbb{P}_n^k)) - \log(T_S(\mathbb{P}_n))| = \infty\} \tag{4.3}$$

with the convention that $\varepsilon^*(T_S, \mathbb{P}_n) = 0$ if $T_S(\mathbb{P}_n) = 0$.

In order to state the theorem for measures of scale corresponding to Theorem 4.2.1 the following definition is required. For any probability measure P put

$$\Delta(P) = \max_x P(\{x\}). \tag{4.4}$$

For an empirical distribution $\mathbb{P}_n$ this becomes

$$\Delta(\mathbb{P}_n) = \max_x \mathbb{P}_n(\{x\}) = \max_x |\{j : x_j = x\}|/n.$$

so that $\Delta(\mathbb{P}_n)$ is the largest proportion of sample values which are the same.

Theorem 4.2.2 *Let T_S be a measure of scale. Then for any sample $\mathbf{x}_n$*

$$\varepsilon^*(T_S, \mathbb{P}_n) \leq \lfloor (n(1 - \Delta(\mathbb{P}_n)) + 1)/2 \rfloor / n.$$

The proof may be found in [42].

The standard deviation functional T_{sd} (2.14) has a breakdown point $\varepsilon^*(T_{\text{sd}}, \mathbb{P}_n) = 1/n$ for any data $\mathbf{x}_n$. A higher breakdown point is attained by the MAD functional defined by

$$T_{\text{MAD}}(P) = T_{\text{med}}(P^A) \tag{4.5}$$

where $A(x) = |x - T_{\text{med}}(P)|$ so that

$$T_{\text{med}}(P^A)(B) = P(\{x : |x - T_{\text{med}}(P)| \in B\}).$$

In terms of the observations

$$T_{\mathrm{MAD}}(\mathbb{P}_n) = \mathrm{median}\{|x_i - \mathrm{median}\{x_1,\ldots,x_n\}|\, i = 1,\ldots,n\}.$$

The finite sample breakdown point of the MAD functional T_{MAD} is given by

$$\varepsilon^*(T_{\mathrm{MAD}},\mathbb{P}_n) = \max\{\lfloor n/2+1\rfloor/n - \Delta(\mathbb{P}_n),0\} \tag{4.6}$$

This is smaller than the upper bound of Theorem 4.2.2. It is not a simple matter to define a scale functional which attains the upper bound but it can be done.

Theorems 4.2.1 and 4.2.2 connect the concepts of breakdown point and equivariance which at first sight are unrelated. However it was argued in [42] equivariance considerations are required in order to derive a non-trivial bound for the breakdown point of a functional.

The concept of breakdown point is easy to understand and important as a pedagogical tool. It can however be misleading if taken to extremes. Consider the two location functionals T_L^1 and T_L^2

$$T_L^1(\mathbb{P}_n) = \frac{\sum_{i=1}^n x_i\{|x_i - T_{\mathrm{med}}(\mathbb{P}_n)| \le 2^{1000}T_{\mathrm{mad}}(\mathbb{P}_n)\}}{\sum_{i=1}^n \{|x_i - T_{\mathrm{med}}(\mathbb{P}_n)| \le 2^{1000}T_{\mathrm{mad}}(\mathbb{P}_n)\}}$$

and

$$T_L^2(\mathbb{P}_n) = T_{\mathrm{med}}(\mathbb{P}_n) + 2^{-1000}T_{\mathrm{ave}}(\mathbb{P}_n).$$

Both are location functionals. The functional T_L^1 has a breakdown point of 0.5 but for all practical purposes behaves like the mean with a breakdown point of $1/n$. The functional T_L^2 has a breakdown point of $1/n$ but for all practical purposes behaves like the median with a breakdown point of $1/2$. The situation is analogous to the use of moments to describe the tail behaviour of random variables. A Cauchy distribution truncated at $\pm 10^{10}$ has all moments but behaves for all practical purposes like the Cauchy distribution. A mixture of the normal distribution and a proportion 10^{-10} of a Cauchy distribution has no moments but behaves for all practical purposes like the normal distribution. This example is attributed to Tukey on page 400 of [103].

Although the breakdown point is an important and simple quantification of the behaviour of a functional a related but more informative one is that of bias. This was already defined in (2.55) of Chapter 2.12 for the special case of the Kolmogorov metric d_{ko} but clearly carries over to any metric d. The finite sample version is

$$b(T,\mathbf{x}_n,k/n) = \sup_{\mathbb{P}_n^k} |T(\mathbb{P}_n) - T(\mathbb{P}_n^k)|.$$

The breakdown point can now be written as

$$\varepsilon^*(T_L,\mathbb{P}_n) = \min\{k/n : b(T_L,\mathbf{x}_n,k/n) = \infty\}.$$

If d is the Kolmogorov metric d_{ko} and F is symmetric and unimodal, it is shown in Chapter 4.2 of [118] that the median T_{med} minimizes the bias amongst all location functionals. This was already noted in Chapter 2.12. It follows that if $\mathbb{P}_n$ is reasonably symmetric and unimodal then the bias of the median will be not too far from the minimum possible bias for location functionals. Thus the concept of bias allows the median to be singled out amongst all location functionals. This is not the case for the breakdown point. The case of scale functionals is not as clear-cut but is treated in Chapter 5.6 of [118] where it is stated that the median absolute deviation 'is a candidate for being "the most robust estimate of scale"'. It is their good bias behaviour rather than their breakdown points which makes the median and MAD appropriate functionals for identifying outliers. Again this was already pointed out in Chapter 2.12 for the case of the median.

4.3 Identifying Outliers and Breakdown

One simple way of identifying outliers is to use a location functional T_L and a scale functional T_S and identify those observations x_i as outliers for which

$$|x_i - T_L(\mathbb{P}_n)| \geq c(n,\alpha)T_S(\mathbb{P}_n) \tag{4.7}$$

for some tuning constants $c(n,\alpha)$. The use of location and scale functionals means that the set of observations which are identified as outliers outliers

$$\{i : |x_i - T_L(\mathbb{P}_n)| \geq c(n,\alpha)T_S(\mathbb{P}_n)\} \tag{4.8}$$

is invariant with respect to affine transformations of the data. In particular it is invariant with respect to the units of measurements.

One simple identifier of the form (4.7) is due to Hampel [101] with $T_L = T_{\text{med}}$, $T_S = T_{\text{MAD}}$ and $c(n,\alpha) = 5.2$ to give

$$|x_i - T_{\text{med}}(\mathbb{P}_n)| \geq 5.2T_{\text{MAD}}. \tag{4.9}$$

The factor $5.2T_{\text{MAD}}(\mathfrak{N}(\mu,\sigma^2)) \approx 3.5\sigma$ which helps in understanding the identifier. In [34] this identifier was erroneously called the Huber identifier, a misnaming that has unfortunately become standard in parts of analytical chemistry.

The values of $c(n,\alpha)$ can be determined by specifying a target distribution P and then requiring that a sample $\mathbf{X}_n(P)$ of size n has no outliers with probability α. This is equivalent to

$$\mathbf{P}\left(\max_i \frac{|X_i(P) - T_L(\mathbb{P}_n)|}{T_S(\mathbb{P}_n)} \leq c(n,\alpha)\right) = \alpha. \tag{4.10}$$

The values of $c(n,\alpha)$ can be obtained by simulations and asymptotics. The values of $c(n,\alpha)$ for the identifier

$$|x_i - T_{\text{med}}(\mathbb{P}_n)| \geq c(n,\alpha)T_{\text{MAD}}(\mathbb{P}_n) \tag{4.11}$$

for $3 \leq n \leq 19$, $\alpha = 0.9, 0.95, 0.99$ and target distribution $P = \mathfrak{N}(0,1)$ are given in Table 4.1. For larger values of n the following simple approximations may be used

n	$\alpha=0.9$	$\alpha=0.95$	$\alpha=0.99$
3	15.40	32.54	158.97
4	7.11	10.19	21.82
5	7.84	11.58	26.42
6	6.18	8.07	13.53
7	6.78	9.17	16.28
8	5.84	7.23	11.98
9	5.99	7.45	12.16
10	5.58	6.77	10.35
11	5.68	6.92	11.06
12	5.37	6.47	9.34
13	5.52	6.58	9.38
14	5.28	6.16	8.46
15	5.44	6.40	8.66
16	5.25	6.12	8.17
17	5.36	6.20	8.42
18	5.21	6.01	7.91
19	5.26	6.02	8.00

Table 4.1 *The values of $c(n,\alpha)$ for the identifier(4.9) with target distribution $\mathfrak{N}(0,1)$.*

(see [32]):

$$1.4826(1+3.42n^{-0.86})\Phi^{-1}\left((1+0.9^{1/n})/2\right), \qquad \alpha=0.9, \tag{4.12}$$

$$1.4826(1+4.65n^{-0.88})\Phi^{-1}\left((1+0.95^{1/n})/2\right), \qquad \alpha=0.95, \tag{4.13}$$

$$1.4826(1+8.10n^{-0.93})\Phi^{-1}\left((1+0.99^{1/n})/2\right), \qquad \alpha=0.99. \tag{4.14}$$

For $n=27$ and $\alpha=0.99$ (4.14) yields the value $c(27,0.99)=7.27$.

Another useful target distribution is the exponential distribution $\mathfrak{E}(1)$ with scale parameter $\lambda=1$. For this a reasonable outlier identifier is (4.7) with $T_L=0$ and $T_S=T_{\text{med}}$. For $n=50$ and $\alpha=0.99$ simulations give $c(100,0.99)=13.94$.

The values of $c(n,\alpha)$ as just described are based on the probability scale specified by α. It is of course possible to define $c(n,\alpha)$ on the linear scale as $c(n)$ or even c without reference to α. The values of $c(n)$ or c can be based on previous experience or on prior knowledge about the data. For many data sets the Hampel choice $c=5.2$ may be perfectly satisfactory.

The breakdown point of an outlier identifier is defined as the smallest proportion of arbitrarily large outliers which are not identified. If the identifier is invariant with respect to affine transformations it is not difficult to see that the highest possible breakdown point is 1/2. This is attained by identifiers of the form (4.7) if T_L and T_S have the highest possible breakdown points of about $1/2$. In particular the identifier (4.10) has the highest possible breakdown point.

One approach to identifying outliers in the frequency paradigm is to test specific observations for being outliers. One such test is the 'two-sided discordancy test' [12] which identifies the smallest observation $x_{(1)}$ as an outlier if

$$|x_{(1)} - T_{\text{ave}}(\mathbb{P}_n)| \geq c(n,\alpha)T_{\text{sd}}(\mathbb{P}_n) \tag{4.15}$$

with the corresponding inequality for the largest observation $x_{(n)}$. If the target distribution is $\mathfrak{N}(0,1)$ it can be shown that for large sample sizes n the breakdown point is less than $\kappa/\log n$ for any $\kappa > 1/2$. For $n = 27$ this suggests a breakdown point of about 0.15. Applied to the copper data of Exercise 1.3.3 with $n = 27$ this translates into four large outliers which may not be identified. Simulations give $c(27, 0.99) = 3.18$. If the first three values of the copper data are replaced by 100 to give an empirical distribution $\mathbb{P}_n^*$ then the mean is 12.89 and the standard deviation is 31.39. The corresponding values for the copper data are 2.016 and 0.116 respectively. A simple calculation gives

$$\frac{|x_{(n)} - T_{\text{ave}}(\mathbb{P}_n^*)|}{T_{\text{sd}}(\mathbb{P}_n^*)} = 2.78 < 3.18$$

so that no outliers are detected. As already noted the value of $c(27, 0.99)$ is for the identifier (4.10) is 7.27. As

$$\frac{|100 - T_{\text{med}}(\mathbb{P}_n^*)|}{T_{\text{MAD}}(\mathbb{P}_n^*)} = 979.7 > 7.27$$

the three outliers are easily identified.

4.4 Outliers in Multivariate Data

The situation changes substantially when moving from univariate to multivariate data. This is mainly due to the standard requirement that the detection of outliers should be affinely invariant which in higher dimensions is a substantially stronger requirement than in one dimension. Affine equivariance and affine invariance were discussed in Chapters 2.4.2 and 2.6.2.1 respectively.

An outlier identification procedure is affine invariant if the same outliers are identified for the transformed data $\mathbf{y}_n = \alpha(\mathbf{x}_n)$ as for the original data $\mathbf{x}_n$. In the location/scale situation an outlier identifier is usually of the form

$$\{i : (x_i - T_L(\mathbb{P}_n))' T_S(\mathbb{P}_n)^{-1}(x_i - T_L(\mathbb{P}_n)) \geq c(n,\alpha)\} \tag{4.16}$$

where T_L is an affinely equivariant location functional (2.17) and T_S an affinely equivariant scale or dispersion functional (2.18). It can be checked that if T_L and T_S are both affinely equivariant then the outlier identifier of (4.10) is affinely invariant. In particular the detection of outliers does not depend on the units.

An example of a multivariate data set is provided by the haematology data given in [182]. The data consist of the following measurements made on 103 car-workers of either West Indian or African descent: (1) haemoglobin concentration, (2) packed cell volume, (3) white blood cell count/100, (4) lymphocyte count, (5) neutrophil count, (6) serum lead concentration. More details of the data are given in [182]. It

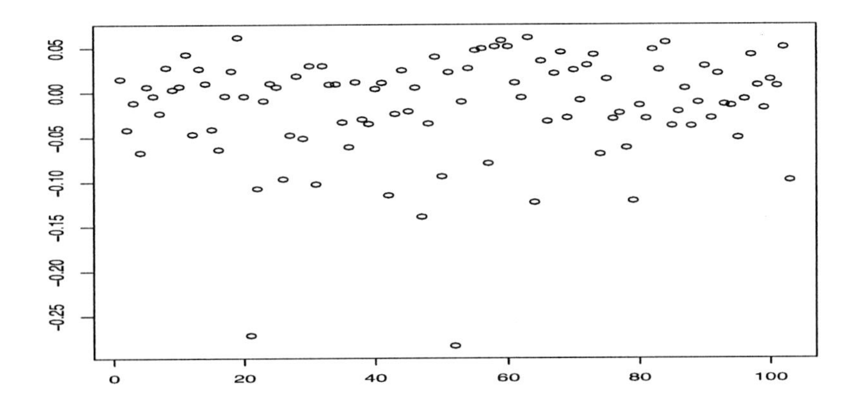

Figure 4.3 *Values of (4.18).*

is apparently standard practice to transform the variables (3)-(6) onto a log-scale [182] and this will be done here. Royston is concerned with adequacy of the normal distribution for multivariate data. This is done in part by testing all pairs and triplets of the variables, the test being based on the squared Mahalanobis distances

$$r_i^2 = (x_i - \bar{x}_n)^t S_n^{-1}(x_i - \bar{x}_n) \tag{4.17}$$

where S_n is the covariance matrix of the variables included. The result for the variables (3), (4) and (5) is a p-value of $6 \cdot 10^{-6}$. This is due to the three observations 21, 47 and 52 which Royston classifies as outliers. After their removal the test yields a p-value of 0.52. None of these observations is an outlier on the marginal plots. They are not even outliers on any of the two dimensional plots. Figure 4.3 shows the plot of the values

$$-0.81(3) + 0.4(4) + 0.42(5) \tag{4.18}$$

where the data have been centred at their mean. The observations (21) and (52) are clearly outliers and observation (47) is the most outlying of the remaining observations.

In the light of the above considerations including the one-dimensional case it seems reasonable to identify outliers using high breakdown affinely equivariant functionals T_L and T_S. There have been many proposals for affinely equivariant functionals which have a breakdown point of almost 1/2 for large samples. The first one due independently to Donoho and Stahel ([191], [57]) involves projecting the data onto all one dimensional subspaces,and then determining a measure of outlyingness for each observation. These measures are then used to weight the observations to provide a high breakdown location and scatter functionals. (see Chapter 6.4 of [148] for the details). It it clearly impossible to project onto all one-dimensional subspaces but the result still holds if, in p dimensions, all subsets of size p of the data are taken and all data sets are projected onto the the one dimensional subspace orthogonal to the $p-1$ dimensional subspace containing the subset. As there are $\binom{n}{p}$ such subsets the computational burden is much too heavy for all but the smallest values of n and p.

A proposal due to Rousseeuw [177] is the Minimum Volume Ellipsoid, MVE, defined as the ellipsoid with the smallest volume which contains half the data points. The centre of the ellipsoid defines the location functional and the symmetric positive definite matrix which describes the shape of the ellipsoid is the scatter functional. It can be seen as a multivariate version of Tukey's shortest half. Again the minimum volume ellipsoid can only be exactly calculated for small values of n and p.

A variant on this idea, also due to Rousseeuw [177], is to choose the covering ellipsoid so as to mimimize the determinant of the covariance matrix of the data points covered by the ellipsoid. This is known as the MCD functional. Although it again can only be calculated for small values of n and p it does have the advantage that there is an iterative algorithm which will improve the ellipsoid from any given starting point [180]. It must be pointed out however that it is not guaranteed to give the solution or to be close to it. Also, as it uses a random search algorithm, it is not in the strict sense of the words affinely equivariant: two applications of the algorithm to the same data set may well yield different values. In spite of these drawbacks it works well for many data sets and is probably the method of choice if highly robust location and scatter functionals for outlier detection are required.

The upper panel of Figure 4.4 shows the results for the transformed haematology data. It gives the squared Mahalanobis distances as in (4.17) but with $\bar{x}_n$ and S_n replaced by the location and scatter functionals respectively which are returned by the default version (*covMcd*) of the MCD functional given in [181]. Based on this it seem reasonable to declare the observation 21 and 52 as outliers. It is not clear whether the observation 47 is an outlier or not, but if it is declared as such then there are also further ones. If an automatic identification procedure is required as in the one-dimensional case then a target model is required. If this is taken to be the multidimensional normal distribution then cut-off points for outliers can be obtained by simulations. For the default version of *covMcd* the 0.99 quantiles of the maximum squared distance is 39.8. Based on this only the observations 21 and 52 are declared as outliers.

The lower panel of Figure 4.4 shows the corresponding results for the raw data without the logarithmic transformation of the variables 3-6. Using the cut-off value 39.8 the observations 52 and 93 are declared to be outliers. The observation 21 is no longer an outlier. Explanations for this cannot be given without expert knowledge of the data.

4.5 Outliers in Linear Regression

Given data (y,x) with y an $n \times 1$ and x an $n \times k$ matrix the standard model for linear regression is

$$Y = x\beta + \sigma\varepsilon \tag{4.19}$$

where $\beta \in \mathbb{R}^k$ and the ε are standard Gaussian white noise. In this version the regressors x are treated as given but they are often treated as random and a conditional

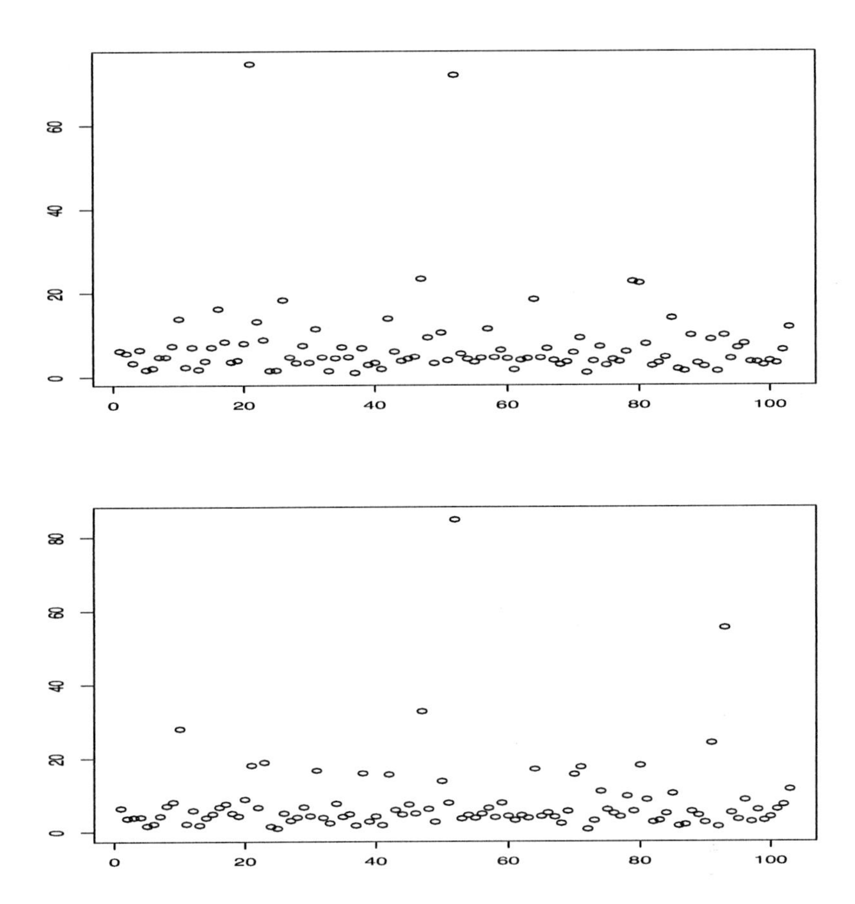

Figure 4.4 *Upper panel: Squared Mahalanobis distances for the transformed haematology data based on the default version (covMcd) of the minimum covariance determinant functional given in [181]. Lower panel: the same for the untransformed data.*

analysis carried out. In general the following group $\mathscr{G}$ of transformations G are allowed defined by

$$G((y,x)) = (\alpha y + x\gamma, xa) \tag{4.20}$$

where $\alpha \neq 0, \gamma \in \mathbb{R}^k$ and a is a $k \times k$ non-singular matrix. The group of transformations $\mathscr{G}$ is not always relevant, an example being the k-way analysis of variance where the x matrix consists of just zeros and ones representing the effects of the factors at the various levels. Here it makes no sense to transform the x matrix. A functional T_R,

$$T_R : \mathscr{P} \to (\mathbb{R}^k, \mathbb{R}_+) \tag{4.21}$$

is called regression equivariant or a regression functional if

$$T_R(P^G) = \big(\alpha a^{-1}(T_{RL}(P) + \gamma), |\alpha| T_{RS}(P)\big) \tag{4.22}$$

where $T_R(P) = (T_{RL}(P), T_{RS}(P))$. The standard least squares functional is regression equivariant.

Just as in multidimensional location/scale situation of the previous section it is not a simple matter to define high breakdown regression equivariant functionals, and those that have been defined suffer from the same computational difficulties. The best known such functional is the 'Least Median of Squares' or LMS. The basic idea again goes back to Tukey's shortest half. It was adapted by Hampel to the regression situation and he gives its breakdown point as being asymtotically 1/2 [100]. The functional was given its name by Rousseeuw who provided the theory and an algorithm for calculating an approximation [177]. For small n and k it can be calculated exactly [198]. It and other high breakdown regression functionals are available in [181].

For an overview of the theory and application of robust equivariance functionals with many examples see [119].

4.6 Outliers in Structured Data

It was argued in [42] that the concept of breakdown point only makes sense when the structure of the data allows a large group of transformations which leave the structure unchanged. Structured data can be defined as data which allow only a small group of such transformations. Although it was never explicitly stated above, the detection of outliers was based on the empirical distribution $\mathbb{P}_n$. This is invariant under every permutation of the data $x_i \to x_{\pi(i)}$ for a permutation π of $1, \ldots, n$. Thus the structure of such data remains invariant under permutations. If the data form a time series or come from some non-parametric regression problem they cannot be permuted as the order of the observations is essential. A discussion of the importance of exchangeability and its relevance for the detection of outliers and robust theory in general is given in Chapter 1.9 of [118].

Another simple example is where the data points represent some Euclidean geometric object. The only transformations which leave the structure unchanged are the orthogonal transformations which leave angles and distances unchanged. A further one was already mentioned, namely the k-way analysis of variance where it makes no sense to consider transformations of the design matrix which consists only of zeros and ones. The methods required for this have nothing in common with the methods standard high breakdown methods for linear regression.

Outliers can occur in structured data but each case must be treated on its own merits. Sometimes there may be different sorts of outliers which require different techniques to identify them. In time series one can distinguish between additive outliers, innovation outliers and shift outliers and possibly even others. The subject of time series is considered in [148].

Chapter 5

The Location-scale Problem

In spite of the title the emphasis in this chapter will be on the location part of the problem as in most situations it is the location parameter which is of interest. If required the scale part can be treated in a similar manner. In [118] the chapter on location is longer by a factor of two than the chapter on scale on the grounds that 'scale usually occurs as a nuisance parameter in robust location'.

5.1 Robustness

In one sense the subject of this chapter belongs to the area of robust statistics. In a second sense the point of view is different.

To take the second sense first. The problem may be seen as providing almost universal measures of location and scale (see [9], [37]), that is functionals which are equivariant with respect to affine transformations, (2.15) for location functionals and (2.16) for scale functionals. The mean functional T_{ave} ((2.12) repeated here for convenience)

$$T_{\text{ave}}(P) = \int x\,dP(x)$$

is the best known functional but it not universal as it is defined only for those distributions P for which

$$\int |x|\,dP(x) < \infty.$$

The second best known location functional is the median T_{med} ((2.54) repeated here for ease of reference)

$$T_{\text{med}}(P) = T_{\text{med}}(F) = (\ell(F) + u(F))/2$$

where

$$\ell(F) = \inf\{x : F(x) \geq 1/2\},\ u(F) = \sup\{x : 1 - F(x) \geq 1/2\}$$

and F is the distribution function of P. The median is universal being well defined, affine equivariant and finite for all distributions P. There are other such ‘universal’ location functionals which are used without direct refrence to a stochastic model. In [196] eleven such functionals are considered and compared with respect to their performance on sets of real data. Although the question of normality is discussed in Section 6 of [196] it plays no role in the comparison. All the functionals were made affine equivariant by including equivariant scale functionals. This was done explicitly to make the results independent of the units of measurement.

There is a corresponding problem for scale. The standard scale functional is the standard deviation T_{sd} of (2.14)

$$T_{\text{sd}}(P) = \sqrt{\int (x - T_{\text{ave}}(P))^2\,dP(x)}$$

but this is even more restrictive than the mean requiring

$$\int x^2\,dP(x) < \infty.$$

The mean absolute deviation

$$\int |x - T_{\text{ave}}(P)|\,dP(x)$$

requires only a first absolute moment and consequently is still not universal. The

MAD functional T_{MAD} of (4.5) is universal but $T_{\text{MAD}}(P) = 0$ for any P with $\Delta(P) \geq 1/2$ where

$$\Delta(P) = \max_x P(\{x\}). \tag{5.1}$$

The 'no name' scale functional T_{nn} defined by

$$T_{\text{nn}}(P) = \inf\{x : P([T_{\text{med}}(P) - x, T_{\text{med}}(P) + x]) \geq (1 + \Delta(P))/2\} \tag{5.2}$$

overcomes this deficit as $0 < T_{\text{nn}}(P) < \infty$ for all P with $\Delta(P) < 1$.

The functionals T_{med} and T_{nn} solve the problem. The median is easily interpretable and is widely used with no reference to any model for the data. It and the functional T_{nn} are however pathologically discontinuous. Any Kolmogorov neighbourhood of any distribution P contains distributions where the median is not continuous. The same applies to all eleven functionals considered in [196] if one takes into account the scale functionals used to standardize the location functionals.

Continuity does not seem to be of any practical importance in statistics. It will not help to construct approximation intervals. For this locally uniform Fréchet differentiability with respect to the Kolmogorov metric is required. This means that in an open neighbourhood of any distribution P the functional has a bounded linear approximation which is uniformly good over the neighbourhood (see [37] and remarks on page 5 of [118] on the importance of uniformity). This enables the application of the central limit theorem to obtain approximation intervals and hence an indication of the variability of the location functional. Although no mention is made of a model the solution to be given below is directed towards data sets such as the copper data where location and scale are important. Other data sets such as waiting times or counts will require other functionals unless the values are so far from the natural origin of zero that they can be treated as a location-scale problem. Such data sets require different functionals but the idea will not be explored further.

In the first sense the chapter can be seen as a contribution to robust statistics as generally understood. In this sense parametric models are important and the idea of robustness is to guarantee stability of analysis within a small neighbourhood of the model. Chapter 1 of [118] contains an excellent discussion of robust statistics. The following desirable features of a statistical procedure are listed:

- Efficiency: It should have reasonably good (optimal or nearly optimal) efficiency at the assumed model.
- Stability: It should be robust in the sense that small deviations from the model should impair the performance only slightly, that is, the latter (described, say, in terms of the asymptotic variance of an estimate, or of the level and power of a test) should be close to the nominal values at the assumed model.
- Breakdown: Somewhat larger deviations from the model should not cause a catastrophe.

Chapter 1 of [103] also gives an discussion of the reasons for robust statistics. It is stated

> *Robust statistics, as a collection of related theories, is the statistics of approximate parametric models.*

The italics are in the original. Similar statements are to be found in [193] and [148].

With the exception of gross outliers (see Chapter 1.2 [103]) there seems to be little attempt to quantify what is meant by 'small' or 'approximate'. The situation is typically modelled by the gross error model

$$Q = (1-\varepsilon)P_\theta + \varepsilon H \tag{5.3}$$

where P_θ is the assumed parametric model and H models the outliers. Realistic values of ε run from 0.01-0.1 for 'routine data' [103].

If the model (5.3) were true it would be clear what the aim of robust statistics is. However one of the premises of robust statistics is not to treat even (5.3) as if it were true. To fix ideas it may help to suppose that the data were generated under a distribution Q with $\{P_\theta : \theta \in \Theta\}$ being the parametric model under consideration and $Q \neq P_\theta$ for all θ. The problem is to make sense of a procedure which proceeds to estimate something, θ, which is not there because it is Q. In Chapter 1.2 of [103] the authors write

> Only parametric models provide the redundancy necessary for a data description which is both simple and complete.

Chapter 8.2d of [103] asks what can actually be estimated. There the authors state

> ..., we may consider entertaining a *simple model* of which we *know that it is wrong*.

where 'know that it is wrong' refers to some goodness-of-fit criterion. Taken together it can be (unfairly) said that the aim of robust statistics is to give simple and complete wrong answers.

An immanent and convincing justification of robust statistics or indeed any method of determining parameters is difficult which can be read as a euphemism for impossible. Any serious data analysis is intended to solve, or help to solve, a problem in the real world. The justification of the procedure will be measured by its ability to solve the problem. In the example of the copper data it is clear what is to be estimated, namely the amount of copper in the sample of drinking water and not the parameter of some hypothetical generating mechanism for the data.

An approximation (confidence) interval for a parameter θ based on a statistic T_n under a model P is a measure of the variability of the statistic. The following is taken from page 145 of [118]:

> 'As a matter of principle, each estimate $T_n = T_n(x_1,\ldots,x_n)$ of any parameter θ should be accompanied by an estimate $D_n = D_n(x_1,\ldots,x_n)$ of its own variability'.

In practice this principle is not taken seriously even in [118]: the reader searches in vain for the phrase 'confidence interval' in the index. Just over one page in [148] is devoted to the subject in the context of the location-scale problem. A similarly casual treatment of the problem is to be found in [103] although the latter emphasize the importance of confidence intervals. It seems that most of the robust literature on the construction of confidence intervals is asymptotic in nature without any attempt to consider the finite sample behaviour of the proposal or its behaviour under deviations from the model. The interpretation of a 95% confidence interval under a given model is clear. Typically robust functionals will be calibrated to be (asymptotically) Fisher consistent at the proposed model and the confidence intervals are based on this. For the copper data it is not a question as to whether the intervals contain the true but unknown parameter but whether they contain the true amount of copper. This can often be decided by analysing prepared samples. The functionals can now be calibrated against real data and the calibration may not be that for the model. The model calibration is best understood as a reasonable default value which can be changed in the light of experience.

The comparisons of the eleven functionals in [196] were all pointwise and in the discussion, with the exception of Pratt, no mention was made of the variability or efficiency of the functionals. The problem of confidence (approximation) intervals was ignored. There are situations where confidence (approximation) intervals are of importance. Interlaboratory tests are a form of quality control. A laboratory whose returned values lie outside appropriate confidence (approximation) intervals may be subject to inspection and run the risk of losing its licence to analyse samples. The one-way table in the analysis of variance to be considered in the next chapter is based on approximation intervals.

If a statistician uses a parametric model for some quantity of interest in the world, then the quantity of interest must be set in some relation to the parameters. In the case of the copper data and any symmetric model, the quantity of copper will almost always be identified with the centre of symmetry. Non-symmetric models are also possible, for example a gamma distribution. It is now no no longer clear which function of the parameters should be identified with the amount of copper. The mean and median are but two possible examples. If a functional is used irrespective of the model, than the functional itself can be identified with the quantity of copper. A location functional can be considered as a differentiable continuation of the centre of symmetry.

5.2 Efficiency and Regularization

The frequentist location-scale problem is the following. It is assumed that the data are an i.i.d. sample $\mathbf{X}_n = \mathbf{X}_n((\mu_0, \sigma_0))$ with common distribution function $F((\cdot - \mu_0)/\sigma_0)$ where F is specified. The problem is to estimate the unknown parameters (μ_0, σ_0) and to give a say 95% confidence interval for μ_0. The confidence interval should be as small as possible given the coverage probability, that is, the estimators for μ_0 and σ_0 should be as efficient as possible given the model. In the case of the Gaussian distribution, $F = \Phi$, the optimal estimators for μ_0 and σ_0 are the mean and the standard deviation of the data. The optimal 95% confidence interval for μ_0 is

$$\left[\overline{\mathbf{X}}_n - \frac{\text{qt}(0.975, n-1)\text{sd}(\mathbf{X}_n)}{\sqrt{n}}, \overline{\mathbf{X}}_n - \frac{\text{qt}(0.975, n-1)\text{sd}(\mathbf{X}_n)}{\sqrt{n}}\right] \tag{5.4}$$

where

$$\overline{\mathbf{X}}_n = \frac{1}{n}\sum_{i=1}^{n} X_i \quad \text{and} \quad \text{sd}(\mathbf{X}_n) = \sqrt{\frac{1}{n-1}\sum_{i=1}^{n}(X_i - \overline{\mathbf{X}}_n)^2}$$

and $\text{qt}(\alpha, k)$ denotes the α-quantile of the t-distribution with k degrees of freedom. For the copper data of Exercise 1.3.3

2.16	2.21	2.15	2.05	2.06	2.04	1.90	2.03	2.06
2.02	2.06	1.92	2.08	2.05	1.88	1.99	2.01	1.86
1.70	1.88	1.99	1.93	2.20	2.02	1.92	2.13	2.13

the mean is 2.016, the standard deviation 0.116 and the 95% confidence interval [1.970, 2.062]. A second example is the student data of Figure 1.6. The logarithm of the data makes the Gaussian model an adequate approximation. The mean is 2.606, the standard deviation 0.226 giving a 95% the confidence interval [2.212, 3.000].

The problems associated with this approach were discussed in Chapter 1.3.6 and Chapter 1.3.7. The conclusion was that the location-scale problem is ill-posed and must be regularized. One form of regularization is to use models with a small Fisher information such as the Gaussian model: the Gaussian distribution minimizes the Fisher information amongst all distributions with a given variance. Other minimum Fisher distributions can be obtained by minimizing over other classes of distributions. One possibility is to minimize over all models in an ε contamination neighbourhood of the $\mathfrak{N}(0,1)$ distribution

$$\mathscr{P}(\mathfrak{N}(0,1), \varepsilon) = \{P : P = (1-\varepsilon)\mathfrak{N}(0,1) + \varepsilon Q, \, Q \in \mathscr{P}(\mathbb{R})\}. \tag{5.5}$$

This gives rise to the Huber distributions with densities of the form

$$f_0(x) = \begin{cases} \frac{1-\varepsilon}{\sqrt{2\pi}} \exp\left(-x^2/2\right), & |x| \le k, \\ \\ \frac{1-\varepsilon}{\sqrt{2\pi}} \exp\left(k^2/2 - k|x|\right), & |x| > k, \end{cases} \tag{5.6}$$

where

$$\frac{2\varphi(k)}{k} - 2\Phi(k) = \frac{\varepsilon}{1-\varepsilon}.$$

Other possibilities are to be found in [118] and [213]. In the latter distributions with a compact support are considered.

Although the minimax strategy is better than accepting free lunches there is no compelling reason to use it. Even under the Gaussian model the statistician is not obliged to use the mean and standard deviation for estimating μ_0 and σ_0 respectively. Apart from the considerations of fit and efficiency there is also the consideration of stability. The statistical analysis should be continuous with respect to small changes in the data, where 'small' can mean large changes in a few observations. The maximum likelihood functional based on the normal distribution or a minimal Fisher distribution of the form (5.6) are not stable in this sense: they have a finite sample breakdown point of $1/n$.

The discussion so far is restricted to models F which are symmetric about $x = 0$ and which have a Lebesgue density. In the case of the student data there is no reason to suppose that the data have a point of symmetry. Nor do the data, as multiples of 1/6, have a Lebesgue density. Once asymmetric models F are allowed the parameters no longer have an interpretation independent of the model.

Finally it may be difficult to find a simple parametric model which is a good approximation to the data. This is the case for some of the data on the lengths of study of German students, for example the data exhibited in Figure 2.11. In such a situation a non-parametric approach may be possible. This was done Chapter 2.13 where the approximation regions of Figure 2.12 were calculated using the mean and standard deviation of the data and relying on the central limit theorem to provide good approximations to their distributions. In order for this to work the approximation region (2.58) includes bounds on the values of the data. Without such bounds the accuracy of the normal approximation would not hold uniformly over the set of approximate models.

The path to be taken in the remainder of this chapter is to replace a specific location-scale model $F((\cdot - \mu)/\sigma)$ by a non-parametric space of models and locally differentiable statistical functionals T_L and T_S [215]. The fact that these functionals are smooth means that they ignore any peculiarities of the model. Data generated under the normal model or the comb model (1.8) with the same mean and variance would give rise to similar values for the functionals. The functionals themselves are a form of regularization.

An approximation interval for μ for a specific location-scale family $F((\cdot - \mu)/\sigma)$ could be defined as

$$\begin{aligned}&\mathscr{A}_n(\mathbf{x}_n, \alpha, \mathscr{M})\\ &\quad= \big\{\mu : \mathrm{qtl}((1-\alpha)/3, n, F) \le \sqrt{n}(\hat{T}_L(\mathbb{P}_n) - \mu)/T_S(\mathbb{P}_n) \le \mathrm{qtl}((2+\alpha)/3, n, F),\\ &\qquad d_{\mathrm{ku}}(\mathbb{P}_n, F((\cdot - \mu)/\sigma) \le \mathrm{qdku}((2+\alpha)/3) \text{ for some } \sigma\big\}. \qquad (5.7)\end{aligned}$$

where $(\hat{T}_L, \hat{T}_S)$ are the maximum likelihood functionals for the model and $\mathrm{qtl}(\alpha, n, F)$ is the α-quantile of $\sqrt{n}(\hat{T}_L(\mathbb{P}_n) - \mu)/T_S(\mathbb{P}_n)$ under the model F. In the non-parametric case the approximation region could be

$$\begin{aligned}&\mathscr{A}_n(\mathbf{x}_n, \alpha, T_L(\mathscr{P}))\\ &\quad= \big\{T_L(P) : \mathrm{qtl}((1-\alpha)/3, n, P) \le \sqrt{n}(T_L(\mathbb{P}_n) - T_L(P))\\ &\qquad \le \mathrm{qtl}((2+\alpha)/3, n, P), d_{\mathrm{ku}}(\mathbb{P}_n, P) \le \mathrm{qdku}((2+\alpha)/3)\big\}. \qquad (5.8)\end{aligned}$$

where $T_L(P)$ is an appropriate functional to be specified. It is to be noted that the approximation interval (5.7) may be empty if the model does not fit. This is not the case for the approximation interval (5.8) as there are always approximating models. In both cases $(1-\alpha)/3$ has been spent on being honest, the remainder $2(1-\alpha)/3$ has been spent on the location functional.

5.3 M-functionals

Given a location-scale model based on a distribution F with density f the maximum likelihood estimates $(\hat{\mu}_n, \hat{\sigma}_n)$ are the solutions of

$$\frac{1}{n}\sum_{i=1}^{n}\frac{-f^{(1)}\left(\frac{x_i-\hat{\mu}_n}{\hat{\sigma}_n}\right)}{f\left(\frac{x_i-\hat{\mu}_n}{\hat{\sigma}_n}\right)} = 0, \tag{5.9}$$

$$\frac{1}{n}\sum_{i=1}^{n}\frac{-f^{(1)}\left(\frac{x_i-\hat{\mu}_n}{\hat{\sigma}_n}\right)}{f\left(\frac{x_i-\hat{\mu}_n}{\hat{\sigma}_n}\right)}\left(\frac{x_i-\hat{\mu}_n}{\hat{\sigma}_n}\right) = 1. \tag{5.10}$$

Functionals derived from maximum likelihood do not necessarily have desirable robustness properties even if the distributions under consideration are minimum Fisher one. This is true for the normal distribution and also for the Huber distributions (5.6). The latter have exponential tails $c\exp(-|x|)$ and consequently finite sample breakdown point of $1/n$. On the other hand there are maximum likelihood functionals which do have a high breakdown point even if they are not minimum Fisher distributions. Under certain conditions the maximum likelihood functionals based on the t-distribution with $\nu > 1$ degrees of freedom has a breakdown point of $1/(1+\nu)$ [77].

Huber in Proposals 1 and 2 of [113] broke the direct connection of estimators of the form (5.11) and (5.12) with maximum likelihood functionals by writing the equations in the form

$$\frac{1}{n}\sum_{i=1}^{n}\psi\left(\frac{x_i-\hat{\mu}_n}{\hat{\sigma}_n}\right) = 0, \tag{5.11}$$

$$\frac{1}{n}\sum_{i=1}^{n}\chi\left(\frac{x_i-\hat{\mu}_n}{\hat{\sigma}_n}\right) = 0. \tag{5.12}$$

In the case of maximum likelihood estimators the ψ- and χ-functions are given by

$$\psi(x) = \frac{-f^{(1)}(x)}{f(x)} \quad \text{and} \quad \chi(x) = x\psi(x) - 1.$$

Huber considered the ψ-function

$$\psi(x) = \psi(k,x) = \begin{cases} x, & |x| \le k, \\ k\cdot \mathrm{sgn}(x), & |x| > k, \end{cases} \tag{5.13}$$

with χ-functions of the form $\chi(x) = \psi^{(1)}(x) - \mathbf{E}(\psi^{(1)}(Z))$ and $\chi(x) = \psi(x)^2 - \mathbf{E}(\psi(Z)^2)$ where $Z \sim \mathfrak{N}(0,1)$. It is not necessary to restrict the χ-function in this manner. If the ψ and χ-functions are given by

$$\psi(x) = \mathrm{sign}(x), \tag{5.14}$$

$$\chi(x) = \mathrm{sign}(|x|-1), \tag{5.15}$$

(Example 6.5 of [118]) then the solutions $\hat{\mu}_n$ and $\hat{\sigma}_n$ are the median and MAD of the data respectively. There is however no location-scale model for which the median and the MAD are the maximum likelihood estimators.

More generally one can define the M-functional $T_{LS}(P) = (T_L(P), T_S(P))$ for any probability distribution P as the solution of the equations

$$\int \psi\left(\frac{x - T_L(P)}{T_S(P)}\right) dP(x) = 0, \tag{5.16}$$

$$\int \chi\left(\frac{x - T_L(P)}{T_S(P)}\right) dP(x) = 0. \tag{5.17}$$

The idea is to decide which desirable properties the functional $T_{LS}(P)$ should have and then to choose the ψ- and χ-functions to achieve them. The following is such a list of desirable properties of $T_{LS}(P)$:

(i) existence and uniqueness for a large class of distributions.

(ii) locally uniform Fréchet differentiability with respect to the Kolmogorov metric.

(iii) a small bias.

(iv) high efficiency for normally distributed data.

The reasons for these properties are as follows.

(i) It is reasonably clear that $T_{LS}(P)$ should be well defined for as large a class of distributions P as possible. The mean functional T_{ave} is defined only for P with $\int |x|\, dP(x) < \infty$. The median functional T_{med} is defined for all P but does not satisfy the next condition. One of the location functionals to be defined below is a compromise lying between T_{ave} and T_{med}.

(ii) Locally uniform Fréchet differentiability is a technical condition but is meant to guarantee the following. Given data $\mathbf{x}_n$ and two models F and G which are close to each other and to the empirical distribution $\mathbb{P}_n$ in the sense of the Kolmogorov metric the approximation region for $T_L(P)$ based on P should be close to the approximation region for $T_L(Q)$. This means stability of analysis. It is to be noted that optimizing with respect to P and Q will lead to a pathologically discontinuous analysis. A functional T is locally uniformly Fréchet differentiable if

$$T(Q) - T(P) = \int I(x, T, P) d(Q(x) - P(x)) + \mathrm{o}(d_{ko}(P, Q)) \tag{5.18}$$

for all P and Q in a ball $B(\varepsilon, P_0) = \{P : d_{ko}(P, P_0) < \varepsilon\}$ where $I(x, T, P)$ is uniformly bounded and continuous on $\mathbb{R} \times B(\varepsilon, P_0)$ and the o-term holds uniformly on $B(\varepsilon, P_0) \times B(\varepsilon, P_0)$. The function $I(x, T, P)$ is the influence function of the functional T (see [103] for an approach to robustness based on the influence function) which is usually defined as

$$I(x, T, P) = \lim_{\varepsilon \to 0} \frac{T((1-\varepsilon)P) + \varepsilon \delta_x}{\varepsilon} \tag{5.19}$$

where δ_x denotes the unit mass in the point x. It is to be noted that (5.19) follows from (5.18) but not conversely. Suppose now that n is large and that $\varepsilon = \varepsilon_n = 4/\sqrt{n}$. Then if $d_{ko}(P, P_0) < 2/\sqrt{n}$ it follows that $d_{ko}(\mathbb{P}_n(P), P) < 4/\sqrt{n}$ and (5.18) is applicable with $Q = \mathbb{P}_n(P)$. This gives

$$\begin{aligned}
\sqrt{n}(T(\mathbb{P}_n(P)) - T(P)) &\approx \frac{1}{\sqrt{n}} \sum_{i=1}^{n} I(X_i(P), T, P) \\
&\stackrel{D}{\Rightarrow} \mathfrak{N}\left(0, \int I(x, T, P)^2 \, dP(x)\right) \qquad (5.20) \\
&\stackrel{D}{\approx} \mathfrak{N}\left(0, \int I(x, T, P_0)^2 \, dP_0(x)\right) \qquad (5.21)
\end{aligned}$$

because of the continuity properties of $I(x, T, P)$. If $P_0 = \mathbb{P}_n$ where $\mathbb{P}_n$ derives from a data set $\mathbf{x}_n$ then the approximation intervals of $T(P)$ based on (5.21) will hold at least for data sets $\mathbf{x}_n$ which are not too strongly bimodal or skewed. In particular this implies that the approximation intervals for $T(P)$ will be approximately the same for all P in $B(\mathbb{P}_n, 2/\sqrt{n})$. This is so because a locally uniformly Fréchet differentiable functional T is so smooth that small changes in P result in only small changes in $T(P)$.

(iii) The requirement of small bias means that the effect of a small percentage of outliers should be small. Typically a small bias can be obtained by choosing the functions ψ and χ so that the breakdown point of the M-functional is high, higher in fact than is necessary: if the data contain a large number of outliers their status should be checked before calculating an approximation interval for $T_L(P)$ rather than relying on the functional to downweight them. A functional with a large breakdown point does not necessarily have a small bias but it is necessary to construct somewhat extreme examples to demonstrate this.

(iv) The normal distribution is a minimum Fisher distribution and therefore, in a certain sense, a worst case distribution. It does not import an increase in 'efficiency' or 'precision', or in Tukey's words, ' a free lunch', through the model. As there is no reason to expect a free lunch in statistics the normal model is a reasonable choice. A high efficiency at the normal model is an attempt to perform reasonably well under the worst possible circumstances: statistical pessimism is always a good guide.

The following conditions will be placed on the ψ- and χ- functions and the probability distribution P. The condition on P is defined in terms of

($\psi 1$) $\psi(-x) = -\psi(x)$ for all $x \in \mathbb{R}$,
($\psi 2$) ψ is strictly increasing,
($\psi 3$) $\lim_{x\to\infty} \psi(x) = 1$,
($\psi 4$) ψ is differentiable with a continuous derivative $\psi^{(1)}$,

($\chi 1$) $\chi(-x) = \chi(x)$ for all $x \in \mathbb{R}$,
($\chi 2$) $\chi : \mathbb{R}_+ \to [-1, 1]$ is strictly increasing,
($\chi 3$) $\chi(0) = -1$,
($\chi 4$) $\lim_{x\to\infty} \chi(x) = 1$,
($\chi 5$) χ is differentiable with a continuous derivative $\chi^{(1)}$,

($\psi\chi 1$) $\chi^{(1)}/\psi^{(1)} : \mathbb{R}_+ \to \mathbb{R}_+$ is strictly increasing,

(P) $\Delta(P) < 1/2$ where $\Delta(P)$ is as in (5.1).

On defining

$$\mathscr{P}_\delta = \{P : \Delta(P) < \delta\} \tag{5.22}$$

the conditions guarantee that the solution $T_{LS}(P) = (T_L(P), T_S(P))$ of (5.16) and (5.17) exists and is unique for all $P \in \mathscr{P}_{1/2}$. Furthermore T_{LS} is uniformly locally Fréchet differentiable in some neighbourhood of any $P \in \mathscr{P}_{1/2}$.

The ψ– and χ–functions to be considered which satisfy the above conditions are of the form

$$\psi(x) = \psi(x,c) = \frac{\exp(cx) - 1}{\exp(cx) + 1}, \tag{5.23}$$

$$\chi(x) = \frac{x^4 - 1}{x^4 + 1}. \tag{5.24}$$

where c is a tuning constant. Large values of c will make the location functional T_L act like the median T_{med} whereas for small values of c it will behave like the mean T_{ave}. The functions fulfil all the conditions for all values of $c > 0$ apart possibly from the condition ($\psi\chi 1$). Direct calculations show that this holds for $c > 2.56149$. This does not exclude the existence and uniqueness if $c \leq 2.56149$ as the condition is sufficient but not known to be necessary.

For empirical data (5.16) and (5.17) become

$$\frac{1}{n}\sum_{i=1}^{n} \psi\left(\frac{x_i - T_L(\mathbb{P}_n)}{T_S(\mathbb{P}_n)}\right) = 0, \tag{5.25}$$

$$\frac{1}{n}\sum_{i=1}^{n} \chi\left(\frac{x_i - T_L(\mathbb{P}_n)}{T_S(\mathbb{P}_n)}\right) = 0, \tag{5.26}$$

which can be solved numerically. The simple, obvious iteration scheme seems to converge but there is no proof. An iteration scheme based on the proof given in [186] of existence and uniqueness of a solution is guaranteed to converge if $c > 2.56149$ but is slightly more complicated. If $c < 2.56149$ there have been rare cases of non-convergence.

5.3.1 *Fisher Consistency*

If T_{LS} is applied to a location-scale model $F(\cdot - \mu)/\sigma)$ with F symmetric about $x = 0$ then

$$T_{LS}(F(\cdot - \mu)/\sigma)) = (\mu, a\sigma)$$

for some constant a. If T_S is replaced by $\tilde{T}_{LS} = (T_L, T_S/a)$ then $\tilde{T}_{LS}$ is Fisher consistent at the model F, that is

$$\tilde{T}_{LS}(F(\cdot - \mu)/\sigma)) = (\mu, \sigma).$$

In particular to make the MAD Fisher consistent at the Gaussian model it must be multiplied by $1/a = 1/\Phi^{-1}(0.75) = 1.4826$. This is the default version of the MAD in R. In the case of the M-functional above with ψ- and χ-functions given by (5.23) with $c = 5$ and (5.24)

$$\tilde{T}_{LS}(\mathfrak{N}(0,1)) = (0, 0.6472)$$

and consequently to make the scale part Fisher consistent at the normal distribution it must be multiplied by 1.5454.

5.3.2 *Breakdown and Bias*

Breakdown of the functional T_{LS} is said to occur if (i) it is no longer well-defined, (ii) if the location part T_L tend to $\pm\infty$ or (iii) the scale part T_S tends to 0 or ∞. It is not easy to calculate the breakdown point with this comprehensive definition of breakdown ([115], [77]). If however only the breakdown of T_L alone is considered this can be explicitly calculated as the solution ε^* of

$$\chi^{-1}\left(-\frac{\varepsilon^*}{1-\varepsilon^*}\right) = \psi^{-1}\left(\frac{\varepsilon^*}{1-\varepsilon^*}\right)$$

(Chapter 6.6 of [118]). For $\psi(x) = \psi(x,c)$ this depends on the value of the tuning parameter c. Some results are given in Table 5.1. The behaviour of the bias of T_L is

c	40	10	5	4	3	2.6	2	1	0.5
ε^*	0.4995	0.489	0.463	0.448	0.428	0.409	0.378	0.279	0.180

Table 5.1 *The breakdown point ε^* as a function of the tuning parameter c.*

shown in the upper panel of Figure 5.1. The bias is calculated for the model

$$P_\varepsilon = (1-\varepsilon)\mathfrak{N}(0,1) + \varepsilon\mathfrak{N}(20,1)$$

with $\varepsilon = 0.005(0.005)0.4$ and with the tuning parameter $c = 2.6, 3(0.5)6$. The black lines from top to bottom are for the values of c in that order. The median functional is included for reference and is shown as a dashed line. It is seen that as the breakdown point increases the bias decreases. For contamination of less than 20% there is not much difference between the curves and the median. The lower panel shows the bias of T_S on a log-scale $\log(T_S(P_\varepsilon)/T_S(P_0))$. The dashed line is the corresponding result for the functional T_{MAD}. Again, for contamination of less than 20% there is not much difference between the curves and the MAD.

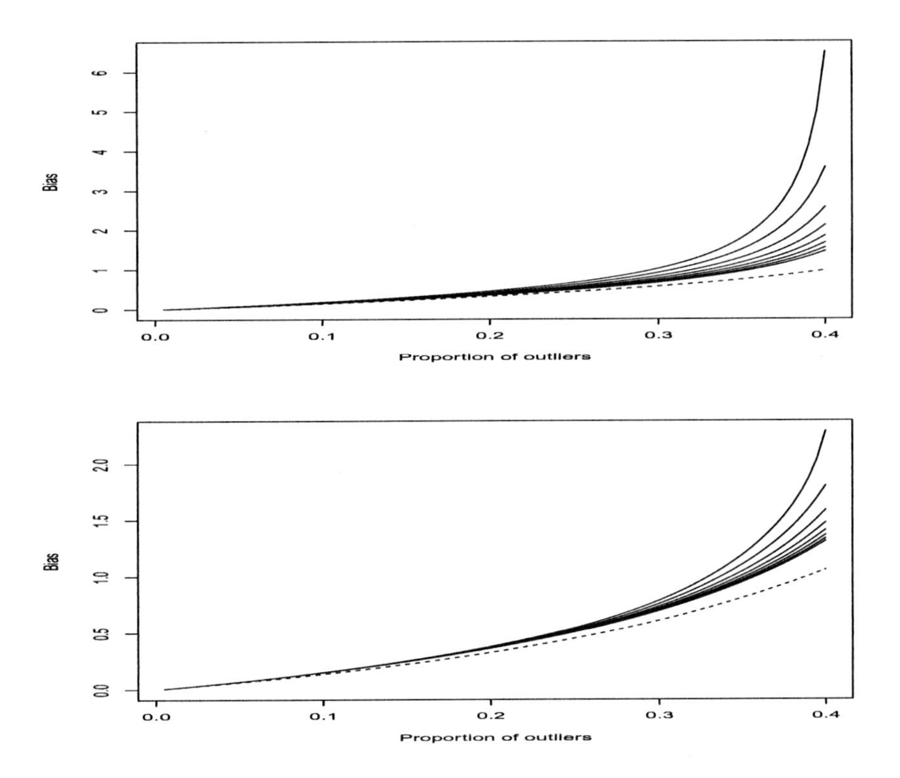

Figure 5.1 *Upper panel:Bias of T_L for various proportions of outliers and various values of c. The dashed line is the median. Lower panel: As for the upper panel but the bias of T_S on a log-scale. The dashed line is the MAD.*

5.3.3 *The Influence Function*

The influence function $I(x,T,P)$ which appears in the definition (5.18) of Fréchet differentiability may be calculated as in (5.19). This extends to the two functionals

T_L and T_S and gives

$$\begin{aligned}
&\psi\left(\frac{x-T_L(P)}{T_S(P)}\right)T_S(P)\\
&= I(x,T_L,P)\int\psi^{(1)}\left(\frac{u-T_L(P)}{T_S(P)}\right)dP(u)+\\
&\quad I(x,T_S,P)\int\left(\frac{u-T_L(P)}{T_S(P)}\right)\psi^{(1)}\left(\frac{u-T_L(P)}{T_S(P)}\right)dP(u) \qquad (5.27)\\
&\chi\left(\frac{x-T_L(P)}{T_S(P)}\right)T_S(P)\\
&= I(x,T_L,P)\int\chi^{(1)}\left(\frac{u-T_L(P)}{T_S(P)}\right)dP(u)+\\
&\quad I(x,T_S,P)\int\left(\frac{u-T_L(P)}{T_S(P)}\right)\chi^{(1)}\left(\frac{u-T_L(P)}{T_S(P)}\right)dP(u). \qquad (5.28)
\end{aligned}$$

The equations can be solved for $I(x,T_L,P)$ and $I(x,T_S,P)$ to give

$$I(x,T_L,P) = T_S(P)\frac{A_4\psi\left(\frac{x-T_L(P)}{T_S(P)}\right)-A_2\chi\left(\frac{x-T_L(P)}{T_S(P)}\right)}{A_1A_4-A_2A_3}, \qquad (5.29)$$

$$I(x,T_S,P) = T_S(P)\frac{A_1\chi\left(\frac{x-T_L(P)}{T_S(P)}\right)-A_3\psi\left(\frac{x-T_L(P)}{T_S(P)}\right)}{A_1A_4-A_2A_3}, \qquad (5.30)$$

(5.31)

where

$$A_1 = \int\psi^{(1)}\left(\frac{u-T_L(P)}{T_S(P)}\right)dP(u), \qquad (5.32)$$

$$A_2 = \int\left(\frac{u-T_L(P)}{T_S(P)}\right)\psi^{(1)}\left(\frac{u-T_L(P)}{T_S(P)}\right)dP(u), \qquad (5.33)$$

$$A_3 = \int\chi^{(1)}\left(\frac{u-T_L(P)}{T_S(P)}\right)dP(u), \qquad (5.34)$$

$$A_4 = \int\left(\frac{u-T_L(P)}{T_S(P)}\right)\chi^{(1)}\left(\frac{u-T_L(P)}{T_S(P)}\right)dP(u). \qquad (5.35)$$

If P is symmetric, then $A_2 = 0$ and $A_3 = 0$, and the expressions simplify to

$$I(x,T_L,P) = T_S(P)\,\psi\left(\frac{x-T_L(P)}{T_S(P)}\right)\Big/A_1, \qquad (5.36)$$

$$I(x,T_S,P) = -T_S(P)\,\chi\left(\frac{x-T_L(P)}{T_S(P)}\right)\Big/A_4. \qquad (5.37)$$

The asymptotics of $\sqrt{n}(T_L(\mathbb{P}_n(P))-T_L(P))$ and $\sqrt{n}(T_S(\mathbb{P}_n(P))-T_S(P))$ follow from (5.20) and (5.27) and (5.28) respectively.

	$c=2.7$	$c=3$	$c=4$	$c=5$
$n=3$	76.5	75.7	74.5	73.9
$n=4$	85.7	85.4	84.7	84.4
$n=5$	82.6	81.4	77.0	76.0
$n=10$	80.9	79.2	77.2	76.2
$n=20$	83.1	81.8	77.4	74.9
$n=50$	84.3	81.6	77.5	75.8
$n=\infty$	83.2	81.7	78.1	75.6

Table 5.2 *Efficiency of T_L for normal samples based on 5000 simulations.*

5.3.4 *Efficiency at the $\mathfrak{N}(0,1)$ Model, Asymptotics and Compromise*

Efficiency at the $\mathfrak{N}(0,1)$ model is number (iv) of the desirable properties. As the mean is the optimal estimator μ of the $\mathfrak{N}(\mu,\sigma^2)$ models it is only necessary to compare the variances of T_L and T_{ave}. The affine equivariance of both functionals means that the comparison can be done on the basis of the standard Gaussian distribution. Table 5.2 shows the relative efficiency of T_L for the tuning parameter values $c=2.7,3,4,5$. The first three lines gives the results for sample sizes $n=5,10,20,50$ based on 5000 simulations. The final line gives the asymptotic efficiency based on (5.20) on noting that in this special case $P=\mathfrak{N}(0,1)$

$$\int I(x,T,P)^2\,dP(x) = \frac{T_S(P)^2 \int \psi\left(\frac{x-T_L(P)}{T_S(P)}\right)^2 dP(x)}{\left(\int \psi^{(1)}\left(\frac{x-T_L(P)}{T_S(P)}\right) dP(x)\right)^2}. \tag{5.38}$$

It is seen that the asymptotic efficiency gives a good approximation to the actual behaviour of the functional T_L down to samples of size $n=5$.

Tables 5.1 and 5.2 suggest that there is a conflict between breakdown point and efficiency at the Gaussian model. This is not the case: it is possible to have a high breakdown point and a high efficiency. As pointed out above, a high breakdown point is usually associated with a small bias and there is indeed a conflict between a small bias (Figure 5.1) and efficiency. This can be seen by noting that the median has the smallest bias in a neighbourhood of a Gaussian distribution but its efficiency is asymptotically $100\cdot 2/\pi \approx 63.7$. Such compromises are common and often necessary in statistics: a procedure which is in one sense optimal may have other properties which are undesirable necessitating a compromise (see [116]). An example is the Huber estimator for location whose ψ function is given by

$$\psi(x) = x\min\left(1, \frac{b}{|x|}\right) \tag{5.39}$$

for some $b>0$. It is optimal B-robust (page 168, Table 4 of Chapter 2.2 of [103] for this and other examples) but is not locally uniformly differentiable. Section 'E for Engineering' of [211] describes the development of a milk bottle where changes were made to the 'optimal' solution of a sphere to give an end product with a flat bottom, a cap, a rounded square cross section to fit in a refrigerator and pushed-in sides for better grip.

5.3.5 *A Two-step Functional*

The smallest value of c which is known to guarantee the existence and uniqueness of $T_{LS} = (T_L, T_S)$ is 2.56149. The asymptotic efficiency of T_L at a Gaussian distribution is about 83% which may be judged to be acceptable or it may be judged to be too small. In the latter case an increase in efficiency can be obtained by a two-step procedure. Having calculated T_{LS} with $c = 5$ (5.23) a second location functional $T_L'(P)$ is defined as the solution of

$$\int \psi\left(\frac{x - T_L'(P)}{T_S(P)}\right) dP(x) = 0. \tag{5.40}$$

with $\psi(x) = \psi(x,c)$ with $c = 1$.

This inherits the breakdown point of T_L with $c = 5$ which is 0.463. The biases for the contamination ε used in Figure 5.1 for T_L' and T_L are as follows:

ε	0.05	0.1	0.15	0.2	0.25	0.3	0.35
T_L'	0.101	0.244	0.380	0.586	0.880	1.344	2.196
T_L	0.068	0.145	0.232	0.335	0.461	0.623	0.857

The influence function of the two-step procedure is given by

$$I(x, T_L', P) = \frac{T_S(P)\psi\left(\frac{x - T_L'(P)}{T_S(P)}\right)}{\int \psi^{(1)}\left(\frac{u - T_L'(P)}{T_S(P)}\right) dP(u)} - I(x, T_S, P)\frac{\int \frac{(u - T_L'(P))}{T_S(P)}\psi\left(\frac{u - T_L'(P)}{T_S(P)}\right) dP(x)}{\int \psi^{(1)}\left(\frac{u - T_L'(P)}{T_S(P)}\right) dP(u)}. \tag{5.41}$$

with $\psi(x) = \psi(x,c)$ with $c = 1$. If P is symmetric this reduces to

$$I(x, T_L', P) = \frac{T_S(P)\psi\left(\frac{x - T_L'(P)}{T_S(P)}\right)}{\int \psi^{(1)}\left(\frac{u - T_L'(P)}{T_S(P)}\right) dP(u)}. \tag{5.42}$$

The asymptotic efficiency of T_L' at a Gaussian model can be directly calculated and is 95.7%. Other choices of c are possible but there is always a trade-off between bias and efficiency. The two-step functional $T_{LS}' = (T_L', T_S)$ with c-values 5 and 1 for the two steps will be the default location-scale functional in this book.

An excellent discussion of of optimization, regularization and asymptotics is to be found in [116].

5.3.6 *Re-descending ψ-functions*

A re-descending ψ-function is one for which $\lim_{|x| \to \infty} \psi(x) = 0$. A more restrictive definition is used in [103] where a re-descending ψ-function is one which has a compact support, that is, it is zero outside a compact interval $[-r, r]$. Such ψ-functions reject large outlying observations completely. They are discussed in detail in Chapter 2.6 of and Chapter 4.8 of [118]. Huber gives 'a word of caution' on their use:

- they are beneficial if there are extreme outliers;
- the improvement is minor and counteracted by an increase in the minimax risk;
- good data screening based on expertise might be more effective;
- they need to be properly tuned.

In [103] it is pointed out that the denominator in the asymptotic variance

$$\left(\int \psi^{(1)} \left(\frac{x - T_L'(P)}{T_S(P)} \right) dP(x) \right)^2$$

with $P = \mathbb{P}_n$ can become zero as the first derivative $\psi^{(1)}$ takes on positive and negative values. This emphasizes the need for proper tuning as in the comment above.

Two different different re-descending ψ-functions will be considered. The first has a compact support and the corresponding location functional is calculated using (5.25) where the scale functional T_S is obtained from a monotone ψ-function. This is a two-step procedure. The second is the ψ function which derives from the maximum likelihood functional based on a t-distribution with $\nu > 1$ degrees of freedom. Surprisingly this functional has the desirable properties (i), (ii) and (iii) above (see [124], [125], [67], [202], [126], [77], [66])

As may be seen from Table 3 of Chapter 2.6 of [103] the actual form of the compact ψ-function does not matter within certain limits. The one to be used here is a piecewise polynomial chosen so that the second derivative is continuous. Given tuning constants $0 < c_1 < c_2 < c_3$ define the polynomial p_1 by

$$p_1(x) = 15(1 - (x/c_1)^2)^2/(8c_1). \tag{5.43}$$

Its first derivative satisfies $p_1^{(1)}(c_1) = 0$. A second polynomial is defined by

$$p_2(x) = \int_0^x p_1(u)\,du = 15(x/c_1 - 2x^3/(3c_1^3) + x^5/(5c_1^5)) \tag{5.44}$$

so that p_2 is an odd function of x with

$$p_2(0) = 0,\ p_2(c_1) = 1,\ p_2^{(1)}(c_1) = p_2^{(2)}(c_1) = 0.$$

A third polynomial is defined by

$$p_3(x) = 15(x^2 - c_2^2)^2(c_3^2 - x^2)^2 \tag{5.45}$$

and satisfies

$$p_3(c_2) = p_3(c_3) = p_3^{(1)}(c_2) = p_3^{(1)}(c_3) = 0.$$

Finally a fourth polynomial is defined by

$$p_4(x) = 1 - c\int_{c_2}^x p_3(u)\,du \tag{5.46}$$

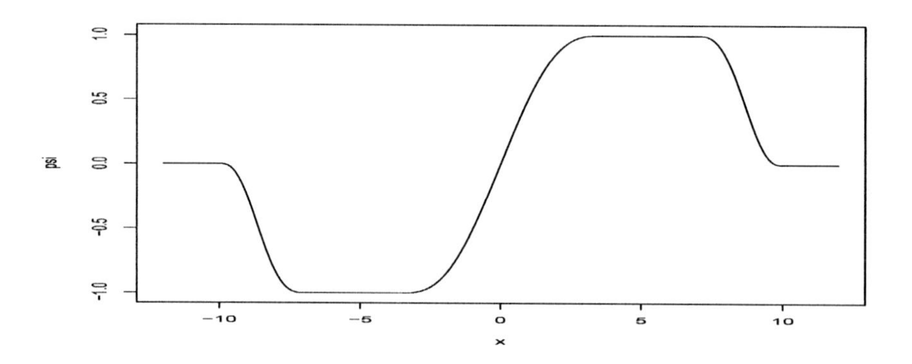

Figure 5.2 *The smooth re-descending ψ–function (5.47) with $c_1 = 3, c_2 = 6, c_3 = 9$.*

where

$$c = 1 \Big/ \int_{c_2}^{c_3} p_3(u)\, du$$

from which it follows that

$$p_4(c_2) = 1,\, p_4(c_3) = 0,\, p_4^{(1)}(c_2) = p_4^{(1)}(c_3) = p_4^{(2)}(c_2) = p_4^{(2)}(c_3) = 0.$$

Finally the resulting re-descending ψ–function is defined by

$$\psi(x) = \begin{cases} p_2(x), & |x| \le c_1, \\ 1, & c_1 < |x| \le c_2, \\ p_4(x), & c_2 < |x| \le c_3, \\ 0, & |x| > c_3. \end{cases} \tag{5.47}$$

The default values which will be used are $c_1 = 3, c_2 = 6, c_3 = 9$. Figure 5.2 shows the resulting function. It can be regarded as a smooth version of the Hampel's three part redescending ψ-function ((2.6.2) of [103]). The location functional T_{rd} based on this ψ-function is defined as that solution of

$$\int \psi\left(\frac{x - T_{\text{rd}}(P)}{T_S(P)}\right) dP(x) = 0 \tag{5.48}$$

which is closest to $T_L(P)$. The reason for this is that (5.48) has multiple solutions. If P is symmetric then $T_{\text{rd}}(P) = T_L(P)$ and both are the point of symmetry. If T_S is locally uniformly Fréchet differentiable at P, so is T_{rd} because of the smoothness of the defining ψ-function. Hampel's three part re-descending M–estimator as given in [103] is not Fréchet differentiable for two reasons. Firstly the ψ–function is piecewise linear and hence not differentiable. Secondly it uses the MAD as an auxiliary measure of scale. As the MAD has points of discontinuity in the neighbourhood of any distribution P so has the resulting location functional.

The second re-descending ψ function is that obtained from the maximum likelihood estimates of a t-distribution with ν degrees of freedom. The estimates $\hat{\mu}_n$ and $\hat{\sigma}_n$ must satisfy (5.9) and (5.10) with

$$f_\nu(x) = \frac{\Gamma((\nu+1)/2)}{\sqrt{\nu\pi}\,\Gamma(\nu/2)}\left(1+\frac{x^2}{\nu}\right)^{-(\nu+1)/2}. \tag{5.49}$$

This results in the ψ- and χ-functions given by

$$\psi(x) = \frac{x}{x^2+\nu}, \tag{5.50}$$

$$\chi(x) = (\nu+1)\frac{x^2}{x^2+\nu} - 1. \tag{5.51}$$

As mentioned above the maximum likelihood estimators for the t-distribution with $\nu > 1$ degrees of freedom is uniquely defined. Furthermore this continues to hold for any distribution P with $\Delta(P) < \nu/(1+\nu)$ (5.1) (see [124], [77], [66]). It follows that the maximum likelihood estimators for a t-distribution with $\nu > 1$ degrees of freedom can be treated as a location-scale functional $T_{LS:t,\nu} = (T_{L:t,\nu}, T_{S:t,\nu})$ defined on $\mathscr{P}_\delta$ (5.22) with $\delta = \nu/(\nu+1)$. The location part with default value $\nu = 1.5$ will be called the Kent-Tyler functional and denoted by T_{kt}. The breakdown point including existence and the scale part $T_{S:t,\nu}$ is treated in [77] for the contamination model $(1-\varepsilon)P + \varepsilon Q$. If $\Delta(P) = 0$ the breakdown point in this sense is $1/(\nu+1)$. Finally it is shown in [66] that the functional $T_{LS:t,\nu}$ is locally uniformly Fréchet differentiable at each $P \in \mathscr{P}_\delta$ with $\delta = \nu/(\nu+1)$ (see also [67]). The default value of ν will be 1.5 giving a breakdown point of 0.4.

The scale part of the Kent-Tyler functional is not Fisher consistent for the $\mathfrak{N}(\mu,\sigma^2)$ family. In fact for $\nu = 1.5$

$$T_{S:t,1.5}(\mathfrak{N}(0,\sigma^2)) = 0.6903\sigma$$

so to make it Fisher consistent it must be multiplied by $1/0.6903 = 1.4486$

The bias behaviour of the location functional T_L with a re-descending ψ function differs from that of location functional T_L with a monotone ψ functional in that the influence of large outliers tends to zero as they move further away from the main body of the data. Figure 5.3 show this effect. The bias is calculated using the model

$$P'_\mu = 0.9\mathfrak{N}(0,1) + 0.1\mathfrak{N}(\mu, 0.01^2), \quad 0 \le \mu \le 20. \tag{5.52}$$

It is seen that the median T_{med} (solid) has the smallest maximum bias attained for even quite small values of μ. The bias of the default version of re-descending functional T_{rd} (dotdash) is zero for $\mu \ge 12.25$ reflecting the fact that the corresponding ψ-function has a compact support. The maximum bias is 0.28 as against 0.14 for the median. The bias of the default version of the Kent-Tyler functional T_{kt} (dashed) is always positive but tends to zero. The largest bias 0.16 is only slightly greater than that of the median.

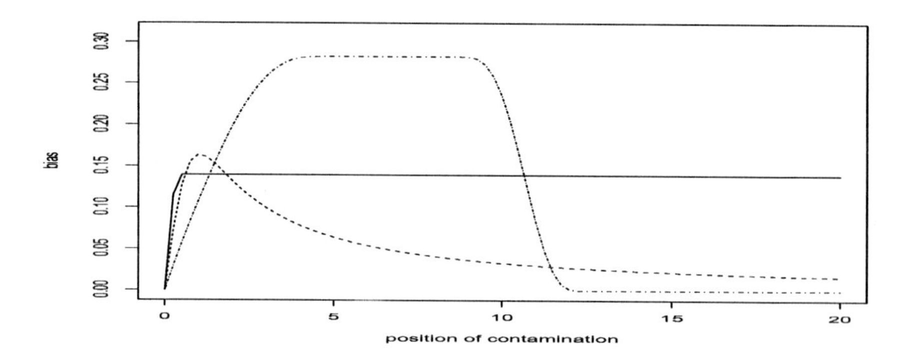

Figure 5.3 *Bias of T_{rd} (dotdash), T_{kl} (dashed) and T_{med} as a function of the position μ of the contamination P'_{μ} of (5.52).*

5.3.7 Relative Efficiencies on Various Test Beds

Table 5.2 gives the efficiency of T_L for normal samples for differing sample sizes and value of the tuning constant c. It is important however to investigate the behaviour of the functionals under different situations. Tukey calls models or real data used to evaluate procedures 'challenges'. Real data can pose a challenge to a procedure as they can exhibit unexpected features which were not initially allowed for when the procedure was specified. In contrast simulated data sets are under the control of the statistician and may be called test beds, just as test beds are under the control of the engineer, at least more rather than less. Test beds are a subclass of 'challenges'.

Questions of efficiency or precision cannot be answered using real data, but only by using simulations where what is to be estimated is known. Heavy tailed symmetric distributions can be modelled by t-distributions with various degrees of freedom and also by the Laplace distribution $\mathfrak{L}(\mu,\sigma)$. All location functionals T_L estimate the same thing for symmetric distribution, namely the centre of symmetry. Asymmetric distributions can be modelled by gamma distributions with various shape parameters and by say by lognormal distributions. Here it is no longer the case that different location functionals estimate the same quantity. For comparison this does not matter as it is possible to quantify the precision with which they estimate whatever they estimate which is that specified by the functional.

Another divergence from normality is bi- or multi-modality. This is in a sense orthogonal to symmetry and asymmetry and is more difficult to treat. The distribution $p\mathfrak{N}(-5,1)+(1-p)\mathfrak{N}(-5,1)$ is bimodal. If $p=1/2$ and it is used as the basic model for the location scale problem all robust functionals will have problems estimating the location parameter μ for all but impossibly large sample sizes. The median will oscillate between $\mu-5$ and $\mu+5$ depending on which component has more observations. The behaviour of the other location functionals will be similar. If $p<0.5$ then the left component will be treated as contamination and the estimators will concentrate on the right component. In other cases the data may be essentially

	t_1	t_2	t_3	t_4	t_5	t_{10}	t_{20}	$\mathfrak{L}(0,1)$	$\mathfrak{N}(0,1)$
T'_L	54	82	98	99	99	100	98	83	95
T_{ave}	0.0	11	54	70	79	95	100	57	100
T_{med}	60	86	85	84	80	74	69	100	65
T_{kt}	100	100	97	96	90	82	77	100	72
T_{rd}	71	86	100	100	100	100	97	84	95

Table 5.3 *Relative efficiencies of of T'_L, T_{ave}, T_{med}, T_{kt} and T_{rd} for test beds given in the first line. The results are based on 1000 simulations.*

multimodal without causing any problems. The upper panel of Figure 3.8 shows the student data on a fine scale where it seems clearly multimodal when compared to the simulated gamma sample in the lower panel. On the coarser scale of distribution functions the student data is essentially unimodal does not cause problems for the location functionals.

The five functionals to be compared are the mean T_{ave}, the median T_{med}, the default two step M-functional T'_L and the two default re-descending functionals T_{rd} and T_{kt}. For each test bed the mean quadratic error is estimated for each functional. The relative efficiency is measured against the best functional for that test bed.

5.3.7.1 *Symmetric Test Beds*

The symmetric distribution used are the t-distributions T_ν with $\nu = 1,2,3,4,5,10$ and 20 degrees of freedom, the standard Laplace distribution $\mathfrak{L}(0,1)$ and the standard normal distribution $\mathfrak{N}(0,1)$. The relative efficiencies of the five functional are given in Table 5.3. It is seen that the re-descending functional T_{rd} essentially dominates the two step functional T'_L for the test beds considered. They perform well for all t-distributions with $\nu \geq 3$ degrees of freedom. In the other direction the Kent-Tyler functional performs well for all t-distributions with $1 \leq \nu \leq 5$ degrees of freedom and for the Laplace distribution. It must be kept in mind that there is a trade off between efficiency and bias behaviour. The table indicates that T_{rd} is a good multi-purpose functional for data whose tails are not too heavy, whilst the Kent-Tyler functional does the same for heavy tailed data.

5.3.7.2 *Non-symmetric Test Beds*

The skewed test beds are the $\Gamma(\nu) = \Gamma(\nu,\lambda)$ distributions with $\lambda = 1$ and $\nu = 1,2,3,4,5,10$, and the lognormal distributions $\mathfrak{LN}(\sigma^2) = \mathfrak{LN}(\mu,\sigma^2)$ for $\mu = 0$ and $\sigma = 1.3, 1.2, 1.1, 1$. The parameters λ and μ are scale parameters and their value has no effect on the results. Although each functional estimates something different they can still be compared by the accuracy with which they estimate whatever they are estimating. This can be seen by considering the problem of quantifying the shift of two samples. At the level of models $T_L(P^\mu) - T_L(P) = \mu$ whatever the location functional T_L where $P^\mu(B) = P(B-\mu)$. The results are given in Table 5.4. On the basis of this table alone the Kent-Tyler functional would seem to be the best. Its only disadvantage is its relatively low asymptotic efficiency of 72% at the normal model which seems to be the appropriate model for most data sets.

	$\Gamma(1)$	$\Gamma(2)$	$\Gamma(3)$	$\Gamma(4)$	$\Gamma(5)$	$\Gamma(10)$	$\mathfrak{LN}(1.3)$	$\mathfrak{LN}(1.2)$	$\mathfrak{LN}(1.1)$	$\mathfrak{LN}(1)$
T_L'	95	100	100	98	98	98	46	50	64	73
T_{ave}	85	95	99	100	100	100	15	20	19	30
T_{med}	87	81	75	71	72	69	80	82	80	81
T_{kt}	100	90	83	80	78	75	100	100	100	100
T_{rd}	88	96	98	96	97	97	58	61	64	71

Table 5.4 *Relative efficiencies of of T_L', T_{ave}, T_{med}, T_{kt} and T_{rd} for test beds given in the first line. The results are based on 1000 simulations.*

5.4 Approximation Intervals, Quantiles and Bootstrapping

5.4.1 *Approximation Intervals*

As mentioned in Chapter 5.1 most of the robust literature on confidence intervals is of an asymptotic nature. Consider a sequence of i.i.d. random variables $(X_i)_{i=1}^{\infty}$ with distribution P and an M-functional T_L. It 'follows' from (5.20) that

$$\left[T_L(\mathbb{P}_n) - 1.96\sqrt{V(T_L,P)}/\sqrt{n}, T_L(\mathbb{P}_n) + 1.96\sqrt{V(T_L,P)}/\sqrt{n}\right]$$

is an asymptotic 0.95-confidence interval for $T_L(P)$ where

$$V(T_L,P) = \int I(x,T_l,P)^2\, dP(x)$$

and $I(x,T_L,P)$ is given by (5.29). Suppose there is a basic model P_0, for example $P_0 = \mathfrak{N}(\mu,\sigma^2)$, but in keeping with one variant of robustness it is required that the asymptotic confidence regions holds for P in a small neighbourhood of P_0. As P is not known standard practice is to substitute $\mathbb{P}_n$ for P to give the 0.95-approximation interval

$$\left[T_L(\mathbb{P}_n) - 1.96\sqrt{V(T_L,\mathbb{P}_n)}/\sqrt{n}, T_L(\mathbb{P}_n) + 1.96\sqrt{V(T_L,\mathbb{P}_n)}/\sqrt{n}\right]$$

with

$$V(T_L,\mathbb{P}_n) = \int I(x,T_L,\mathbb{P}_n)^2\, d\mathbb{P}_n(x) = \frac{1}{n}\sum_{i=1,n} I(x_i,T_L,\mathbb{P}_n)^2.$$

This is the recommendation of page 42 of [148] where the neighbourhood is a gross error neighbourhood.

To make this work assumptions must be placed on the M-functional and it was argued in [37] (see above) that locally uniformly Fréchet differentiability is sufficient. The result is precise and the statement correct but this is making life too simple. In practice n is not infinite and there may be no basic model. This requires investigating the behaviour for small n under various models. The task is a thankless but necessary one and will be undertaken in this section.

One result is that for small samples the efficiency of the location part can be misleading. The efficiencies of T_L' at the standard normal distribution for $n = 3$, $n = 4$

and $n = 6$ are 0.94, 0.99 and 0.99 respectively. The efficiencies as measured by the squared lengths of the confidence interval with $\alpha = 0.95$ are 53.3, 4.75 and 20.5 respectively.

Given data $\mathbf{x}_n$ and a location-scale functional $T_{LS} = (T_L, T_S)$ a non-parametric approximation interval is given by (5.8). In view of the results of the previous section this is now amended to

$$\begin{aligned}\mathscr{A}_n(\mathbf{x}_n, \alpha, T_L(\mathscr{P}_\delta)) &= \{T_L(P) : \mathrm{qtl}((1-\alpha)/3, n, P) \le \sqrt{n}(T_L(\mathbb{P}_n) - T_L(P)) \\ &\quad \le \mathrm{qtl}((2+\alpha)/3, n, P),\, d_{\mathrm{ku}}(\mathbb{P}_n, P) \le \mathrm{qdku}((2+\alpha)/3), \\ &\quad \Delta(P) < \delta\} \end{aligned} \tag{5.53}$$

for some $\delta < 1/2$.

As it stands it is not possible to calculate the approximation region (5.53) as this requires knowledge of the quantiles $\mathrm{qtl}(\alpha, n, P)$ of $\sqrt{n}(T_L(\mathbb{P}_n(P)) - T_L(P))$ for all P satisfying $d_{\mathrm{ku}}(\mathbb{P}_n, P) \le \mathrm{qdku}((2+\alpha)/3)$. However as T_L is locally uniformly Fréchet differentiable the

$$\sqrt{n}(T_L(\mathbb{P}_n(P)) - T_L(P)) \overset{D}{\Rightarrow} \mathfrak{N}(0, V(P)^2) \tag{5.54}$$

with

$$V(P)^2 = \int I(x, T_L, P)^2\, dP(x) \tag{5.55}$$

where $I(x, T_L, P)$ is the influence function of T_L given either by (5.29) or (5.41) depending on the choice of T_L. Furthermore for any given P_0 with $\Delta(P_0) < 1/2$ there exists $\varepsilon(P_0) > 0$ such that the convergence in (5.54) is uniform for all P in the Kuiper ball $B_{\varepsilon(P_0)}(P_0)$ of radius $\varepsilon(P_0)$ centred at P_0 . Moreover for sufficiently small $\varepsilon(P_0)$ the locally uniform Fréchet differentiability implies

$$\begin{aligned} V(P)^2 &= \int I(x, T_L, P)^2\, dP(x) \\ &\approx \int I(x, T_L, P_0)^2\, dP_0(x) \\ &= V(P_0)^2 \end{aligned}$$

again uniformly for all $P \in B_{\varepsilon(P_0)}(P_0)$. Combining this gives

$$\sqrt{n}(T_L(\mathbb{P}_n(P)) - T_L(P)) \overset{D}{\approx} \mathfrak{N}(0, V(P_0)^2) \tag{5.56}$$

with the approximation being uniformly accurate for $P \in B_{\varepsilon(P_0)}(P_0)$ and $n \ge n(P_0)$ for sufficiently small $\varepsilon(P_0)$ and sufficiently large $n(P_0)$.

In (5.53) the role of P_0 is played by the empirical distribution of the data $\mathbb{P}_n$. It follows that the approximation interval can be calculated with sufficient accuracy if $\varepsilon(\mathbb{P}_n) < \mathrm{qdku}((2+\alpha)/3)$ and $n \ge n(\mathbb{P}_n)$. Everything is known and therefore given a specified degree of accuracy the values of $\varepsilon(\mathbb{P}_n)$ and $n(\mathbb{P}_n)$ can in principle be calculated using the Berry-Esseen inequality ([18], [84]). The calculations are not simple and will probably lead to values which are much too pessimistic. This will accentuated by the use of the Berry-Esseen inequality which is also pessimistic [109]. An analytical approach will therefore probably lead to values of $\varepsilon(\mathbb{P}_n)$ and $n(\mathbb{P}_n)$ which are much too small and large respectively for them to be usable only for very large sample sizes n. The alternative is to use simulations to give a guide as to the shape and size of empirical $\mathbb{P}_n$ for which the approximation will be acceptable. This will be done in the next section.

5.4.2 *Approximating the Quantiles*

The above discussion suggests approximating qtl$((2+\alpha)/3,n,P)$ by qnorm$((2+\alpha)/3)\sqrt{V(\mathbb{P}_n)}$ with $V(P)$ given by (5.55). This requires some idea of how large n should be for the asymptotic approximation to be sufficiently accurate. Moreover it is also desirable to have good approximations for smaller values of n to extend the range of applicability of the procedure. Towards this end consider a model P with $\Delta(P) < 0.5$. It follows from (5.54) that

$$\sqrt{n}\,T_L^t(P) \overset{D}{\Longrightarrow} \mathfrak{N}(0,1)\,. \tag{5.57}$$

where

$$T_L^t(P) = (T_L(\mathbb{P}_n(P)) - T_L(P))/\sqrt{V(\mathbb{P}_n(P))} \tag{5.58}$$

Consider now a suite of models $\mathscr{M} = \{P_1,\ldots,P_k\}$ for which the convergence in (5.57) is uniform and for which the approximation is judged to be sufficiently accurate for $n \geq n_1$ with n_1 not too large. For this it is not necessary for the P_i to be close to one another with respect to the Kuiper metric. As $T_L^t(P)$ is invariant with respect to affine transformations the suit can be extended to

$$\mathscr{M}^{\mathscr{A}} = \{P^A : P \in \mathscr{M}, A \text{ affine }\}.$$

Table 5.5 shows the results of a simulation for $T_L = T_L'$ for the suite

$P_1 = \mathfrak{N}(0,1)$ $\quad P_2 = \mathfrak{U}(0,1)$ $\quad P_3 = \mathfrak{C}(0,1)$

$P_4 = \mathfrak{LN}(0,1)$ $\quad P_5 = \mathfrak{LN}(0,0.75^2)$ $\quad P_6 = \mathfrak{LN}(0,0.5^2)$

$P_7 = \Gamma(1)$ $\quad P_8 = \Gamma(2)$ $\quad P_9 = \Gamma(3)$

$P_{10} = 0.5\mathfrak{N}(0,1) + 0.5\mathfrak{N}(20,1)$ $\quad P_{11} = 0.7\mathfrak{N}(0,1) + 0.3\mathfrak{N}(20,1)$

and for the sample sizes $n = 50, 100, 300, 500$.

Table 5.5 shows that the normal approximation is already good for samples of size $n = 50$ for the symmetric distributions P_1, P_2, P_3. For samples of size $n = 500$ it

	P_1	P_2	P_3	P_4	P_5	P_6	P_7	P_8	P_9	P_{10}	P_{11}
50	8	7	19	35	30	21	26	18	19	334	189
100	7	6	13	22	17	16	17	17	12	348	143
300	6	7	6	14	12	8	12	7	8	257	65
500	5	5	8	12	8	7	9	9	8	95	5

Table 5.5 *Kuiper distances* $\times 10^3$ *of the distribution of (5.58) under the models* $P_1,\ldots,P_{11}$ *from the* $\mathfrak{N}(0,1)$ *distribution for the sample sizes* $n = 50, 100, 300, 500$.

is very good for all models apart from the strongly bimodal model P_{10}. On the basis of this the strategy of Chapter 2.9 will be followed to give finite sample approximations to the distribution of (5.58) using the model $P = P_1 = \mathfrak{N}(0,1)$. The distribution will be approximated by the t distribution with $n-1$ degrees of freedom. More precisely the the empirical estimate $\hat{\mathrm{q}}(\beta,n,T_L^t(\mathfrak{N}(0,1)))$ of $\mathrm{q}(\beta,n,T_L^t(\mathfrak{N}(0,1)))$ obtained obtained from $5 \cdot 10^5$ simulations will be approximated in turn by a cubic function of $\mathrm{qt}(\beta,n-1)$ which is the β-quantile of the t-distribution with $n-1$ degrees of freedom. The approximation is for β in the range $0.8 \le \beta \le 0.9999$

$$\hat{\mathrm{q}}(\beta,n,T_L^t(\mathfrak{N}(0,1))) \approx \sum_{k=0}^{3} a_k(n)\mathrm{qt}(\beta,n-1)^k$$

The values of the constants $a_k(n), k = 0,1,2,3$ are determined by least squares for each $n, 3 \le n \le 50$. This strategy works for $n \ge 9$. For $3 \le n \le 7$ the simulated values of $\hat{\mathrm{q}}(\beta,n,T_L^t(\mathfrak{N}(0,1)))$ are stored and used. For $n \ge 50$ the simple approximation

$$\hat{\mathrm{q}}(\beta,n,T_L^t(\mathfrak{N}(0,1))) \approx \mathrm{qt}(\beta,n-1)$$

is sufficient for all practical purposes.

There are two other possibilities. One is to use

$$\frac{\sqrt{n}(T_L(\mathbb{P}_n(P)) - T_L(P))}{T_S(\mathbb{P}_n)} \tag{5.59}$$

instead of $T_L^t(P)$ of (5.58) and to calibrate it with the $\mathfrak{N}(0,1)$ model. The second is to bootstrap $T_L(\mathbb{P}_n(P)) - T_L(P)$. Simulations indicate that both are inferior to using (5.58): the expression (5.59) does not adjust to the shape of the samples and the sample sizes required for bootstrapping are such that the normal approximation to the distribution of (5.58) is then very good. They will not be considered further.

	$\mathfrak{N}(0,1)$				$\mathfrak{S}(0,1)$				$\mathfrak{E}(1)$			
	5	10	50	100	5	10	50	100	5	10	50	100
T_L^t	94.8	94.7	94.8	94.7	96.7	97.3	96.1	95.8	90.9	91.7	93.7	94.4
	6.87	4.76	4.10	4.04	25.9	15.2	11.3	10.9	6.14	4.39	3.76	3.74
T_{rd}	95.5	94.7	94.8	94.7	95.9	96.1	95.7	95.6	90.1	89.8	93.6	94.5
	10.2	6.26	4.11	4.05	54.3	21.9	10.5	10.0	10.1	7.04	4.15	3.95
T_{kt}	92.3	91.5	94.1	94.5	95.4	94.6	94.2	95.3	91.0	89.3	93.3	94.1
	8.41	5.71	4.81	4.71	25.9	12.4	9.13	8.87	6.95	4.53	3.70	3.64

Table 5.6 *Covering probabilities of the approximation regions on the* $\mathfrak{N}(0,1)$, $\mathfrak{S}(0,1)$ *and the* $\mathfrak{E}(1)$ *test beds.*

Table 5.6 gives the covering probabilities for three different functionals (T_L^t, T_{rd} and T_{kt}), three different test beds (the standard normal $\mathfrak{N}(0,1)$, the standard slash $\mathfrak{S}(0,1) = \mathfrak{N}(0,1)/\mathfrak{U}(0,1)$ and the standard exponential $\mathfrak{E}(1)$) and four different sample sizes ($n = 5, 10, 50, 100$). The first lines gives the covering probabilities, the second lines give the average lengths of the intervals. The results are based on 5000 simulations. The results for the log-normal model $\mathfrak{LN}(0,1)$ and for the absolute Cauchy model $|\mathfrak{C}(0,1)|$ are the same as for the exponential model but with somewhat longer intervals. Overall the two-step functional T_L' and the redescending functional T_{rd} have the better coverage behaviour and of the two T_L' has the shorter intervals. On the basis of this T_L' would seem to be the best default functional.

The resulting approximation to the approximation region $\mathscr{A}_n(\mathbf{x}_n, \alpha, T_L(\mathscr{P}))$ of (5.53) without the 'honesty' term

$$d_{\text{ku}}(\mathbb{P}_n, P) \leq \text{qdku}((2+\alpha)/3)$$

is, for $9 \leq n \leq 50$,

$$\begin{aligned} &\mathscr{A}_n(\mathbf{x}_n, \alpha, T_L(\mathscr{P})) \\ &\approx \Big[T_L(\mathbb{P}_n) - \Big(\sum_{k=0}^{3} a_k(n)\text{qt}((5+\alpha)/6, n-1)^k\Big)\sqrt{V(\mathbb{P}_n)}/\sqrt{n}, \\ &\qquad T_L(\mathbb{P}_n) + \Big(\sum_{k=0}^{3} a_k(n)\text{qt}((5+\alpha)/6, n-1)^k\Big)\sqrt{V(\mathbb{P}_n)}/\sqrt{n}\Big] \end{aligned} \tag{5.60}$$

where with $a_0(n)$, $a_1(n)$ and $a_3(n)$ are as described above. For $3 \leq n \leq 8$ the simulated quantiles are used and for $n \geq 51$ the approximation with $a_1(n) = 1$ and the remainder zero is sufficient.

Although honesty condition has been omitted in (5.60) it still hovers in the background as it was used to justify the approximation. It also leaves its mark in (5.60) which looks at first glance like a $(2+\alpha)/3$-approximation region rather than the α-approximation region it was defined as. If the honesty term in (5.53) is omitted, that is all the $1-\alpha$ is spent on the location term the approximate approximation interval is as in (5.60) but with $(1+\alpha)/2$ in place of $(5+\alpha)/6$. The accuracy of the approximation (5.60) depends on the size n and shape of the sample $\mathbf{x}_n$. In the next section an attempt will be made to specify conditions on the sample $\mathbf{x}_n$ so that the approximation (5.60) is satisfactory.

5.4.3 *Covering Efficiency and $n = 4$*

So far the efficiency of a location functional on a particular test bed has been defined as the ratio of its variance and the variance of the optimal location functional or variance of the best of the functionals under consideration. An example of the former is Table 5.2 and of the latter Table 5.3. In the robustness literature emphasis is placed on asymptotic efficiency which is defined as the ratio of the asymptotic variances. This can be supplemented by finite sample simulations as in Tables 5.2 and 5.3. Very often, as in Table 5.2, the finite efficiencies are well described by the asymptotic efficiency. For $n = 4$ the efficiency of T_L when compared to that of the mean is about 85%. For T_L' this increase to about 95% and certainly gives no hint of any problem.

The argument for efficiency is that it leads to small confidence or approximation intervals and hence to an increase in precision. It follows that a better measure of efficiency is the comparison of the lengths of the confidence intervals for a given

	T'_L				Median/MAD				Best		
	0.9	0.95	0.99	asymp	0.9	0.95	0.99	asymp	0.9	0.95	0.99
$n=3$	72.4	53.3	35.6	93.9	10.1	5.1	0.97	73.9	75.7	53.8	33.0
$n=4$	4.93	4.75	4.34	96.2	42.6	33.5	19.3	84.0	86.4	82.3	70.5
$n=5$	84.0	80.5	68.2	97.1	28.9	21.0	9.74	69.7	84.9	81.9	70.9
$n=6$	22.1	21.7	20.5	97.7	46.1	39.5	25.9	78.4	91.1	87.8	77.9
$n=7$	83.1	81.8	77.4	97.4	39.3	32.2	20.0	67.9	89.3	86.1	77.4
$n=8$	87.1	86.3	84.1	98.3	51.5	45.6	34.1	75.1	92.4	90.0	83.6

Table 5.7 *Covering efficiencies of T'_L and median/MAD for normal samples based on 10000 simulations.*

coverage probability. More precisely the relative efficiency of two location functionals for a given sample size can be defined as the ratio of the average of the squared lengths of the interval. Asymptotically the two measures will be the same.

Table 5.7 gives the efficiencies with respect to the standard confidence interval based on the mean and standard deviation. It is seen that for $n = 4$ the efficiency of T'_L measured in this sense is about 5% which bears no relationship to the 95% using the normal measure of efficiency. The interval based on the median and MAD are much better for $n = 4$ but for $n = 3$ and $\alpha = 0.99$ the efficiency is as low as 1%. For $n \geq 9$ the covering efficiencies of T'_L are all greater than 90% and can be regarded as satisfactory. One of the normal samples generated in the course of the simulations was

$$\mathbf{x}_4 = (0.07034408, 0.80260169, -0.33693350, 0.07510794)$$

giving $T'_L(\mathbf{x}_4) = 0.07776037$ and $T_S(\mathbf{x}_4) = 0.1119544$. The normalized sample $(\mathbf{x}_4 - T'_L(\mathbf{x}_4)/T_S(\mathbf{x}_4))$ is

$$(-0.06624455, 6.47450558, -3.70417313, -0.02369229)$$

which gives rise to the following values of the A_i of (5.32) to (5.35):

$$A_1 = 1.21,\ A_2 = 5.48 \cdot 10^{-5},\ A_3 = -4.18 \cdot 10^{-6},\ A_4 = 2.00 \cdot 10^{-4}$$

and

$$A_1 A_4 - A_2 A_3 = 2.43 \cdot 10^{-4}$$

leading to $V(\mathbb{P}_4) = 229.94$. The quantile for $n = 4$ and $\alpha = 0.9$ is 4.251 and results in a confidence interval of length $2 \cdot 4.251 \cdot \sqrt{V(\mathbb{P}_4)/2} = 64.46$. The length of the corresponding confidence interval based on the mean and standard deviation is 1.12. For this sample the default R-version of the MAD gives 0.305.

The problem is that $T_S(\mathbf{x}_4)$ is largely determined by the two values 0.07034408 and 0.07510794 which are very close. The MAD on the other hand depends also on -0.3369335. This can be seen more clearly for the sample

$$\mathbf{x}'_4 = (0.07034408, 10.80260169, -10.33693350, 0.07510794)$$

with $T'_S(\mathbf{x}'_4) = 0.336$ but $MAD(\mathbf{x}'_4) = 7.718$. For this data $V(\mathbb{P}'_4) = 722.7$. The same reasoning applies for $n = 3$ but with four observations the chance that two are very close together is higher than for three observations. This explains the fact the the efficiency for $n = 3$ is higher than for $n = 4$. For $n = 5$ and $n = 6$ the same phenomenon would occur if three observations were very close together which is much more unlikely. Nevertheless this can explain the drop in efficiency going from $n = 5$ to $n = 6$.

Ar first glance it may seem that the problem could be avoided by studentizing with $T_S'(\mathbb{P}_4)$ rather than $\sqrt{V(\mathbb{P}_4)}$. The resulting covering efficiencies for $\alpha = 0.95$ are

$$7.52, 21.2, 35.8, 49.9, 54.4, 62.3$$

which can be compared with the corresponding column of Table 5.7 for T_L'. There has been some improvement for $n = 4$ and $n = 6$ but for the other values of n there has been a significant drop in efficiency.

A second possibility is to try an emulate the behaviour of the MAD. This can be done by choosing χ to be close to the χ-function of the MAD, namely

$$\chi_{\text{MAD}} = \begin{cases} -1, & |x| \leq 1, \\ 1, & |x| > 1. \end{cases}$$

It can be achieved by replacing $(x^4 - 1)/(x^4 + 1)$ by, for example, $(x^8 - 1)/(x^8 + 1)$ or some higher power of $|x|$. Calculations show that if a power of 8 is chosen the default ψ function must now be at least 6 so that the condition $\psi\chi 1$ is satisfied. Increasing the power of $|x|$ leads to an increase in the constant c so the (T_L, T_S) will converge to $(T_{\text{med}}, T_{\text{MAD}})$. As the median/MAD performance for other n values ranges from very bad to just acceptable (Table 5.7) this does not solve the problem either.

A third possibility is to replace the χ-function $(x^4 - 1)/(x^4 + 1)$ by $(x^4 - \gamma)/(x^4 + 1)$ for some $\gamma > 0$ and with different choices of γ for different n. This works to some extent. For example for $n = 4$ and $\gamma = 0.37$ the efficiencies for $\alpha = 0.9, 0.95, 0.99$ are $81.2, 66.9, 51.4$ respectively. In order to accommodate one large observation γ must be at least 0.348 but this leads to poor bias behaviour.

The fourth possibility is to break the connection between the location and scale functionals. The two most common location functionals, the mean and the median, are both defined independently of a scale functional. In the opposite direction this is not common: both the standard deviation and the MAD rely on a location functional. For a scale functional the link can be broken by symmetrizing the sample by considering all differences $x_i - x_j$ and then calculating a scale for the symmetrized data (see [179]). More precisely and more generally a scale functional T_S^s can be defined by

$$\int\int \chi\left(\frac{u - v}{T_S^s(P)}\right) dP(u)\,dP(v) = 0. \tag{5.61}$$

Having calculated $T_S^s(P)$ the corresponding location functional T_L'' is defined by

$$\int \psi\left(\frac{x - T_L''(P)}{T_S^s(P)}\right) dP(x) = 0 \tag{5.62}$$

where the function $\psi(,x,c)$ of (5.23) will be used with $c = 1$. It turns out that for a suitable choice of χ this strategy works to a large extent. Put

$$\chi_\gamma(x) = \frac{x^2 - \gamma}{x^2 + 1} \tag{5.63}$$

for some $\gamma > 0$. The value of γ controls the finite sample breakdown point. For $n = 3$ and $n = 4$ the functional must be able to withstand one outlier. This implies $\gamma > 0.8$ for $n = 3$ and $\gamma > 0.6$ for $n = 4$. The values chosen are 0.81 and 0.61 respectively. The reasoning can be repeated for the other values of n. For $\alpha = 0.95$ the resulting coverage efficiencies are

$$64.6, 83.6, 78.9, 88.7, 85.5, 90.0.$$

This has lead to improvement apart for $n = 3$ and $n = 5$.

The result of the above considerations is to studentize using $\sqrt{V(\mathbb{P}_n)}$ as in (5.58) for all n except for $n = 4, 6, 7, 8$. For these values of n the location functional is standardized using $T_S^s(\mathbb{P}_n)$. This will be the default choice.

5.5 Stigler's Comparison of Eleven Location Functionals

In [196] Stigler compared the following eleven location functionals: the mean, the median, the 10%, 15% and 20% trimmed means, Huber's P15, Andrew's AMT, Tukey's biweight, the Edgeworth functional, the outmean and Hogg's T_1 (see [196] for the definitions). The data sets used were Short's determinations of the parallax of the sun, Cavendish's determinations of the mean density of the earth and Michelson's and Newcomb's measurements of the velocity of light (see [196] for the data sets and original references). For each data set the 'true' value was taken to be the modern value. This is particularly questionable for the Michelson data as 96 of the 100 measurements are larger than the true value. There is clearly a systematic error which no statistical analysis can correct for. This and other comments and criticisms were made by the discussants of the paper and replied to by Stigler.

The absolute error for each functional and data set was calculated and then converted to a relative error. The functionals were ordered according to the average relative error and average rank over the data sets, the results for small and large data sets being given separately. The smallest relative error for small data sets was the 10% trimmed mean with an error of 0.916. For large data sets the smallest relative error was that of the mean with an error of 0.924. In terms of rank the best functional both for the small and large data sets was the 10% trimmed mean with an average rank of 4.6 for the small samples and 4.5 for the large samples. The default two-step functional T_L' has an average relative error over the twenty-four data sets of 0.882. The redescending functional T_{rd} was marginally better with an average relative error of 0.876. The Kent-Tyler functional T_{kt} has an average relative error of 1.126. This is just slightly better than that of the median (1.149 for small samples and 1.152 for large samples) which was the worst performing functional in the comparison. The average rank of the two-step functional was 4.77 which was the smallest value. It is even smaller than all the values given by Stigler with the exception of the 10% trimmed mean in spite of the fact that now twelve functionals are being compared and not eleven.

5.6 An Attempt at an Automatic Procedure

The approximation region (5.60) can be regarded as a procedure for specifying a range of reasonable values of $T_L(P)$ based on a data set $\mathbf{x}_n$. Its accuracy depends on the data $\mathbf{x}_n$. Ideally it should therefore be accompanied by a (necessarily vague) description of the data sets where it can be expected to perform reasonably well such as the following:

(a) the sample size is not too small;

(b) the data are not too skewed depending on the sample size;

(c) the tails are not too heavy;

(d) the data have not too many one-sided outliers;

(e) the data have no large atoms;

(f) not too discrete;

(g) the data are not too bimodal.

Some of these properties can be quantified and incorporated into an automatic procedure with the following outputs:

(a) the sample size;

(b) the Kuiper distances from a suite of model;

(c) the number of outliers based on the closest model;

(d) the size of the largest atom;

(e) the number of different values.

It seems difficult to give some good measure of bimodality, but it can possibly be done using the taught string method for densities [46]. The examples in mind are the two models P_{10} and P_{11} of Table 5.5. The suite of models are not to be thought of as models for the data but to given an indication of whether the asymptotic normality approximation for T_L^t of (5.58) is likely to be accurate. The skewed models $P_4 - P_9$ in Table 5.5 give similar results which opens up the possibility of some adaptivity of the quantiles of T_L^t for skewed distributions, based for example on the model $P_5 = \mathfrak{LN}(0,1)$.

Once a procedure has been defined it can be evaluated on some well-defined test beds, that is by using simulations, where the experimental conditions are under the full control of the statistician (see 'C for challenges' in [211]). An example is given by the covering probabilities of Table 5.6 for samples generated with different distributions. Other possibilities are to include one- and two-sided outliers of different sizes.

A procedure should also be evaluated using real data sets. The use of real data may reveal weaknesses that were not suspected when the procedure was defined. A disadvantage is that the conditions are not under the control of the statistician and so there be no obvious way of evaluating the performance it can sometimes be done (see [196] and Chapter 5.5).

One situation in which a direct evaluation of a procedure is possible is the analysis of interlaboratory tests. Samples are prepared with a given amount of

39.0	48.0	51.0	55.0	57.0	60.0	61.8	62.0	62.5	63.0	63.0	65.0
65.2	65.9	66.0	66.0	66.0	66.0	66.5	69.0	69.0	70.0	76.0	82.1

Table 5.8 *The amount of cadmium in a sample of drinking water in units of* 10^{-5} *milligrammes per litre.*

	copper data		cadmium data		student data	
T_{ave}	2.016	[1.970,2.062]	63.12	[59.41,66.84]	13.59	[13.09,14.08]
T_L'	2.021	[1.975,2.067]	63.97	[61.36,66.59]	13.22	[12.74,13.71]
T_{rd}	2.021	[1.975,2.068]	64.20	[61.22,67.17]	13.19	[12.70,13.67]
T_{kt}	2.029	[1.987,2.072]	64.63	[62.56,66.70]	12.79	[12.33,13.26]

Table 5.9 *Point values and 0.95-approximation regions without the honesty term for the copper (Exercise 1.3.3), cadmium (Table 5.8) and student study (Figure 2.11) data sets and the four functionals* T_{ave}, T_L', T_{rd} *and* T_{kt}.

contamination which can then be taken as known, even though there may be inaccuracies due to evaporation and/or interaction with other chemicals in the prepared sample. In the discussions leading to DIN 38402-45:2003-09 [1] one experienced analytical chemist reported that in his experience the mean after elimination of outliers was better than both the mean and the median. The mean after the elimination of outliers downweights the influence of the outliers to zero. The median is not as radical in that it bounds the influence of outliers but does not downweight it to zero. The final version of DIN 38402-45:2003-09 uses Hampel's piecewise linear redescending ψ-function ((2.6.2) of [103] and [8]) which downweights the influence of outliers to zero but in a smoother manner than simply eliminating them.

There can be very practical considerations which limit the choice of procedure. In the case of the lengths of study of German students the data played a role in changes made in the German higher education system, more specifically the replacement of the 'Diplom' by the 'Bachelor'. Any measure of the length of study had to be one that was immediate and easy to understand. This reduced the field to two candidates, the mean and the median. There was a decided preference for the median which was always smaller giving perhaps the impression that things were not quite as bad as they would otherwise appear. One argument in favour of the mean is that it perhaps more accurately reflects the cost of long study times than does the median.

The copper data of Exercise 1.3.3 have been used several times. Another example from analytic chemistry is given in Table 5.8. It gives 24 readings of the amount of cadmium in 10^{-5} milligrammes per litre in a sample of drinking water. Table 5.9 gives point values and 0.95-approximation intervals without an honesty term for the copper, cadmium and student study times (Figure 2.11) for the four functionals $T_{\text{ave}}, T_L', T_{\text{rd}}$ and T_{kt}.

5.7 Multidimensional M-functionals

The M-functional $T_{LS} = (T_L, T_S)$ defined as the solution of (5.16) and (5.17) can be extended to higher dimensions by defining it as the solution of

$$\int (x - T_L(P))\psi((x - T_L(P))^t T_S(P)^{-1}(x - T_L(P)))\,dP(x) = 0 \tag{5.64}$$

$$\int \chi((x - t_L(P))^t T_S(P)^{-1}(x - T_L(P)))(x - T_L(P))(x - T_L(P))^t\,dP(x) = T_S(P) \tag{5.65}$$

where P is a probability distribution on (the Borel sets of) $\mathbb{R}^k$ with $k \geq 2$. Here $T_L(P) \in \mathbb{R}^k$ and $T_S(P) \in \mathscr{M}_k$ where $\mathscr{M}_k$ denotes the set of symmetric, positive definite $k \times k$ matrices.

The situation changes radically when moving from one-dimensional to higher dimensional M-functionals. Whereas in one dimension the only restriction on the measure P was that its largest atom be less than 1/2 in size the restrictions in higher dimensions are much more severe. They often include the existence of a density with certain monotonic properties (see [147], [33], [125], [202] and [118]). A class of M-functionals of scatter which does not require the assumption of a density was given in [212] but requires knowledge of the location parameter although this can be avoided by considering the symmetrized version as in [67]. A detailed treatement of the breakdown properties is given in [77]. The results given there also apply to the multivariate Kent-Tyler functional $T_{kt,\nu} = (T_{L:kt;\nu}, T_{S:kt;\nu})$ [124] which can be defined in the first instance as the maximum likelihood estimates for the multivariate t-distribution with $\nu \geq 1$ degrees of freedom:

$$T_{kt,\nu} = \text{argmax}_{\mu,\Sigma} \sum_{i=1}^{n} \rho((x_i - \mu)^t \Sigma^{-1}(x_i - \mu)) + \frac{1}{2} n \log |\Sigma| \tag{5.66}$$

where

$$\rho(s) = \frac{1}{2}(\nu + k)\log(1 + s/\nu). \tag{5.67}$$

In the present context $T_{kt,\nu} = (T_{L:kt;\nu}, T_{S:kt;\nu})$ loses its interpretation as a maximum likelihood estimator and becomes an M functional with the functions ψ and χ given by

$$\psi(s) = (\nu + k)/(\nu + s), \quad \chi(s) = \psi(s) \tag{5.68}$$

and solution $(T_L(P), T_S(P)) = T_{kt,\nu}$. Existence and uniqueness for $k = 1$ was proved for $\nu = 1$ in [29] and for $\nu > 1$ in [146]. Existence and uniqueness for general k was proved in [124] under the condition that the total mass on any ℓ dimensional hyperplane does not exceed $(\nu + \ell)/(\nu + k)$. This corresponds to a restriction on the size of the largest atom in the one-dimensional case. Moreover a convergent algorithm for calculating the solution of (5.66) was also provided in [124]. The results of [124] were extended to general measures P in [67] and [66] where it was also shown that $T_{kt,\nu}$ locally uniformly Fréchet differentiable. Finally the breakdown point was considered in [77]: for atomless P it is $1/(k+\nu)$ with a somewhat complicated expression for the general case. In all the Kent-Tyler functional $T_{kt,\nu}$ has the desired properties apart from a high breakdown point.

Theorem 7 (b) of [66] shows that

$$\sqrt{n}\,(T_{L:kt,\nu}(P_n(P)) - T_{L:kt,\nu}(P)) \Rightarrow \mathfrak{N}(0,\Sigma_\nu(P)) \tag{5.69}$$

for some $\Sigma_\nu(P)$. Using this an α-approximation region for $T_{L:kt,\nu}(P)$ is given by

$$\begin{aligned}&\mathscr{A}_n(\mathbf{x}_n,\alpha,T_{L:kt,\nu}(\mathscr{P})) = \\ &\{T_{L:kt,\nu}(P) : n\|\Sigma_\nu(P)^{-1/2}(T_{L:kt,\nu}(P) - T_{L:kt,\nu}(P_n))\|^2 \le \mathrm{q}((1+\alpha)/2,k,P,n), \\ &\qquad d_{\mathscr{H}}(\mathbb{P}_n,P) \le \mathrm{q}_{\mathscr{H}}((1+\alpha)/2,P,n)\}\end{aligned} \tag{5.70}$$

where $\mathrm{q}(\alpha,k,P,n)$ and $\mathrm{q}_{\mathscr{H}}(\alpha,P,n)$ are the α-quantiles of

$$\|\Sigma_\nu(P)^{-1/2}(T_{L:kt,\nu}(P) - T_{L:kt,\nu}(P_n(P)))\|^2$$

$d_{\mathscr{H}}(\mathbb{P}_n(P),P)$ respectively.

As it stands the approximation region is not calculable for the same reason that the approximation region (5.53) is not calculable. It was however possible to approximate the approximation region (5.53) to give the approximate approximating region (5.60) where the 'honesty' term was omitted but can be included if desired. Similar reasoning can be applied to (5.70).

Firstly because of the locally uniform differentiability combined with $\mathrm{q}_{\mathscr{H}}((1+\alpha)/2,P,n) = O(1/\sqrt{n})$ it follows that asymptotically $\Sigma_\nu(P)$ can be replaced by $\Sigma_\nu(\mathbb{P}_n)$. Further $\mathrm{q}((1+\alpha)/2,k,P,n)$ can be replaced by the $(1+\alpha)/2$ quantile of the chi-squared distribution with k degrees of freedom qchisq$((1+\alpha)/2,k)$. If desired the honesty term can be included, and again asymptotically $\mathrm{q}_{\mathscr{H}}(\alpha,P,n)$ can be replaced by $\mathrm{q}_{\mathscr{H}}(\alpha,\mathbb{P}_n,n)$ where the latter can be estimated using bootstrap (see [14]). All this holds uniformly in a $d_{\mathscr{H}}$ ball centred at $\mathbb{P}_n$. These approximations lead to

$$\begin{aligned}&\mathscr{A}_n(\mathbf{x}_n,\alpha,T_{L:kt,\nu}(\mathscr{P})) \approx \\ &\{T_{L:kt,\nu}(P) : n\|\Sigma_\nu(\mathbb{P}_n)^{-1/2}(T_{L:kt,\nu}(P) - T_{L:kt,\nu}(P_n))\|^2 \le \mathrm{qchisq}((1+\alpha)/2,k), \\ &\qquad d_{\mathscr{H}}(\mathbb{P}_n,P) \le \mathrm{q}_{\mathscr{H}}((1+\alpha)/2,\mathbb{P}_n,n)\}\end{aligned} \tag{5.71}$$

for large n.

The easiest method to calculate $\Sigma_\nu(\mathbb{P}_n)$ is to calculate the influence function of $T_{kt,\nu}$ using (5.64), (5.64) and (5.68). Because of the equivariance properties it is sufficient to consider a distribution with $T_{kt,\nu}(P) = (0,I_k)$. Put $T_{kt,\nu}((1-\varepsilon)P+\varepsilon\delta_x) = \varepsilon(\eta,\Delta)$ where the dependence of η and Δ on x has been suppressed. The equations

(5.64), (5.64) give

$$\begin{aligned}(1-\varepsilon)\int \psi((y-\varepsilon\eta)^t(I_k+\varepsilon\Delta)^{-1}(y-\varepsilon\eta))(y-\varepsilon\eta)\,dP(y) \\ +\varepsilon\psi((x-\varepsilon\eta)^t(I_k+\varepsilon\Delta)^{-1}(x-\varepsilon\eta))(x-\varepsilon\eta) &= 0\\ (1-\varepsilon)\int \psi((y-\varepsilon\eta)^t(I_k+\varepsilon\Delta)^{-1}(y-\varepsilon\eta))(y-\varepsilon\eta)(y-\varepsilon\eta)^t\,dP(y) \\ +\varepsilon\psi((x-\varepsilon\eta)^t(I_k+\varepsilon\Delta)^{-1}(x-\varepsilon\eta))(x-\varepsilon\eta)(I_k+\varepsilon\Delta)^{-1}(x-\varepsilon\eta)^t &= I_k+\varepsilon\Delta.\end{aligned}$$

On noting that $(I_k+\varepsilon\Delta)^{-1}=I_k-\varepsilon\Delta$ to the first order of ε and retaining only terms of the first order of ε the above simplifies to

$$\begin{aligned}-\varepsilon\eta\int\psi(y^ty)\,dP(y)-\varepsilon\int\psi^{(1)}(y^ty)(2\eta^ty+y^t\Delta y)y\,dP(y) \\ +\varepsilon\psi(x^tx)x &= O(\varepsilon^2)\\ -\varepsilon\eta\int\psi(y^ty)y^t\,dP(y)-\varepsilon\int\psi^{(1)}(y^ty)(2\eta^ty+y^t\Delta y)yy^t\,dP(y) \\ +\varepsilon\psi(x^tx)xx^t &= \varepsilon\Delta+O(\varepsilon^2).\end{aligned}$$

This leads to the following set of linear equations in η and Δ

$$\begin{aligned}\eta\int\psi(y^ty)\,dP(y)+2\int\psi^{(1)}(y^ty)(\eta^ty)y\,dP(y) \\ +\int\psi^{(1)}(y^ty)(y^t\Delta y)y\,dP(y) &= \psi(x^tx)x \qquad (5.72)\\ \eta\int\psi(y^ty)y^t\,dP(y)+2\int\psi^{(1)}(y^ty)(\eta^ty)yy^t\,dP(y) \\ +\Delta+\int\psi^{(1)}(y^ty)(y^t\Delta y)yy^t\,dP(y) &= \psi(x^tx)xx^t \qquad (5.73)\end{aligned}$$

which can be solved for each x to give

$$I(x,T_{L:kt,\nu},P)=\eta(x),\quad I(x,T_{S:kt,\nu},P)=\Delta(x).$$

If now $Y_n, n=1,2,\ldots$ are independently distributed with common distribution P it follows that

$$\sqrt{n}\,T_{L:kt,\nu}(\mathbb{P}_n(P))\overset{D}{\Rightarrow}\mathfrak{N}\left(0,\int\eta(y)\eta(y)^t\,dP(y)\right).$$

If $X_n, n=1,2,\ldots$ are independently distributed with common distribution P where $T_{kt,\nu}(P)=(\mu,\Sigma)$ the affine equivariance of $T_{kt,\nu}$ implies that

$$\begin{aligned}&\sqrt{n}\,\Sigma^{-1/2}(T_{L:kt,\nu}(\mathbb{P}_n(P))-\mu)\overset{D}{\Rightarrow}\\ &\qquad\mathfrak{N}\left(0,\int\eta\left(\Sigma^{-1/2}(x-\mu)\right)\eta\left(\Sigma^{-1/2}(x-\mu)\right)^t dP(x)\right)\end{aligned}$$

and the general case follows. If P is an elliptical transformation of a radial distribution the calculations simplify and $\Sigma_\nu(P)$ is simply a multiple of $T_{S:kt,\nu}(P)$ (see

[148] (6.22) and (6.109)). For data generated under such a model $T_{L:kt,v}(\mathbb{P}_n(P))$ and $T_{S:kt,v}(\mathbb{P}_n(P))$ are asymptotically independently distributed. However once the equations (5.72) and (5.73) have been solved and programmed there seems to be no advantage in using the simplification.

Another possibility is simply to use the bootstrap which works for locally uniformly Fréchet differentiable functionals. Both methods are asymptotic and just as in the case $k = 1$ finite sample approximations can be attained. These must be based on a model and as almost always the Gaussian model is the default choice. If the ball of radius $q_{\mathscr{H}}((1+\alpha)/2, \mathbb{P}_n, n)$ centred at $\mathbb{P}_n$ contains a normal distribution then the assumption holds. This can be 'checked' using the bootstrap method in [14].

The results for the different methods of calculating the covariance matrix for the haematology data of [182] are given in (5.74) - (5.76). Solving (5.72) and (5.73) results in the matrix of (5.74), the bootstrap method gives (5.75) whereas the simplified calculation assuming almost elliptically transformed radial data gives (5.76). It is seen that the first two are close but the simplified method leads to a somewhat reduced dispersion. The data are skewed and it is presumably this which leads to the difference between (5.76) and the other two.

$$\Sigma_{1.5}(\mathbb{P}_n) = \begin{pmatrix} 1.09 & 2.2 & 1.6 & 1.8 & -0.5 & 0.47 \\ 2.25 & 7.9 & 3.5 & 4.7 & -2.2 & 2.00 \\ 1.57 & 3.5 & 189.6 & 103.1 & 64.1 & -1.45 \\ 1.84 & 4.7 & 103.1 & 99.6 & -7.7 & 3.22 \\ -0.50 & -2.2 & 64.1 & -7.7 & 68.4 & -5.76 \\ 0.47 & 2.0 & -1.4 & 3.2 & -5.8 & 24.09 \end{pmatrix} \tag{5.74}$$

$$\Sigma_{\text{bootstrap}}(\mathbb{P}_n) = \begin{pmatrix} 1.20 & 2.6 & 1.4 & 1.6 & -0.5 & 0.62 \\ 2.62 & 9.1 & 3.7 & 4.6 & -2.5 & 2.49 \\ 1.41 & 3.7 & 193.2 & 105.3 & 64.2 & 0.43 \\ 1.60 & 4.6 & 105.3 & 106.5 & -12.1 & 4.05 \\ -0.48 & -2.5 & 64.2 & -12.1 & 72.1 & -5.00 \\ 0.62 & 2.5 & 0.4 & 4.1 & -5.0 & 23.46 \end{pmatrix} \tag{5.75}$$

$$\Sigma_{\text{symmetric},1.5}(\mathbb{P}_n) = \begin{pmatrix} 0.92 & 2.0 & 1.0 & 1.2 & -0.4 & 0.37 \\ 2.00 & 7.4 & 3.3 & 4.4 & -1.8 & 1.63 \\ 0.97 & 3.3 & 168.3 & 94.8 & 54.2 & -0.03 \\ 1.24 & 4.4 & 94.8 & 86.3 & -0.1 & 2.01 \\ -0.42 & -1.8 & 54.2 & -0.1 & 51.3 & -2.64 \\ 0.37 & 1.6 & 0.0 & 2.0 & -2.6 & 17.29 \end{pmatrix} \tag{5.76}$$

One disadvantage of the Kent-Tyler functional which it shares with other M-functionals is that if the model P is a product measure $P = P_1 \otimes P_2$ then this is not reflected in the structure of the scatter matrix which ideally should be a block matrix. The problem can be overcome by symmetrization. For this and further analyses of M-functionals see [75]: for a much more efficient way of calculating the Kent-Tyler functional see [74].

One problem with multidimensional approximation region is their representation. In k dimensions $k(k+3)/2$ numbers are involved. In one and two dimensions this is not a problem but becomes one at the latest when $k = 4$ and 14 numbers are involved. One possibility would be to visualize all two dimension projections of the ellipsoid but even this becomes tiresome: there are ten such projections for $k = 5$ and 45 for $k = 10$. The large number of parameters also means that large sample sizes are required to attain an accuracy comparable to that of the one-dimensional case.

Chapter 6

The Analysis of Variance

The analysis of variance goes back to R. A. Fisher who developed the idea whilst working at the Rothamsted Experimental Station in the 1920s. The term 'analysis of variance' refers to Fisher's decomposition of the variance into the sum of variances within agricultural blocks and between agricultural blocks. There are many variations

Control	5.37	5.80	4.70	5.70	3.40	7.48	5.77	7.15	6.49	4.09	5.94
	6.38	9.24	5.66	4.53	6.51	7.00	6.20	7.04	4.82	6.73	
Active	3.96	3.04	5.28	3.40	4.10	6.16	3.22	7.48	3.87	4.27	4.05
	2.40	5.81	4.29	2.77	4.40						
Inactive	5.37	10.60	5.02	14.30	4.27	5.75	5.03	5.74	7.85	6.82	7.90
	8.36	5.72	6.00	4.75	5.83	7.30	7.52	5.32	6.05	5.68	7.57
	5.68	8.91	5.39	4.40	7.13						

Table 6.1 *Measurements of plasma bradykininogen levels in three different groups of patients.*

on this theme but the idea of a decomposition of variances is still present in much modern work.

The results of such an analysis are often displayed in an analysis of variance table which specifies for each factor the amount of variance which can be explained by this factor. Not all questions of interest can be answered by such a decomposition. In the one-way table the decomposition can answer the question as to whether there are significant differences between the various samples but it cannot of itself specify which these samples are. Methods specifically developed to answer this latter question are Tukey's studentized range [209] for all pairwise comparisons and Scheffé's method for considering all contrasts ([184, 185]).

One seductive property of the decomposition of the total sum of squares is its mathematical elegance and, in more abstract settings, the applicability of Hilbert space theory. Mathematical elegance alone is no argument in favour of a particular statistical methodology. The approach taken to the analysis of variance detailed in this chapter makes no use of a decomposition into sums of squares. To this extent the name 'analysis of variance' is a misnomer. It is to be understood as a reference to a set of problems that are modelled by the same models as in the analysis of variance. The analysis is different.

6.1 The One-Way Table

6.1.1 *The General Case*

This section is based largely on [38]. The one-way table is concerned with statements about the locations of $k \geq 2$ one-dimensional data sets

$$\mathbf{x}_{in_i} = (x_{i1}, x_{i2}, \ldots, x_{in_i}),\ i = 1, \ldots, k.$$

Three examples of such data sets are the following. The first is taken page 114 of [159] and shown in Table 6.1. It gives the measurements of plasma bradykininogen levels in control patients, patients with active Hodgkin's disease and patients with inactive Hodgkin's disease. The second data set given in Table 6.2 is taken from page 387 of [176] and are the results of an interlaboratory test. The reader is referred to [176] for details. The third data set (Table 6.3) and contains 187 measurements of the velocity of light in kilometres per second grouped according to the years when the measurement were made.

Lab 1	Lab 2	Lab 3	Lab 4	Lab 5	Lab 6	Lab 7
4.13	3.86	4.00	3.88	4.02	4.02	4.00
4.07	3.85	4.02	3.88	3.95	3.86	4.02
4.04	4.08	4.01	3.91	4.02	3.96	4.03
4.07	4.11	4.01	3.95	3.89	3.97	4.04
4.05	4.08	4.04	3.92	3.91	4.00	4.10
4.04	4.01	3.99	3.97	4.01	3.82	3.81
4.02	4.02	4.03	3.92	3.89	3.98	3.91
4.06	4.04	3.97	3.90	3.89	3.99	3.96
4.10	3.97	3.98	3.97	3.99	4.02	4.05
4.04	3.95	3.98	3.90	4.00	3.93	4.06

Table 6.2 *Results of an inter-laboratory tests given on page 397 of [176]*

Year	Velocity of light								
1675-1783	320000	300000	292000	317700	307500	352000	303430	301416	299289
	303320	300650	300313	303434	300093	299406	303440	302220	300460
1841-1899	300340	300270	300250	299980	315300	313300	305650	298000	310700
	300030	300050	300096	299904	298000	299870	301280	299660	284000
	280900	299980	298500	299810	289700	300400	299990	299900	299921
	300011	300140	298400	296700	296000	299990	299910	299900	299660
	301382	295500	299700	299000	299810	299780	287000	299860	299853
	299850	299850	296400	301880	301500	300570	292000	300500	300000
	300460	300100	300100	299600	300560	302200	300920	300750	300390
	299870	299850	299850	299850	300320	300250	297200	300430	300395
	300390	300390	300100	299795	300150	302590	299730	300390	299600
	300900	300330	299195	300032	300000				
1900-1949	299700	299880	299760	299660	299901	299860	299590	300480	300080
	299710	299710	299803	299760	299650	299670	299780	300090	299590
	299850	299690	299630	299660	299650	299840	299940	299870	299750
	299780	299700	299795	299980	299802	299800	299796	299890	299798
	300090	299778	299786	299630	299774	300420	299790	299920	299771
	299771	299768	299771	299776	299798	299792	299687	299695	299701
	299796	299792.4	299794.3	299793.1					
1950-1967	299780.0	299775.0	299794.2	299793.1	299792.6	299776.0	299792.85	299795.1	299792.75
	299789.8	299792.4	299792.0	299792.9	299792.7	299792.4	299792.2	299791.9	299792.6
	299792.5	299792.6	299792.44	299792.56	299792.5				

Table 6.3 *187 measurements of the velocity of light in kilometres per second.*

Such samples are typically modelled as

$$X_{ij} = \mu_i + \sigma_i \varepsilon_{ij}, \; j = 1, \ldots, n_i, \, i = 1, \ldots, k \tag{6.1}$$

where the ε_{ij} are i.i.d. with a common specified distribution P_0 symmetric about zero, for example $P_0 = \mathfrak{N}(0,1)$.

The analysis of the data sets $\mathbf{x}_{in_i}$ will be based on the approximation intervals of the last chapter. Although calibrated using a specific P_0 the procedure is intended to work well under more general models than that described by (6.1), namely where the distribution P_i of the ε_{ij} can vary from sample to sample. It is not even necessary for the P_is to be symmetric, they may all be skewed to some extent. However a precise description of the models under which the procedure will work well is not possible (Chapter 2.15).

The procedure is the following. Given one of the locally uniformly Fréchet differentiable location-scale functionals $T_{LS} = (T_L, T_S)$ of the previous chapter and α, an approximation interval $[a_i, b_i]$ for μ_i is calculated for each sample $\mathbf{x}_{in_i}$ with $\alpha_k = \alpha^{1/k}$. The choice of α is to guarantee that under the model

$$\mathbf{P}(\mu_i \in [a_i, b_i], \, i = 1, \ldots, k) \geq \alpha \,. \tag{6.2}$$

Data	$T'_L(\mathbb{P}_n)$	$T_S(\mathbb{P}_n)$	Interval
Hodgkin control	6.00	0.72	[5.29, 6.71]
Hodgkin active	4.14	0.66	[3.23, 5.04]
Hodgkin inactive	6.36	0.95	[5.41, 7.30]
Lab 1	4.06	0.0164	[4.02, 4.10]
Lab 2	4.00	0.0578	[3.89, 4.12]
Lab 3	4.00	0.0175	[3.97, 4.03]
Lab 4	3.92	0.0204	[3.87, 3.97]
Lab 5	3.96	0.0487	[3.88, 4.04]
Lab 6	3.97	0.0295	[3.88, 4.05]
Lab 7	4.01	0.0334	[3.91, 4.11]
1675-1783	302344.4	1964.17	[299233.4, 305455.4]
1841-1899	300001.0	337.30	[299805.2, 300196.8]
1900-1949	299792.2	72.85	[299745.1, 299839.3]
1950-1967	299792.4	0.48	[299791.8, 299793.0]

Table 6.4 *Empirical values of* $(T'_L(\mathbb{P}_n), T_S(\mathbb{P}_n))$ *and the approximation intervals for the data of Tables 6.1, 6.2 and 6.3 with* $\alpha = 0.95$.

This assumes the independence of the samples under the model. If this seems unreasonable the replacing $\alpha^{1/k}$ by $1-(1-\alpha)/k$ will guarantee (6.2) even if the samples are dependent. For α close to 1 the difference is negligible.

The functional used for the calculations in the following is the two-step functional $T'_L(P)$ of Chapter 5.3.5. Figure 6.1 shows the simultaneous approximation intervals for the data sets PTT times (top panel), the interlaboratory test (centre panel) and the velocity of light (bottom panel).

All questions concerning the parameters μ_i and the relationships between them can be answered using the approximation intervals. The null hypothesis $H_0 : \mu_1 = \mu_2$ in the standard approach corresponds to the question as whether the two data sets $\mathbf{x}_{1n_1}$ and $\mathbf{x}_{2n_2}$ can be adequately approximated by models with $\mu_1 = \mu_2$. This will be the case if the intersection of the approximation intervals $[a_1, b_1] \cap [a_2, b_2]$ is non-empty. The intersection gives the approximation region for the joint value. The approximation interval for patients with active Hodgkin's disease is disjoint from the other two intervals which can be interpreted as the value for patients with the active disease is significantly lower than for the controls patients or those where the disease is inactive. The interval for modern measurements of the velocity of light (1950-1967) is disjoint from interval for the years 1841-1899 but, so to speak, just disjoint. If α is increased to 0.99 then the interval for the modern readings is contained in all the other intervals. In this sense all earlier measurements are consistent with the modern ones.

Approximation intervals for the differences $\mu_i - \mu_j$ are given by $[a_i - b_j, b_i - a_j]$. Such an interval covers the origin if and only if the two intervals $[a_i, b_i]$ and $[a_j, b_j]$ are not disjoint. The approximation intervals for the Hodgkin data are

$$(1,2)\ [0.1553.418], \quad (1,3)\ [-2.134, 1.196], \quad (2,3)\ [-4.140, -0.372]. \qquad (6.3)$$

Contrasts can be treated in the same manner as can more complicated relationships such as $\mu_1 = \mu_2 \log(1+\mu_3^2)$. Ordering properties such as $\mu_1 \le \ldots \le \mu_k$ are

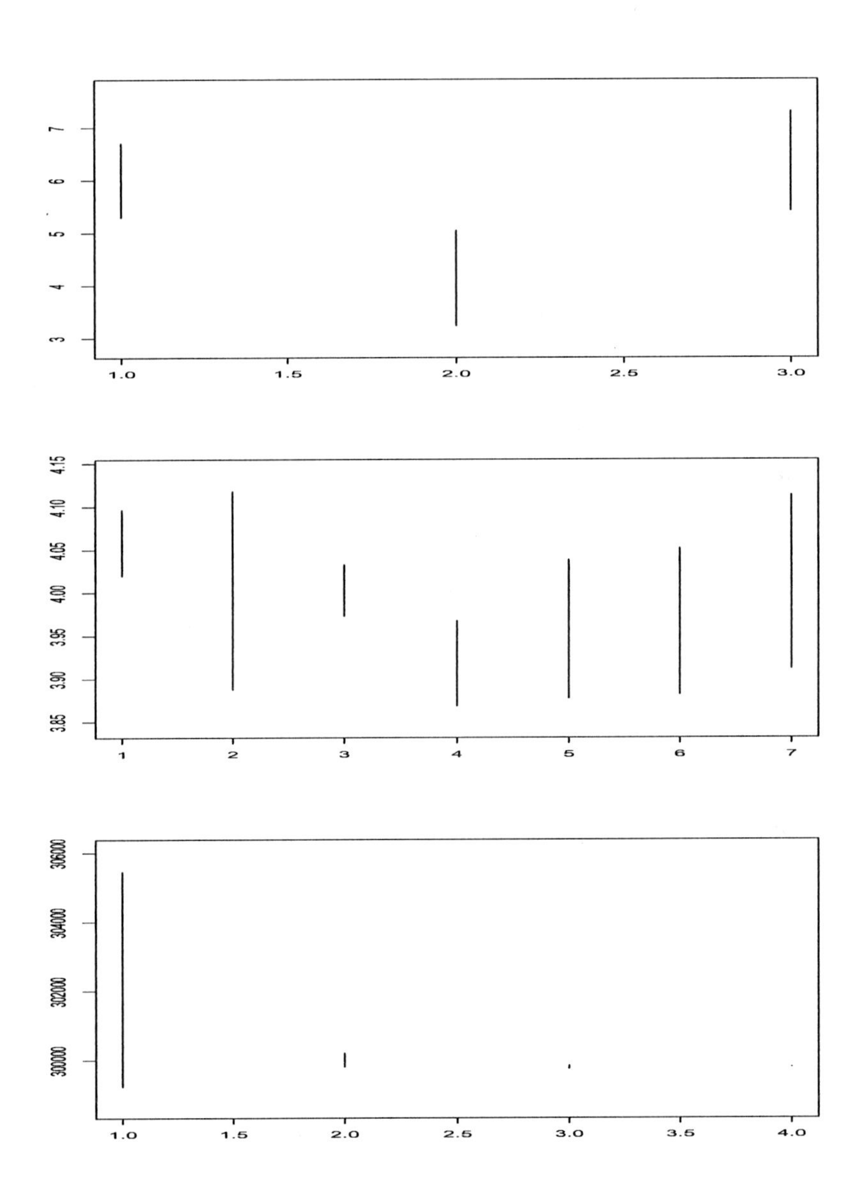

Figure 6.1 *Top: approximation intervals for Table 6.1. Centre: Approximations intervals for Table 6.2. Bottom: Approximations intervals for Table 6.3.*

judged to be consistent with the data if and only if there exist $m_i \in [a_i, b_i]$ with $m_1 \leq \ldots \leq m_k$. If such m_i exist then monotone upper and lower bounds for consistent values of the μ_i can be derived. It is to be noted that for data generated under the model all approximation (confidence) regions for whatever functions of the μ_is are considered hold simultaneously, there is no need for a Bonferroni adjustment. This follows from (6.2).

6.1.2 *Borrowing Strength*

The procedure expounded above allows for different scales for the different samples. This is seen most clearly in the case of the data on the velocity of light where the final approximation interval is so small compared to the first as to be barely visible in Figure 6.1. Other procedures such as the standard decomposition table, Tukey's studentized range and Scheffé's method derive in the first instance from a Gaussian model where the variance is the same for all samples. This justifies, at least at the level of the model, 'borrowing strength' by estimating the common variance using the data from all samples. If 'borrowing strength' is judged to be justified on the basis of the data or knowledge about the mechanism giving rise to the data then it can be incorporated into the approximation interval method. This can be done in several ways. The one chosen here is the following. For each sample the asymptotic variance for this sample (5.55) is calculated $V(\mathbb{P}_{i,n_i})$ and a weighted average of these is used

$$V(\mathbb{P}_n) = \Big(\sum_{i=1}^{k} V(\mathbb{P}_{i,n_i})n_i\Big)/n, \quad n = \sum_{i=1}^{k} n_i. \tag{6.4}$$

The approximation interval for the ith sample is

$$\big[T_L(\mathbb{P}_{i,n_i})) - \sqrt{V(\mathbb{P}_n)}\,\mathrm{qbs}(\beta_k,n,i)/\sqrt{n_i},\, T_L(\mathbb{P}_{i,n_i})) + \sqrt{V(\mathbb{P}_n)}\,\mathrm{qbs}(\beta_k,n,i)/\sqrt{n_i}\big] \tag{6.5}$$

where $\beta_k = (1+\alpha^{1/k})/2$ and $\mathrm{qbs}(\beta_k,n,i)$ are the corresponding quantiles. These will depend on the design $n = (n_1,\ldots,n_k)$. It turns out that it is sufficient to regress $\log(\mathrm{qbs}(\alpha,n,i))$ on $\log(\mathrm{qt}(\alpha,(n_i-1)k))$ using simulations for each design. This results in an approximation of the form

$$\mathrm{qbs}(\alpha,n,i) \approx \exp(\mathrm{fac}(n,i;1))\mathrm{qt}(\alpha,(n_i-1)k)^{\mathrm{fac}(n,i;2)}$$

A library can be created for further use. Some examples are

$$\begin{aligned}
\mathrm{fac}(3,5) &= ((0.268,1.146),(0.020,\ 1.144))\\
\mathrm{fac}(3,3,3) &= (0.006\ 1.062)\\
\mathrm{fac}(3,5,10) &= ((0.615\ 1.075),(0.351,1.086),(0.026,\ 1.073))\\
\mathrm{fac}(16,21,27) &= ((0.275,\ 1.008),(0.140,0.995),(0.005,1.013))\\
\mathrm{fac}(10,10,10,10,10,10,10) &= ((0.006,1.004)).
\end{aligned} \tag{6.6}$$

Figure 6.2 shows the approximation intervals for the data sets of Tables 6.1 and 6.2 when the scale is calculated by borrowing strength: it would be clearly misleading to do this for the velocity of light data. On comparing Figure 6.2 with the first two

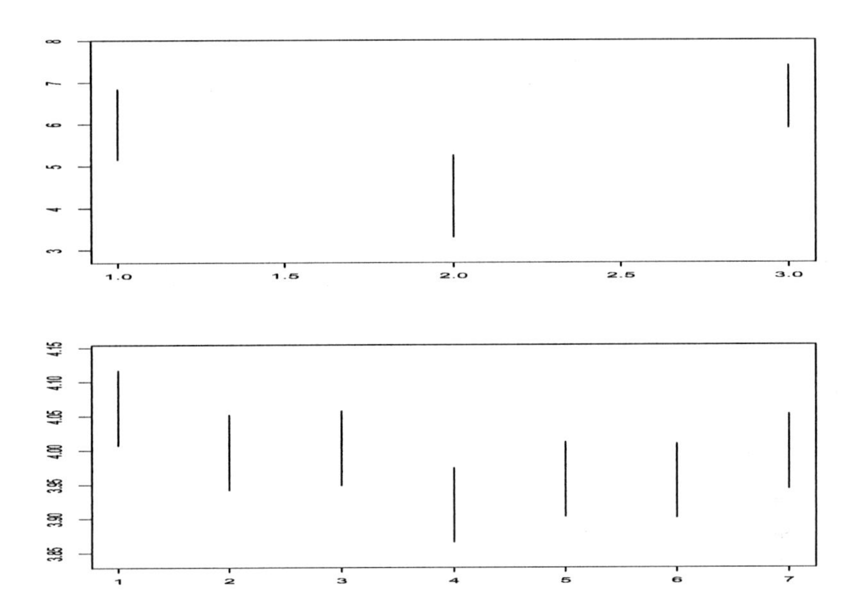

Figure 6.2 *Upper: approximation intervals for Table 6.1 using borrowing strength. Lower: the same for for Table 6.2.*

panels of Figure 6.1 it is seen that borrowing strength for the plasma bradykininogen levels data does not change the picture. In the case of the interlaboratory data there are large changes in the intervals for all the laboratories except for laboratory 4.

6.1.3 *Paired Comparisons: Borrowing Strength*

This section is concerned with robustifying Tukey's studentized range or honest significant difference test (HSD). Given k samples $\mathbf{x}_{in_i}, i = 1,\ldots,k$ simultaneous α-confidence intervals for the differences in location functionals are given by

$$\text{ave}(\mathbf{x}_{in_i}) - \text{ave}(\mathbf{x}_{jn_j}) \pm \sqrt{\frac{\sum_{\ell=1}^{k}(n_\ell - 1)\text{sd}(\mathbf{x}_{\ell n_\ell})^2}{n-k}}\,\text{qtukey}(\alpha,k,n-k)\sqrt{\frac{1}{2}\left(\frac{1}{n_i}+\frac{1}{n_j}\right)} \tag{6.7}$$

where $\text{qtukey}(\alpha,k,df)$ is the α-quantile of the distribution of Tukey's studentized range for k samples with df degrees of freedom. The the robustified version of this is

$$T_L(\mathbb{P}(\mathbf{x}_{1n_1})) - T_L(\mathbb{P}(\mathbf{x}_{2n_2})) \pm \sqrt{V(\mathbb{P}_n)}\,\text{fac1}(\mathbf{n})\text{qtukey}(\alpha,k,n-k)^{\text{fac2}(\mathbf{n})}\sqrt{\frac{1}{2}\left(\frac{1}{n_i}+\frac{1}{n_j}\right)} \tag{6.8}$$

where n and $V(\mathbb{P}_n)$ are as in (6.4) and $\text{fac1}(\mathbf{n})$ and $\text{fac2}(\mathbf{n})$ are correction factors depending only on $\mathbf{n} = (n_1,\ldots,n_k)$.

The correction factors fac1($\mathbf{n}$) and fac2($\mathbf{n}$) are obtained by simulations and a linear regression of the logarithms of the empirical quantiles on the logarithm of the quantiles qtukey$(\alpha, k, n-k)$ for $0.8 \le \alpha \le 0.999$. The results for each design can be saved and a library built up. The results for the designs of (6.6) are

$$\begin{aligned} \text{fac(3,3,3)} &= \text{(-0.035, 1.069)} \\ \text{fac(3,5,10)} &= \text{(-0.002, 1.021)} \\ \text{fac(3,20,20)} &= \text{(-0.009, 1.009)} \\ \text{fac(16,21,27)} &= \text{(0.003, 1.008)} \\ \text{fac(10,10,10,10,10,10,10)} &= \text{(0.009, 1.001)}. \end{aligned} \tag{6.9}$$

6.1.4 Paired Comparisons: Heterogeneous Scales

The problem is to compare the location parameters of two samples with different scales and sizes. Given two samples $\mathbf{x}_{in_i}, i = 1,2$ the direct comparison of the two locations parameters will be based on

$$\frac{T_L(\mathbb{P}(\mathbf{x}_{1n_1})) - T_L(\mathbb{P}(\mathbf{x}_{2n_2}))}{\left(\frac{V(\mathbb{P}(\mathbf{x}_{1n_1}))}{n12} + \frac{V(\mathbb{P}(\mathbf{x}_{2n_2}))}{n_2}\right)^{1/2}} \tag{6.10}$$

as in the classical treatment of this problem, the Behrens-Fisher problem. The distribution of the statistic will be simulated under the model $\mathbf{X}_{in_i} \stackrel{D}{=} \mathfrak{N}(\mu_i, \sigma_i^2), i = 1,\ldots,2$. Equivariance considerations show that it is only necessary to do this for the models $(\mu_1, \sigma_1^2) = (0,1)$ and $(\mu_2, \sigma_2^2) = (0, \sigma^2)$. In the general case σ^2 corresponds to the ratio of the variances σ_2^2/σ_1^2. The procedure is to obtain lower and upper bounds for $\sigma^2 = \sigma_2^2/\sigma_1^2$. This is done by simulating the ratio $V(\mathbb{P}(X_{1n_1}))/V(\mathbb{P}(X_{2n_2}))$ and approximating the quantiles by the corresponding Fisher distribution. For each value of σ in this interval the distribution of the statistic will be approximated by a t-distribution with the Welch degrees of freedom (see [218]).

$$\nu = \frac{\left(\frac{1}{n_1} + \frac{\sigma^2}{n_2}\right)^2}{\left(\frac{1}{n_1^2(n_1-1)} + \frac{\sigma^4}{n_2^2(n_2-1)}\right)}.$$

More precisely, for $3 \le n_1 \le n_2 \le 50$ the logarithm of the quantiles $q(p)$ for $0.8 \le p \le 0.9999$ of the statistic will be approximated by a cubic function of the logarithm of the corresponding quantiles of the t-distribution with the Welch degrees of freedom. This is done for a grid of values of σ^2. The one used in the simulations is $\sigma^2 = 2^j, j = -14, -13, \ldots, 14$.

For small and large values of σ^2 the values for 2^{-14} and 2^{14} are used respectively. For other values of σ^2 linear interpolation is used. The largest value of the quantile for σ^2 in the interval is used as in [16]. This guarantees that under the model the confidence interval is honest.

For a given value of α an amount $\alpha/2$ will be spent on the ration σ_1^2/σ_2^2 and $\alpha/2$ on the statistic (6.10) with none spent on an honesty component for the distributions

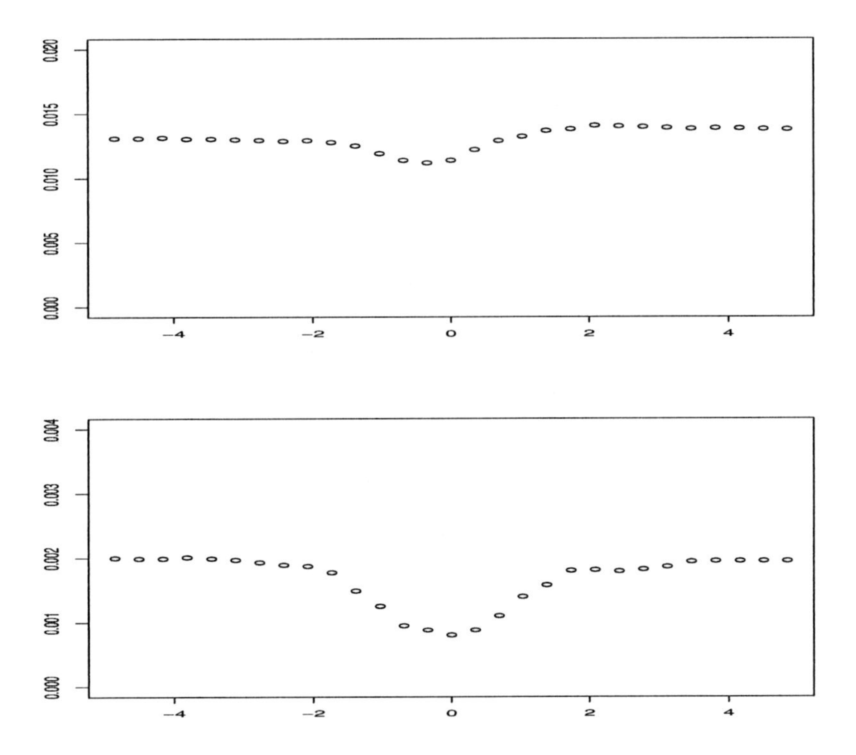

Figure 6.3 *Attained* $(1-\alpha)$ *values plotted against the logarithm of the ratio of the scales* σ. *Top:* $\alpha = 0.96^{1/3} = 0.986$ *with* $n_1 = 16, n_2 = 21$ *as in the data of Table 6.1. Bottom:* $\alpha = 0.96^{1/21} = 0.998$ *with* $n_1 = n_2 = 10$ *as in the data of Table 6.2.*

of $\mathbf{x}_{1n_1}$ and $\mathbf{x}_{2n_2}$. The upper panel of Figure 6.3 shows the simulated $1-\alpha$ values for $\alpha = 0.9^{1/3}$ plotted against $\log(\sigma_1^2/\sigma_2^2)$ under the model and the range $(\sigma_1^2/\sigma^2 = 2^j, j = -14, -13, \ldots, 14$ and for $(n_1, n_2) = (16, 21)$ (control-active for Table 6.1). The lower panel shows the results for $\alpha = 0.9^{1/3}$ plotted against $\log(\sigma_1^2/\sigma_2^2)$ under the model and the range $\sigma_1^2/\sigma_2^2 = 2^j, j = -14, \ldots, 14$ and for $(n_1, n_2) = (16, 21)$ (control-active for Table 6.1). The lower panel shows the results for $\alpha = 0.96^{1/21} = 0.998$ with $n_1 = n_2 = 10$ as in Table 6.2.

6.1.5 *Results for the Tables 6.1 and 6.2*

The following procedures

(1) interval based as in (6.3),

(2) interval based borrowing strength,

(3) paired comparisons heterogeneous scales,

(4) paired comparisons borrowing strength,

(5) Tukey's studentized range.

are compared for the data sets of Tables 6.1 and 6.2. In all cases $\alpha = 0.95$ where 0.01 is spent on the ratio of the scales and 0.04 on the intervals. The results are as follows:

Table 6.1

Procedure	Significant differences
(1),(3),(4),(5)	control - active, active - inactive
(2)	active - inactive

Table 6.2

Procedure	Significant differences
(1)	1-4,1-5,3-4
(2)	1-4
(3)	1-3,1-4,1-5,1-6,3-4
(4)	1-4,1-5,1-6
(5)	1-4,1-5,1-6,3-4

6.2 The Two-Way Table

6.2.1 L_0, L_1 and L_2 Interactions

This section is based on [205], [204], [206] and [40].

The standard model for the two-way table is

$$X_{ijk} = \mu + a_i + b_j + c_{ij} + \sigma\varepsilon_{ijk}, k = 1,\ldots,K_{ij}, j = 1,\ldots,J, i = 1,\ldots,I \tag{6.11}$$

where the ε_{ijk} are standard Gaussian white noise. As the model stands the parameters μ, $a = (a_i)$, $b = (b_j)$ and $c = (c_{ij})$ are not identifiable. To guarantee identifiability the following constraints are usually placed on the parameters

$$\sum_i a_i = \sum_j b_j = 0, \quad \sum_i c_{ij} = 0, \forall j, \quad \sum_j c_{ij} = 0, \forall i. \tag{6.12}$$

One consequence of the constraints on the interactions c_{ij} is that the model now excludes a single interaction in one cell. The minimum number of non-zero interactions is four but there is no convincing argument why this should be so. The problem concerns the very definition of what is meant by an interaction and is best analysed for exact data with one observation per cell:

$$x_{ij} = a_i + b_j + c_{ij}, j = 1,\ldots,J, i = 1,\ldots,I. \tag{6.13}$$

The term μ has been dropped and no constraints are placed on the row and column effects as they play no role in the following.

The simplest exact data with a non-zero interaction are

$$x_1 = \begin{pmatrix} 1 & 0 & 0 \\ 0 & 0 & 0 \\ 0 & 0 & 0 \end{pmatrix}. \tag{6.14}$$

A standard least squares analysis

$$(a^{(2)}, b^{(2)}) = \text{argmin}_{a,b} \sum_{i,j} (x_{ij} - a_i - b_j)^2 . \tag{6.15}$$

results in the interactions

$$x_1 - a^{(2)} - b^{(2)} = c^{(2)} = \begin{pmatrix} 4/9 & -2/9 & -2/9 \\ -2/9 & 1/9 & 1/9 \\ -2/9 & 1/9 & 1/9 \end{pmatrix} . \tag{6.16}$$

This is the $4-2-1$ pattern of [31]. If the L_1-norm

$$(a^{(1)}, b^{(1)}) = \text{argmin}_{a,b} \sum_{i,j} |x_{ij} - a_i - b_j| \tag{6.17}$$

is used instead of the L_2-norm the the resulting interactions c are

$$x_1 - a^{(1)} - b^{(1)} = c^{(1)} = \begin{pmatrix} 1 & 0 & 0 \\ 0 & 0 & 0 \\ 0 & 0 & 0 \end{pmatrix} .$$

that is the x_1 of (6.14) itself: the single imposed interaction has been identified. This is also the case for the L_0-'norm'

$$(a^{(0)}, b^{(0)}) = \text{argmin}_{a,b} \ |\{(i,j) : |x_{ij} - a_i - b_j)| > 0\}| \tag{6.18}$$

where $|S|$ denotes the number of elements in the set S. Again the x_1 of (6.14) is returned as the interactions:

$$x_1 - a^{(0)} - b^{(0)} = c^{(0)} = \begin{pmatrix} 1 & 0 & 0 \\ 0 & 0 & 0 \\ 0 & 0 & 0 \end{pmatrix} .$$

Here are two further examples. The first is

$$x_2 = \begin{pmatrix} 1 & 0 & 0 \\ 0 & 1 & 0 \\ 0 & 0 & 0 \end{pmatrix} . \tag{6.19}$$

The interactions returned by the L_2 method are

$$c^{(2)} = \begin{pmatrix} 5/9 & -4/9 & -1/9 \\ -4/9 & 5/9 & -1/9 \\ -1/9 & -1/9 & 2/9 \end{pmatrix} \tag{6.20}$$

whereas both the L_1 and L_0 methods again return the original data. The second example is

$$x_3 = \begin{pmatrix} 1 & 0 & 0 \\ 0 & -1 & 0 \\ 0 & 0 & 0 \end{pmatrix} . \tag{6.21}$$

The interactions returned by the L_2 method are

$$c^{(2)} = \begin{pmatrix} 1/3 & 0 & -1/3 \\ 0 & -1/3 & 1/3 \\ -1/3 & 1/3 & 0 \end{pmatrix}. \tag{6.22}$$

For this data set the L_1 and L_0 methods do not result in a unique set of interactions. The original data are one set of possible interactions as are

$$c_1^{(1)} = c_1^{(0)} = \begin{pmatrix} 0 & 0 & -1 \\ 0 & 0 & 0 \\ 0 & 1 & 0 \end{pmatrix}. \tag{6.23}$$

and

$$c_2^{(1)} = c_2^{(0)} = \begin{pmatrix} 0 & 0 & 0 \\ 0 & 0 & 1 \\ -1 & 0 & 0 \end{pmatrix}. \tag{6.24}$$

So far the L_1 and L_0 interactions coincide. That this is not always the case is shown by the example

$$x_4 = \begin{pmatrix} 0 & 0 & 1 & 2 & 3 \\ 0 & 0 & 0 & 0 & 0 \\ 0 & 0 & 0 & 0 & 0 \end{pmatrix}. \tag{6.25}$$

The unique L_0 solution is x_4 itself but the unique L_1 solution is

$$c_4^{(1)} = \begin{pmatrix} -1 & -1 & 0 & 1 & 2 \\ 0 & 0 & 0 & 0 & 0 \\ 0 & 0 & 0 & 0 & 0 \end{pmatrix}. \tag{6.26}$$

To investigate the problem of lack of uniqueness of the L_0 method [206] introduced the concept of an interaction pattern. An interaction pattern is a matrix ι with entries either 0 or 1 where 1 denotes the presence of an interaction. The interaction pattern of interactions c will be denoted by $\iota(c)$ The interaction pattern of the data x_1 of (6.14) is

$$\iota_1 = \iota(x_1) = \begin{pmatrix} 1 & 0 & 0 \\ 0 & 0 & 0 \\ 0 & 0 & 0 \end{pmatrix}. \tag{6.27}$$

The data x_2 and x_3 of (6.19) and (6.21) respectively both have the interaction pattern

$$\iota_2 = \begin{pmatrix} 1 & 0 & 0 \\ 0 & 1 & 0 \\ 0 & 0 & 0 \end{pmatrix}. \tag{6.28}$$

An interaction pattern ι is called unconditionally identifiable if the L_0 interactions are unique irrespective of the values of the interactions. The interaction pattern ι_1 of (6.27) is unconditionally identifiable but the interaction pattern ι_2 of (6.28) is not, as has been shown. The complete theory of unconditionally identifiable interactions patterns was given in [206] (see Theorem 6.2.1 below). In particular it was shown that the L_0 and L_1 solutions coincide for such patterns so that the L_0 solution can be obtained from the L_1 solution.

6.2.2 *Unconditionally Identifiable Interaction Patterns*

The following theorem was proved in [206].

Theorem 6.2.1

(a) Let ι be an interaction pattern and let ι' be any other interaction pattern obtainable from ι by adding rows and columns of ones mod 2. *If there is no $\iota' \neq \iota$ with the same number or fewer ones, then ι is unconditionally identifiable.*

(b) Let c be a matrix of interactions with an unconditionally identifiable interaction pattern $\iota(c)$. Then the L_1 problem has a unique solution and the L_1 interactions coincide with c.

An example of the application of (a) of the theorem is the interaction matrix

$$\iota_1 = \begin{pmatrix} 1 & 0 & 0 \\ 0 & 1 & 0 \\ 0 & 0 & 0 \end{pmatrix}.$$

Adding a row of ones mod 2 to the first row results in the matrix

$$\begin{pmatrix} 0 & 1 & 1 \\ 0 & 1 & 0 \\ 0 & 0 & 0 \end{pmatrix}$$

and on adding a column of ones mod 2 to the second column results in the matrix

$$\begin{pmatrix} 0 & 0 & 1 \\ 0 & 0 & 0 \\ 0 & 1 & 0 \end{pmatrix}.$$

This matrix has the same number of ones as ι_1 but in different cells. It follows that the interaction pattern ι_1 is not unconditionally identifiable. This was already shown using the x_4 of (6.21). It satisfies $\iota(x_4) = \iota_1$ and has two L_1 and L_0 solutions given by (6.23) and (6.24).

It is a simple matter to programme (a) of the theorem to check any given interaction pattern. The complexity is $2^{\min(I,J)}$. It can be used to show that the interaction pattern

$$\begin{pmatrix} 1 & 0 & 0 & 0 & 0 \\ 0 & 1 & 0 & 0 & 0 \\ 0 & 0 & 1 & 0 & 0 \\ 0 & 0 & 0 & 1 & 0 \\ 0 & 0 & 0 & 1 & 1 \end{pmatrix} \tag{6.29}$$

is unconditionally identifiable as are all patterns obtainable by permuting both the rows and the columns.

The theorem suggests the following: calculate the L_1 solution and then check using part (a) of the theorem whether the interaction pattern of the residuals is unconditionally identifiable. If this is the case the L_1 solution is also the sparsity solution.

Simpler sufficient conditions for unconditional identifiability are available. Two such are the following. An interaction pattern is unconditionally identifiable if (i) less than half the rows and less than half the columns contain ones or (ii) less than one quarter of the cells in each row and less than one quarter of the cells in each column are ones.

Finally, if ι is an unconditionally identifiable interaction matrix and ι' any other interaction matrix with $\iota' \leq \iota$ cell wise, then ι' is also unconditionally identifiable.

The above theory does not cover the interactions $c = x_4$ of (6.25) that are identifiable although the corresponding interaction pattern

$$\iota(x_4) = \begin{pmatrix} 0 & 0 & 1 & 1 & 1 \\ 0 & 0 & 0 & 0 & 0 \\ 0 & 0 & 0 & 0 & 0 \end{pmatrix}$$

is not unconditionally identifiable. This can be seen by adding a row of ones $\mod 2$ to the first row to give the interaction pattern

$$\begin{pmatrix} 1 & 1 & 0 & 0 & 0 \\ 0 & 0 & 0 & 0 & 0 \\ 0 & 0 & 0 & 0 & 0 \end{pmatrix}.$$

with fewer ones.

6.2.3 *An Algorithm for Identifiability of Interactions*

The following is based on [40].

Interactions which are identifiable but whose interaction pattern is not unconditionally identifiable are not covered by the results of [206]. One such set of interactions is x_2 of (6.19)

$$c = \begin{pmatrix} 1 & 0 & 0 \\ 0 & 1 & 0 \\ 0 & 0 & 0 \end{pmatrix} \tag{6.30}$$

which are identifiable. This was claimed but not proved in the last section.

To show that the interactions are identifiable it is necessary to show that there are no row or column operations (adding a row or column) which result in an interaction matrix $c' \neq c$ with the same number or fewer interactions. If it were possible to obtain such an interaction matrix c' then either one or both of the interactions of c must be set to zero using row and column operations. On indexing the cells from 1 to 9 starting with the first column, the first interaction of c is in cell 1 and the second in cell 5. Suppose firstly that the interaction in cell 1 is set to zero whilst that in cell 5 remains non-zero. If the interaction in cell 1 is set to zero then not all the entries in the cells 2, 7 and 8 marked by stars

$$\begin{pmatrix} 1 & 0 & \star \\ \star & 1 & \star \\ 0 & 0 & 0 \end{pmatrix}. \tag{6.31}$$

can remain zero: at least one of them must be non-zero. The same holds for each of the following sets of three cells $\{2,7,8\},\{3,4,6\},\{3,7,9\}$. This is only possible if at least two cells now have non-zero values, for example the cells $\{3,7\}$. Setting the interaction in cell one to zero creates at least two new interactions in cells at present without interactions. The same applies to the interaction in cell five. It follows that both interactions must be set to zero. The sets of three cells each of which must contain at least one new interaction are now

$$\{2,7,8\},\{3,4,6\},\{3,7,9\},\{2,3,6\},\{4,7,8\},\{6,8,9\}.$$

The problem is to choose a set containing at most two numbers such that the intersection with each of the above six sets is non-empty. This is accomplished by choosing $\{3,8\}$ or $\{6,7\}$ and these are the only possibilities.

The argument can be strengthened as follows. Let $*$ denote a cell without an interaction whose entry was initially zero but has been chosen as one of the cells which now have a non-zero entry and consider the pattern

$$\begin{pmatrix} 0 & * \\ 0 & 0 \end{pmatrix}. \tag{6.32}$$

The same argument as given above shows that such a pattern is impossible. This places further constraints on the cells chosen to be non-zero. It can happen that there are no constellations such as (6.31) in which case there are initially no constraints on the cells which are not set to zero. The restrictions (6.32) remain however in force and may still be active. This part of the problem will be referred to as the intersection problem. It is to be noted that the intersection problem concerns only the interaction pattern of the interactions and not their values. From this it follows that if the intersection problem is not solvable then the interaction pattern $\iota(c)$ of the interactions is unconditionally identifiable.

The final part of the problem is to decide whether there exist row and column effects which set the interactions in cells 1 and 5 to zero whilst leaving the zero values in the cells $\{2,3,4,8,9\}$ unchanged. This gives rise to a system of linear equations and their consistency can be checked by performing a linear regression. For the values 1 and 1 the remaining sum of squares is non-zero and the equations are consequently inconsistent. The interactions of (6.30) are therefore identifiable.

If the interaction in cell 5 is set to -1 to give x_3 of (6.21) then the first part of the analysis is as before. This time however the set of linear equations is consistent and gives rise to interactions in the cells 6 and 7 or in the cells 3 and 8. These correspond to the solutions given in (6.23) and (6.24) respectively. The interactions are therefore not identifiable.

The intersection problem has no solutions for the interactions

$$\begin{pmatrix} 1 & 0 & 0 & 0 & 0 \\ 0 & 2 & 0 & 0 & 0 \\ 0 & 0 & 3 & 0 & 0 \\ 0 & 0 & 0 & 4 & 0 \\ 0 & 0 & 0 & 5 & 6 \end{pmatrix} \tag{6.33}$$

which implies that the corresponding interaction pattern is unconditionally identifiable. This also follows from (a) of Theorem 6.2.1 and demonstrates that the intersection method is quite powerful. The method shows that the interactions

$$\begin{pmatrix} 2 & 0 & 0 & 0 & 0 \\ 2 & 2 & 0 & 0 & 0 \\ 0 & 0 & 2 & 0 & 0 \\ 0 & 0 & 0 & 2 & 0 \\ 0 & 0 & 0 & 2 & 2 \end{pmatrix} \tag{6.34}$$

are identifiable but not the interactions

$$\begin{pmatrix} 2 & 0 & 0 & 0 & 0 \\ 2 & 2 & 0 & 0 & 0 \\ 0 & 0 & 2 & 0 & 0 \\ 0 & 0 & 0 & 2 & 0 \\ 0 & 0 & 0 & -2 & -2 \end{pmatrix}. \tag{6.35}$$

For each of the interactions (6.34) and (6.35) it was only necessary to test eight sets of linear equations for consistency. All other possibilities were ruled out by the intersection criterion.

6.2.4 *L_0 and L_1*

The following is based on [40].

There is a relationship between the L_0 and L_1 as shown by Theorem 6.2.1: the example (6.19) shows that the two can coincide and be unique even if the interaction pattern is not unconditionally identifiable and example (6.21) shows that they can coincide but not be unique. Finally they can be different as is shown by the example x_4 of (6.25).

The above suggests the following strategy of obtaining the L_0 solution from an L_1 solution. Consider a given cell (i_0, j_0) and calculate the L_1 solution without this cell

$$(\tilde{a}_{(i_0,j_0)}, \tilde{b}_{(i_0,j_0)}) = \text{argmin} \sum_{(i,j)\neq(i_0,j_0)} |x_{i,j} - a_i, b_j| \tag{6.36}$$

and then check whether the residuals

$$\tilde{r}_{(i_0,j_0)} = x - \tilde{a}_{(i_0,j_0)} - \tilde{b}_{(i_0,j_0)}$$

are an L_0 solution. More generally given a subset $\mathscr{I} \subset \{(i,j) : 1 \le i \le I, 1 \le j \le J\}$ consider the reduced L_1 solution

$$(\tilde{a}_{\mathscr{I}}, \tilde{b}_{\mathscr{I}}) = \text{argmin} \sum_{(i,j) \notin \mathscr{I}} |x_{i,j} - a_i, b_j| \tag{6.37}$$

and the corresponding residuals

$$\tilde{r}_{\mathscr{I}} = x - \tilde{a}_{\mathscr{I}} - \tilde{b}_{\mathscr{I}}.$$

Eventually this strategy will provide all L_0 solutions but the hope is that an L_0 solution will be found when $\mathscr{I}$ consists of only a few, more precisely at most three cells.

For the example (6.25) the strategy works when $\mathscr{I} = \{(1,4),(1,5,)\}$. This results in

$$\tilde{a}_{\mathscr{I}} = (0,0,0), \tilde{b}_{\mathscr{I}} = (0,0,0,0,0)$$

with

$$\tilde{r}_{\mathscr{I}} = x_4 - \tilde{a}_{\mathscr{I}} - \tilde{b}_{\mathscr{I}} = x_4$$

which is the L_0 solution.

6.2.5 *Interactions, Outliers and Optimal Breakdown Behaviour*

The results of this section are based on [206].

In the analysis of variance an interaction can also be interpreted as an outlier: Tukey uses the word 'exotic' to cover both cases. Whether an exotic observation is an interaction or an outlier can only be decided in the context of the data. There is in consequence a link to robust statistics and, in particular, to that area of robustness concerned with breakdown. The simplest quantification of breakdown is the finite sample breakdown point due to [61]. The replacement version is as follows. For a given data set x_n let x_n^k denote any other data set derivable from x_n by altering at most k of the values of x_n. For a statistic T the finite sample breakdown point $\varepsilon(x_n, T)$ is defined by

$$\varepsilon(x_n, T) = \min\left\{k : \sup_{x_n^k} \|T(x_n^k)\| = \infty\right\}/n. \tag{6.38}$$

It was convincingly argued in [42] that the concept of breakdown point only makes sense for statistics that are equivariant with respect to a group operation. The group operations for the two-way table are permutations of rows, permutations of columns, addition of row and column effects and multiplication by a non-zero number. The L_0, L_1 and L_2 estimators are all equivariant with respect to this group. It follows that the maximum finite sample breakdown point for any equivariant estimator T is

$$\varepsilon^*(x_n) = \lfloor \min\{I,J\} + 1 \rfloor / n \tag{6.39}$$

(see for example [148, 118]). If $I = J = 5$ this implies a maximum breakdown point of $3/25 = 0.12$. As pointed out in [118] this is unduly pessimistic as it requires that all three outliers be either in the same row or in the same column. To overcome this pessimism [118] suggested considering the random placement of outliers.

Given the correspondence between interactions and outliers, an interaction matrix can also be interpreted as an outlier matrix where a '1' denotes the presence of an outlier. As is normal in breakdown analysis the outliers are regarded as being very large, 'clear distant outliers, not of marginal cases' in the words of [103]. This can be modelled by letting the noise tend to zero, that is, by exact data.

In robust statistics it is standard practice to try and fit the model to the majority of the data. Data points with large deviations from this model are identified as outliers. It is this which leads in unstructured models to the ubiquitous value of $1/2$ as the largest possible breakdown point. Fitting as many observations to the model as possible leads to choosing the row and column effects to minimize the number of outliers. This is also the principle adopted in [103]. Again in breakdown analysis it is assumed that the outliers are malicious: the worst possible case is considered just as in (6.38). If the outliers matrix corresponds to an unconditionally identifiable interaction matrix, then the outliers will be correctly identified. If the matrix is not unconditionally identifiable then row and column effects exist which reduce the number of outliers, or will find the same number but at other locations. In either case one cell (i,j) which initially contained an outlier will now have a value of zero. Consequently either the a_i or the b_j is now non-zero which corresponds to an arbitrarily large value of one of the parameters. The method has broken down. The best any method can do is to identify all outliers with an unconditionally identifiable matrix. The L_1 method does this in the two-way table and therefore has the best possible breakdown behaviour for this table.

Suppose outliers are distributed at random over the table as suggested by [118]). Three outliers can cause a breakdown in the 5×5 two-way table only if they are all in the same row or column. The probability of this is $1/23 \approx 0.042$. For four outliers the result is $41/253 \approx 0.162$. For the $I \times J$ two-way table the maximum number of interactions of an unconditionally identifiable interaction matrix is at most

$$\min\left\{\left(J - \left\lfloor \frac{J-1}{2} \right\rfloor\right) \left\lfloor \frac{I-2}{2} \right\rfloor, \left(I - \left\lfloor \frac{I-1}{2} \right\rfloor\right) \left\lfloor \frac{J-2}{2} \right\rfloor\right\} + \left\lfloor \frac{I-1}{2} \right\rfloor \left\lfloor \frac{J-1}{2} \right\rfloor . \tag{6.40}$$

(see [206]). For $I = J = 5$ the upper bound is seven. The unconditionally identifiable interaction matrix of (6.29) has six and it can be checked that this is the maximum. The upper bound (6.40) is therefore not in general sharp. It is of course possible for the outliers to be identifiable without their interaction matrix being unconditionally identifiable. The L_1 method is then not guaranteed to find them.

Outliers or interactions can often be identified using Tukey's median polish [108]. At each sweep the L_1 norm of the residuals either remains the same or decreases. It may happen that the median polish will find the L_1 solution but this is not guaranteed (see [148] and [118]).

6.3 The Three-Way and Higher Tables

The following is based on the unpublished manuscript [204] and on [40].

The exact k-factor model with one observation per cell is given by

$$x_{i_1\ldots i_k} = \sum_{j=1}^{k} a_{ji_j} + c_{i_1\ldots i_k},\ 1 \le i_j \le I_j,\ 1 \le j \le k \tag{6.41}$$

which can be written more compactly as

$$x_i = \ell(a_i) + c_i,\ i \in I, \quad \text{or} \quad x = \ell(a) + c, \tag{6.42}$$

depending on the context. The interactions c are not decomposed into two-way, three-way interactions etc. Whether such a decomposition is of advantage remains to be seen.

An interaction pattern ι is an array of the form

$$\iota_i \in \{0,1\},\ i \in I. \tag{6.43}$$

Given interactions c the interaction pattern $\iota(c)$ associated with c is defined by

$$\iota(c)_i = \begin{cases} 1, & c_i \neq 0, \\ 0, & \text{otherwise}, \end{cases} \quad i \in I. \tag{6.44}$$

A summation pattern σ is a tensor of the form

$$\sigma_i \in \{0,1\},\ i \in I. \tag{6.45}$$

for which there exist effects a such that

$$\sigma_i = \begin{cases} 1, & \ell(a_i) \neq 0, \\ 0, & \text{otherwise}, \end{cases} \quad i \in I. \tag{6.46}$$

For any array x the number of non-zero elements is given by

$$sp(x) = \left|\{i : x_i \neq 0, i \in I\}\right|. \tag{6.47}$$

6.3.1 *Unconditionally Identifiable Interactions*

An interaction pattern ι is called unconditionally identifiable if the L_0 problem is uniquely solvable for c with solution c for all c with $\iota(c) = \iota$. The following theorem is due to Terbeck [204] and corresponds to (a) of Theorem 6.2.1 for the two-way table.

Theorem 6.3.1 *An interaction pattern ι is unconditionally identifiable if and only if there exists no summation pattern σ not identically zero such that $sp(\iota + \sigma) \le sp(\iota)$ where the addition in $\iota + \sigma$ is* $\bmod (2)$.

The proof of the theorem is similar to the proof of (a) of Theorem 6.2.1 given in [206]. If $k \geq 3$ however there does not seem to be a simple way of generating all the summation patterns. Moreover part (b) of Theorem 6.2.1 no longer holds if $k \geq 3$ as is shown by the following example given in [204]. For the $3^2 4^1$ design the interactions

$$\left(\begin{array}{rrr|rrr|rrr} -3 & 0 & 0 & 0 & 0 & 0 & 0 & 0 & 0 \\ 0 & -1 & 0 & 0 & 0 & 0 & 0 & 0 & 4 \\ 0 & 0 & 3 & 0 & 0 & 4 & 0 & 4 & 4 \\ 0 & 0 & 0 & 0 & 2 & 0 & 0 & 2 & 6 \end{array}\right) \tag{6.48}$$

have an unconditionally identifiable interaction pattern and an L_1 norm of 33. Setting the effects a to

$$a_{1i_1} = (0,0,-2,-1),\ a_{2i_2} = (1,0,-1),\ a_{3i_3} = (1,0,-1)$$

for all relevant i_j results in the interactions

$$\left(\begin{array}{rrr|rrr|rrr} -1 & 1 & 0 & 1 & 0 & -1 & 0 & -1 & -2 \\ 2 & 0 & 0 & 1 & 0 & -1 & 0 & -1 & 2 \\ 0 & -1 & 1 & -1 & -2 & 1 & -2 & 1 & 0 \\ 1 & 0 & -1 & 0 & 1 & -2 & -1 & 0 & 3 \end{array}\right) \tag{6.49}$$

with an L_1 norm of 32. This is also the L_1 optimal solution.

The proof given by Terbeck that the interactions (6.48) have an unconditionally identifiable interaction pattern makes use of Theorem 6.3.1. It can also be shown using the intersection method described in Section 6.2.3. The intersection method can also be used in principle to reconstruct the interactions (6.48) from the L_1 solution (6.49). In fact however the computational complexity is too great for this to be carried out in practice. There are twenty-four interactions in (6.48) and consequently $2^{24} \approx 1.7 \cdot 10^7$ subsets which may have to be checked. Many of these result in only a few restrictions on the cells which are not set to zero which increases the algorithmic complexity further. If however the strategy described in Section 6.2.4 is applied to the L_1 solution (6.49) then the interactions (6.48) are identified immediately. Including an interaction term for the cell $(1,1,1)$ gives the following L_1 residuals

$$\left(\begin{array}{rrr|rrr|rrr} 0 & 0 & 0 & 0 & 0 & 0 & 0 & 0 & 0 \\ 0 & -1 & 0 & 0 & 0 & 0 & 0 & 0 & 4 \\ 0 & 0 & 3 & 0 & 0 & 4 & 0 & 4 & 4 \\ 0 & 0 & 0 & 0 & 2 & 0 & 0 & 2 & 6 \end{array}\right) \tag{6.50}$$

with a value for the interaction of -3. This is precisely (6.48).

In spite of the example (6.48) there are many cases in which the L_0 and L_1 solutions coincide and are unique for interactions with an unconditionally identifiable interaction pattern. Simple sufficient conditions are given in [204] and [207]. In practice however it is better to adapt the intersection method for the two-way table.

6.3.2 *Identifiable Interactions*

Just as in the two-way table interactions c may still be identifiable even if their interaction pattern $\iota(c)$ is not unconditionally identifiable and they may still be L_1 optimal. The intersection method of Chapter 6.2.3 carries over to higher tables as follows. Consider a cell indexed by i where the interaction is to be set to zero. Choose a pair j_1 and j_2 and keep indices $i_j, j \neq j_1, j_2$ fixed. Letting the indices i_{j_1} and i_{j_2} vary gives a two-way table to which the intersection method can be applied. This is done for every choice of $1 \leq j_1 < j_2 \leq k$. If for any given interactions c the intersection conditions cannot be fulfilled then the interaction pattern $\iota(c)$ is unconditionally identifiable. If the intersection conditions can be fulfilled the method then either shows that the interactions are nevertheless identifiable or it provides all other sets of interactions which can be obtained by choosing different effects a. The result of applying the intersection method to (6.48) is that the intersection criterion cannot be satisfied and the interaction pattern is therefore, as claimed, unconditionally identifiable.

6.3.3 *Interactions, Outliers and Optimal Breakdown Behaviour*

Just as in the two-way table interactions in exact data can be interpreted as arbitrarily large outliers in noisy data. The limits of breakdown for equivariant functionals are the unconditionally identifiable interaction patterns, again just as in the two-way table, but now with the restriction that the L_1 functional is not guaranteed to find all such interactions. The L_1 functional breaks down for the data set (6.48) as is shown by the L_1 solution (6.49). The problem of defining an equivariant functional with the optimal breakdown behaviour for three and higher tables remains open.

6.4 Interactions in the Presence of Noise

The following is based on [40].

So far the analysis has been for exact data. The presence of noise makes the identification of interactions more difficult in practice but the main idea is clear: if the L_1 solution is unique then this will not be destroyed by adding a small amount of noise. This is the basis of the proof given in [206] where it was shown that if the interaction pattern is unconditionally identifiable and therefore L_1 optimal, then the interactions will be identified by the L_1 solution as long as they are sufficiently large compared to the noise. This result carries over to any interaction pattern which is L_1 optimal and identifiable. The first step is to quantify the noise and here it is necessary to distinguish between data with and without replications. If the data are replicated the noise can be quantified from the replications which simplifies the problem considerably. If only one observation per cell is available the situation is more difficult. A solution was offered in [206] but the following is much superior.

6.4.1 One Observation per Cell

If only one observation per cell is available the quantification of the noise must be based on the residuals after the linear effects have been removed by some method. The obvious method to choose is L_1 minimization but this has the following disadvantage mentioned in [206]. Firstly the L_1 norm itself cannot be used to quantify the noise as this may well be influenced by the very interactions which are to be detected. The result is the well-known masking effect by which the scale is so affected by the interactions which are to be detected that they are not detected. To overcome this the L_1 norm can be replaced by the MAD of the residuals. This allows for about 50% interactions, a number which is more than sufficient for applications. The problem now is that that the L_1 solution for noise is in general not unique (see [206]) and that although the residuals of all the different L_1 solutions have the same L_1 norm, their MADs vary considerably. To overcome this the L_1 minimization can be replaced by minimizing

$$\sum_i \rho_\lambda(x_i - \ell(a_i)) \tag{6.51}$$

over the effects a and where

$$\rho_\lambda(x) = \int_0^x \psi_\lambda(u)\,du \tag{6.52}$$

with

$$\psi_\lambda(x) = \frac{\exp(\lambda x) - 1}{\exp(\lambda x) + 1}. \tag{6.53}$$

The function ρ_λ is strictly convex so that the minimization problem (6.51) has a unique solution. As

$$\lim_{\lambda \to \infty} \rho_\lambda(x) = |x|$$

the parameter λ can be so chosen that the L_1 norm of the residuals $r_i = x_i - \ell(\hat{a}_i)$ after the minimization of (6.51) is within a certain factor, say $1 + 10^{-3}$, of the L_1 norm of the residuals of an L_1 solution. The result is an approximate L_1 solution (see for example [136]). The scale can now be quantified by the median of the absolute residuals $|x_i - \ell(\hat{a}_i)|$. In fact simulations indicate that a further improvement can be obtained as follows. As an L_1 simplex solution puts $\sum_{j=1}^k I_j - k + 1$ residuals to zero the median of the absolute residuals is replaced by the mth quantile of the absolute residuals where

$$m = \sum_{j=1}^k I_j - k + 1 + \left\lfloor \frac{\prod_{j=1}^k I_j - \sum_{j=1}^k I_j + k - 1}{2} \right\rfloor .$$

On denoting this by $\mathfrak{s}_{(m)}$ the noise level is finally quantified by

$$\mathfrak{s} = \left(\frac{1}{m} \sum_i r_i^2 \{|r_i| \le \mathfrak{s}_{(m)}\} \right)^{1/2} .$$

The classification of a cell i as one with an interaction is based on the ideas of [32] (see also [101]),who considered the case of one-dimensional data. Put

$$\mathfrak{O}_i = \frac{|r_i|}{\mathfrak{s}}.$$

and given $\alpha, 0 < \alpha < 1$, define $\tau(\alpha)$ by

$$\mathbf{P}\left(\max_i \mathfrak{O}_i \geq \tau(\alpha)\right) = \alpha\,. \tag{6.54}$$

The cell i is classified as one with an interaction if $\mathfrak{O}_i \geq \tau(\alpha)$. To evaluate $\tau(\alpha)$ a model used for the error term ε is required. The default model, and the one used here, is that of homoscedastic Gaussian white noise. The value of $\tau(\alpha)$ also depends on the design but its value can be obtained with sufficient accuracy from simulations. In general 1000 simulations suffice and take but a few seconds of computer time. There remains the choice of α. The value $\alpha = 0.05$ would be a standard choice (see [32]) but here the identification of interactions based on $\mathfrak{O}_i$ is only regarded as provisional. It is suggested that the final decision be based on an L_2 regression where interaction terms are explicitly included for cells with high values of $\mathfrak{O}_i$. With this in mind a choice of $\alpha = 0.5$ is quite reasonable and is the one used here. Of course there is much to be said in favour of the statistician making a subjective informed choice based on particular knowledge of the data.

As it stands the method is sensitive to the scale of the observations. It can be made numerically more stable by starting with the L_1 or L_2 residuals normalized by their MAD or standard deviation. Numerically it seems not to make any difference. If this is done then it is sufficient to use a large value of λ rather than increasing λ to obtain a sufficiently close approximation to the L_1 norm. The default value of λ used is $\lambda = 128$.

The method is demonstrated using a $4^1 3^2$ design. Simulations give $\tau(0.5) \approx 7.6$. The following interactions are as in (6.48) but without the interaction in cell (4,3,3):

$$\left(\begin{array}{ccc|ccc|ccc} -3 & 0 & 0 & 0 & 0 & 0 & 0 & 0 & 0 \\ 0 & -1 & 0 & 0 & 0 & 0 & 0 & 0 & 4 \\ 0 & 0 & 3 & 0 & 0 & 4 & 0 & 4 & 4 \\ 0 & 0 & 0 & 0 & 2 & 0 & 0 & 2 & 0 \end{array}\right). \tag{6.55}$$

If white noise with $\sigma = 0.3$ is added then percentage of times the interactions are identified are

$$72.6, 2.4, 72.0, 42.0, 90.4, 90.8, 29.3, 96.8, 86.3$$

so that on average 58.3% of the interactions are found. In [40] the percentages are given correctly but the mean (!) was incorrectly calculated resulting in a false claim of 64.8% of the interactions being identified.

The situation changes if the interaction 6 in cell 36 is included. Without noise $\mathfrak{s} = 0.674$ and $\max|r_i|/\mathfrak{s} = 4.450 < 7.6$ so that no interactions are detected: the L_1-functional has broken down. The residuals of the L_0-functional give $\mathfrak{s} = 0$ and all the interactions are correctly identified. The problem is that the method of obtaining the L_0-functional from L_1-functionals described above cannot be used in the presence of noise as it relies on counting the number of residuals which are zero. The obvious solution is to replace 'zero' by 'small'. The following proposal is taken from [40]. It is formulated for the case of ignoring one cell but the extension to a collection $\mathscr{I}$ of cells is canonical.

(i) Calculate the L_1 solution ignoring the cell i_0

$$a(i_0) = \text{argmin}_a \sum_{i \neq i_0} |x_i - \ell(a_i)|$$

(ii) Calulate all residuals $r_i(i_0) = x_i - \ell(a_i(i_0))$ and the $\mathfrak{s}(i_0)$.

(iii) Repeat (i) and (ii) for all cells i_0 and put $\tilde{\mathfrak{s}} = \min_{i_0} \mathfrak{s}(i_0)$

(iv) Choose the solution $r_i(i_0)$ for which minimizes the number of cells i with $|r_i(i_0)| \geq 3\tilde{\mathfrak{s}}$. This is motivated by the L_0 criterion.

Given $r_i(i_0)$ the interactions are located in the cells i for which $|r_i(i_0)| \geq \mathfrak{s}(i_0)\tau$ where the value of τ is obtained from simulations. For the $3^2 4^1$ design the values are 7.99 and 8.28 for the leave one out and leave two out procedures respectively. For the leave one out procedure and a noise level $\sigma = 0.3$ the percentage of times the interactions of (6.48) are found are

$$43.9,\ 0.4,\ 53.3,\ 15.3,\ 76.3,\ 76.2, 8.1,\ 80.7,\ 64.0,\ 95.0\,.$$

so that on average 51.3% of the interactions are identified. This increases to 68.3% for the leave two out method. Even if the L_0 and L_1 solutions coincide the procedure can result in an improvement. For example in the case of the interactions (6.55) the percentage of correctly identified interactions increases from 58.3% to 77.2% (leave one out) and to 81.3% (leave two out). The reason for this is not clear.

6.4.2 *Several Observations per Cell*

There is no canonical method with replications but any reasonable method will work which allows for outliers in the replications (see the seat-belt data below). The one adopted here is to use the median of the values in each cell as the measure of location and the MAD as a measure of dispersion. If the design is balanced and the dispersions are more or less equal the median of the MADs is used as the global dispersion value $\mathfrak{s}$. The value of the threshold τ can be obtained using simulations: for a 3^3 design with three replications per cell the value of $\tau(0.5)$ is approximately 4.34.

6.5 Examples

The following is based on [40].

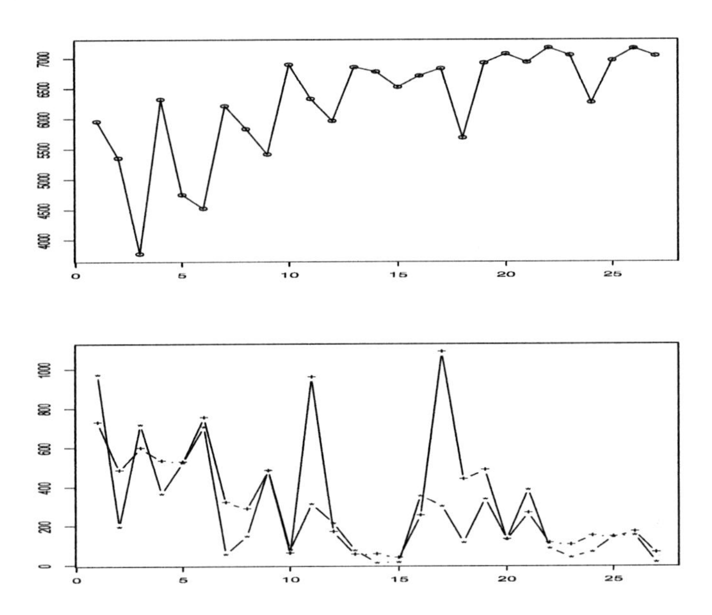

Figure 6.4 *Upper panel: locations m_i of the cells of the seat-belt data with factor C increasing. Lower panel: standard deviations (+) and MADs (∗).*

6.5.1 The Seat-Belt Data

The data are given in Table 6.2 on page 269 of [222]. The response is the pull strength of truck seat belts. Only the first three factors A pressure, B die flat and C crimp length will be considered giving a 3^3-table with three observations per cell. The levels will be 1-3 as against 0-2 in [222]. The analysis of such a data set cannot be complete as this would require a discussion and interpretation of the results with the person who provided the data. This is not possible. Nevertheless an attempt will be made to give an analysis which goes beyond the decomposition into sums of squares, their sources, degrees of freedom and their F-values. The upper panel of Figure 6.4 shows a plot of the median of the three observations in each cell where the the factor A increases from level 1 to level 3. It seems clear that this factor is relevant. The relevance of the factor C also seems clear. The lower panel shows the corresponding plot for the standard deviations and the MADs. A certain degree of heteroscedasticity is apparent. The cells $(2,1,2)$ and $(2,3,2)$ which correspond to 11 and 17 in the figure exhibit large differences between the standard deviations and the MADs. The individual measurements for these cells are (6538,6328,4784) and (6832,7034, 5057) respectively.

In a first analysis both the heteroscedasticity and the deviant observations in the cells 11 and 17 will be ignored to allow a direct comparison with a standard analysis. The results of a standard analysis are given in Table 6.5 which is Table 6.3 of [222].

Source	Degrees of Freedom	Sum of Squares	Mean Squares	F	p-value
A	2	34621746	17310873	85.58	0.000
B	2	938539	469270	2.32	0.108
C	2	9549481	4774741	23.61	0.000
$A \times B$	4	3298246	824561	4.08	0.006
$A \times C$	4	3872179	968045	4.79	0.002
$B \times C$	4	448348	112087	0.55	0.697
$A \times B \times C$	8	5206919	650865	3.22	0.005
residual	54	10922599	202270		
total	80	68858056			

Table 6.5 *Decomposition of the sums of squares for an L_2 analysis of the seat-belt data taken from Table 6.3 of [222].*

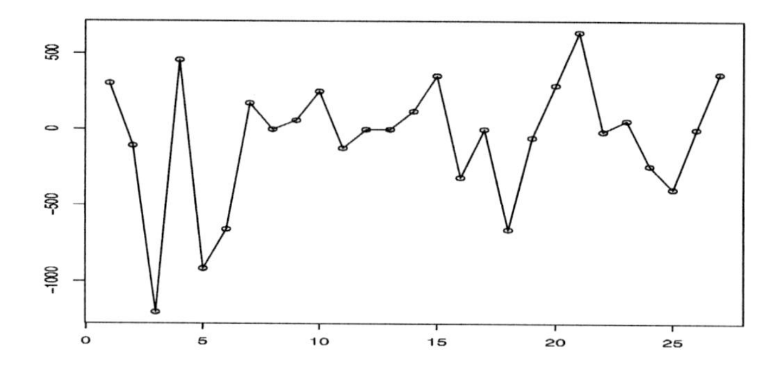

Figure 6.5 *The approximate L_1 residuals calculated as described in Section 6.4.2.*

The total sum of squares is given in Table 6.5 as 68858056 which differs slightly from the sum 68858058 calculated using R. The sum of squares due to the effects is 45109766 which again differs slightly from the value of 45109767 calculated using R.

The results of an L_1 analysis as described in Section 6.4.2 are as follows. The approximate L_1 residuals are given in Figure 6.5. The value of $\mathfrak{s}$ calculated as in Section 6.4.2 is 108. The threshold $\tau(0.5) = 4.34$. On this basis the cells 3, 4, 5, 6, 15, 18, 21, 25 and 27 are candidates for cells with interactions. If terms for the factor effects and interactions in these cells are included and an L_2 regression performed the R output is the following;

```
Coefficients:
      Estimate Std. Error t value Pr(>|t|)
xxx0  5890.748    159.630  36.902  < 2e-16 ***
xx    -224.076    134.459  -1.667  0.10043
xx    -755.361    178.046  -4.242 7.16e-05 ***
xx     249.892    160.018   1.562  0.12323
xx     182.147    147.326   1.236  0.22078
xx     676.420    160.018   4.227 7.55e-05 ***
xx    1103.380    178.090   6.196 4.45e-08 ***
xx   -1435.387    323.050  -4.443 3.53e-05 ***
xx       3.694    328.225   0.011  0.99106
xx   -1052.897    333.837  -3.154  0.00244 **
xx    -685.279    342.623  -2.000  0.04967 *
xx     466.301    323.050   1.443  0.15370
xx    -522.954    325.056  -1.609  0.11250
xx     692.567    325.056   2.131  0.03691 *
xx    -183.941    317.876  -0.579  0.56482
xx     599.086    334.948   1.789  0.07834 .
```

The non-identifiability of the effects has been broken by setting the first level of each factor to zero.

On this basis the cells corresponding to the numbers 3, 5, 6 and 21 have significant interactions at the 5% level; the cell 27 is significant at the 10% level. If just these cells are included with interaction terms and the regression run again the R output is

```
Coefficients:
      Estimate Std. Error t value Pr(>|t|)
xxx0  5910.56     145.49  40.625  < 2e-16 ***
xx    -188.96     127.89  -1.477 0.144095
xx    -745.40     149.92  -4.972 4.64e-06 ***
xx     330.76     143.09   2.311 0.023800 *
xx      90.06     136.50   0.660 0.511563
xx     639.02     141.77   4.508 2.61e-05 ***
xx    1032.47     149.85   6.890 2.10e-09 ***
xx   -1465.16     319.53  -4.585 1.96e-05 ***
xx   -1188.69     313.55  -3.791 0.000318 ***
xx    -795.92     318.44  -2.499 0.014823 *
xx     733.70     320.34   2.290 0.025060 *
xx     732.31     313.79   2.334 0.022530 *
```

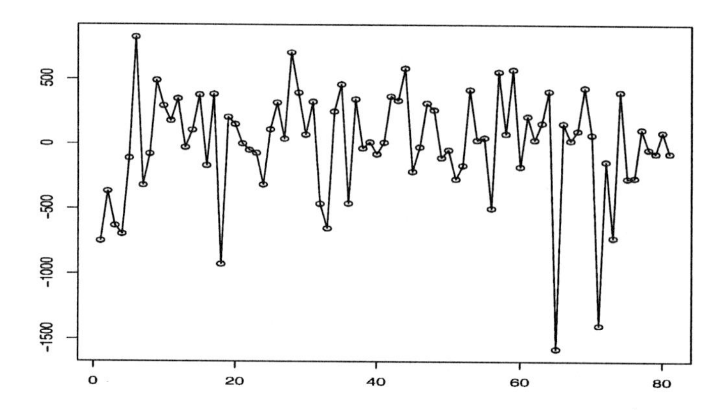

Figure 6.6 *The L_2 residuals with terms included for the interactions in the cells 3, 5, 6, 21 and 27.*

The interactions are

$$\left(\begin{array}{rrr|rrr|rrr} 0 & 0 & 0 & 0 & 0 & 0 & 0 & 0 & 0 \\ 0 & -1189 & 0 & 0 & 0 & 0 & 0 & 0 & 0 \\ -1465 & -796 & 0 & 0 & 0 & 0 & 734 & 0 & 732 \end{array}\right). \tag{6.56}$$

The algorithm of Section 6.2.3 shows that the interaction pattern is unconditionally identifiable so that the solution is sparse. The results should of course be discussed with the provider of the data to decide to what extent if any they are of significance. Statistical significance is not identical with practical significance and the latter always has priority. It is perhaps worth noting that four of the five interactions are at level 3 of factor C.

The residual sum of squares is 14105541. The values of all eighty-one residuals are shown in Figure 6.6. The two deviant values mentioned above are clearly visible.

It is not possible to discuss the status of the two observations with the person who provided the data. Nevertheless it seems reasonable to assess their effect, if any, on the analysis. There are many ways of doing this. The one chosen here is to replace the observations 4784 and 5057 by 6784 and 7057 respectively as if they were due to transcription errors. The candidate cells for interactions are now 3, 4, 5, 6, 15, 16, 18, 21, 25 and 27. If the above analysis is repeated the cells 3, 5, 6, 18, 21 and 27 have p-values below 0.1 are are included in a second regression. This results in the following interactions.

$$\left(\begin{array}{rrr|rrr|rrr} 0 & 0 & 0 & 0 & 0 & 0 & 0 & 0 & 0 \\ 0 & -1144 & 0 & 0 & 0 & 0 & 0 & 0 & 0 \\ -1460 & -698 & 0 & 0 & 0 & -764 & 705 & 0 & 608 \end{array}\right). \tag{6.57}$$

The algorithm of Section 6.2.3 shows that the interactions are identifiable so that the solution is sparse. However the pattern is not unconditionally identifiable as can be seen from the last line of (6.57). It is notable that five of the six interactions are associated with level 3 of factor C. The residual sum of squares is 7873978 which is considerably less than the 14105541 of the first analysis. This is almost entirely due to the two altered values in the cells 11 and 17. It can be compared with the residual sum of squares 10922599 of the standard L_2 analysis which includes all interaction terms. The values of all 81 residuals are shown in Figure 6.7.

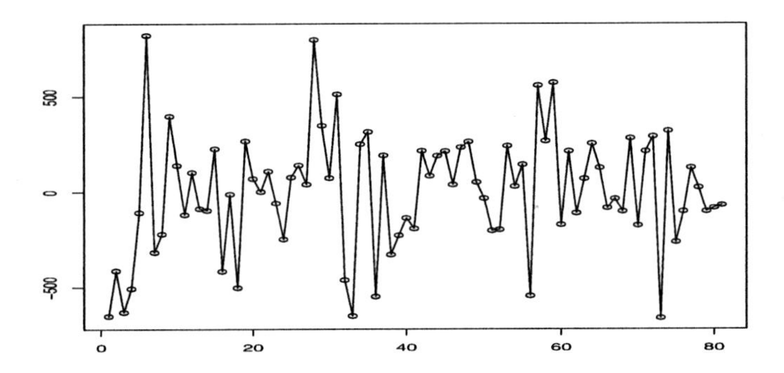

Figure 6.7 *The L_2 residuals with altered values for the observations in the cells indexed 11 and 17 and including interactions in the cells indexed 3, 5, 6, 18, 21 and 27.*

Finally the heteroscedasticity of the measurements can be taken into account. There are many ways of doing this, either using an automatic method, or by specifying the observations with the same scale, or by specifying the scales themselves. The second alternative seem the appropriate one in this case as the large variability is associated with the first level of factor A. The following analysis does this and continues to use the altered values in the cells indexed by 11 and 17. In fact, because of the local nature of the noise more deviant observations can be found. The most deviant of these is the observation 6220 in the cell indexed 19. One does not have to alter it but its deviant value becomes apparent in a plot of the residuals just as did the deviant values in the cells 11 and 17. In the following it is altered to 7000. Heteroscedasticity will be taken into account by allowing for two different scales. The first is for those observations where the level of the first factor A is 1. The second is for the remaining observations. The calculated values are 264.4 and 51.0 respectively. The observations are now weighted with the inverse of these values and the analysis performed as before. Simulations give a threshold of 5.84 which results in the observations 15, 18, 21 and 27 being candidates for interactions. The linear regression

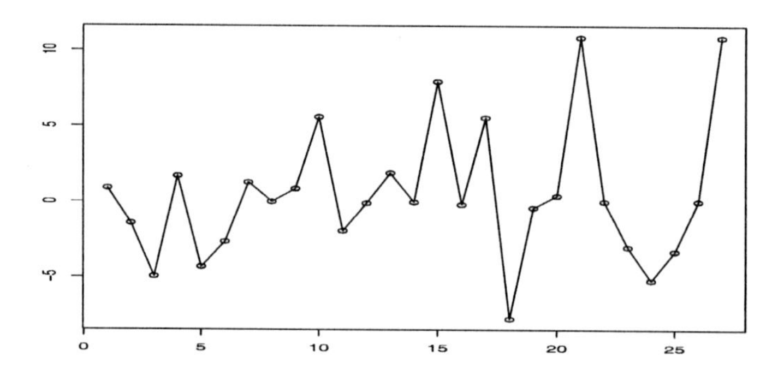

Figure 6.8 *The normalized approximate L_1 residuals corresponding to Figure 6.5 but allowing for heteroscedasticity.*

results in all four cells as having interactions which are given by

$$\left(\begin{array}{ccc|ccc|ccc} 0 & 0 & 0 & 0 & 0 & 0 & 0 & 0 & 0 \\ 0 & 0 & 0 & 0 & 0 & 0 & 0 & 0 & 0 \\ 0 & 0 & 0 & 0 & 488 & -573 & 702 & 0 & 705 \end{array}\right). \tag{6.58}$$

The interaction pattern is unconditionally identifiable. The interactions are concentrated at level 3 of factor C. The normalized L_2 residuals are shown in Figure 6.9.

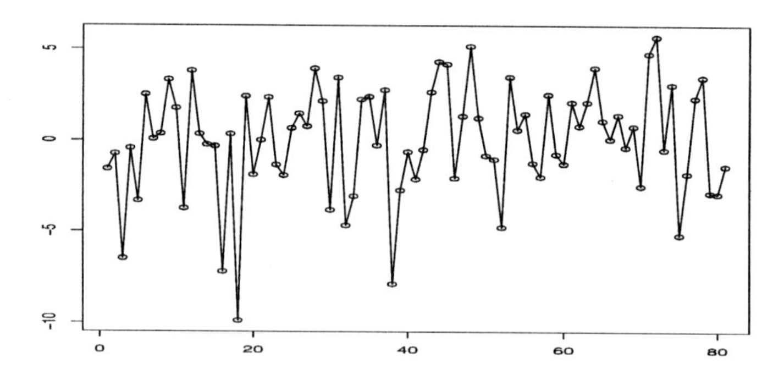

Figure 6.9 *The normalized L_2 residuals corresponding to Figure 6.7 but allowing for heteroscedasticity.*

Source	df	Mean Square
(1)	1	65118387
D	4	54394
C	2	298808
G	7	31477
DC	8	32930
DG	28	7458
CG	14	14984
DCG	56	9969

Table 6.6 *Decomposition of the sums of squares for a standard L_2 analysis of the dental data taken from Table 2 of [189].*

6.5.2 *The Dental Data*

The data are taken from [189] which contains a long and detailed description and analysis of the data. It is a $8 \times 5 \times 3$-table with one observation per cell. The first factor is the gold alloy used, the second is the dentist who prepared the filling and the third is the method of condensation. The final results are a measure of the hardness of a filling obtained from surface indentation measured in millimetres. It is a difficult data set with possible interactions/outliers, heteroscedasticity and a high degree of discreteness: four of the first values in the first column are 792 and three of the values in the second column are 803. This calls into question the appropriateness of the Gaussian model as mentioned by [189]. Two analyses will be given. The first is the standard analysis described in the main body of the paper. In the second analysis the threshold for interactions is reduced. Table 6.6 is taken from [189] and shows the results for a standard L_2 analysis. As there is only one observation per cell the third order interactions are also the residuals.

Figure 6.10 shows the L_1 residuals for the dental data calculated as described in Section 6.4.1. The non-standard appearance of the residuals is perhaps to some extent explained by the large number of repeated values in the data. The value of $\mathfrak{s}$ is 26.7 and the threshold for interactions $\tau(0.5)$ is 8.37. Thus all cells with absolute residuals of greater than $26.7 \cdot 8.37 = 223.5$ are candidates for interactions. These are the cells 14, 90, 93, 96, 103, 119 and 120. If now an ordinary least squares regression is performed as described above then all of these observations are identified as interactions.

For reasons of space the interaction tables given below are such that the columns correspond to the dentists, the rows to the alloys and the three blocks to the method of condensation. To recover a cell from its number the three blocks must be imagined

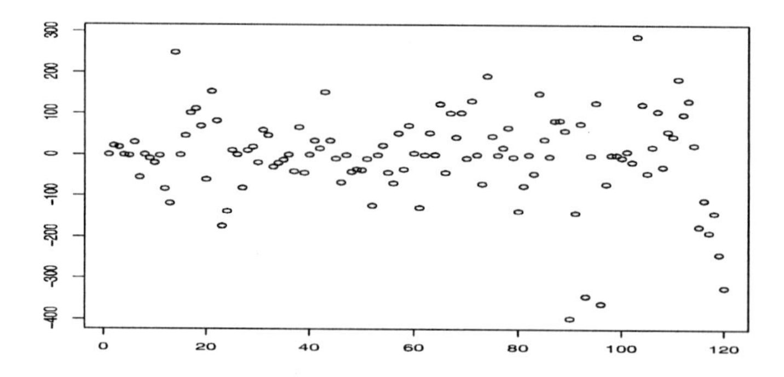

Figure 6.10 *The L_1 residuals calculated as described in Section 6.4.1 for the dental data.*

as being aligned vertically rather than horizontally. The interactions are

$$\left(\begin{array}{ccccc|ccccc|ccccc}
0 & 0 & 0 & 0 & 0 & 0 & 0 & 0 & 0 & 0 & 0 & 0 & 0 & 0 & 0 \\
0 & 0 & 0 & 0 & 0 & 0 & 0 & 0 & 0 & 0 & 0 & 0 & 0 & -435 & 0 \\
0 & 0 & 0 & 0 & 0 & 0 & 0 & 0 & 0 & 0 & 0 & 0 & 0 & 0 & 0 \\
0 & 0 & 0 & 0 & 0 & 0 & 0 & 0 & 0 & 0 & 0 & 0 & 0 & 0 & 0 \\
0 & 0 & 0 & 0 & 0 & 0 & 0 & 0 & 0 & 0 & 0 & 0 & 0 & -360 & 0 \\
0 & 0 & 0 & 0 & 0 & 250 & 0 & 0 & 0 & 0 & 0 & 0 & 0 & 0 & 0 \\
0 & 0 & 0 & 0 & 279 & 0 & 0 & 0 & 0 & 0 & 0 & 0 & 0 & 0 & -266 \\
0 & 0 & 0 & 0 & 0 & 0 & 0 & 0 & 0 & 0 & 0 & 0 & 0 & -385 & -326
\end{array}\right) \tag{6.59}$$

and the corresponding interaction pattern is unconditionally identifiable.

If instead of the threshold of 223 a much lower one of 150 is chosen then there are fourteen candidate cells 14, 21, 23, 43, 74, 90, 93, 96, 103, 111, 115, 117, 119 and 120. Including terms for interactions in these cells in a standard L_2 regression results in all interactions except that in cell 43 being classified as significant. As 43 is borderline it was still included giving rise to the interactions

$$\left(\begin{array}{ccccc|ccccc|ccccc}
0 & 0 & 0 & 0 & 0 & 0 & 0 & 0 & 0 & 0 & 0 & 0 & 0 & 0 & 0 \\
0 & 0 & 0 & 195 & 0 & 0 & 0 & 0 & 0 & 0 & 0 & 0 & 0 & -422 & 0 \\
0 & 0 & 0 & 0 & 0 & 0 & 0 & 0 & 0 & 0 & 0 & 135 & 0 & 0 & -210 \\
0 & 0 & 0 & 0 & 0 & 0 & 0 & 0 & 0 & 0 & 0 & 0 & 0 & 0 & 0 \\
0 & 0 & 0 & 0 & 0 & 0 & 0 & 0 & 0 & 0 & 144 & 0 & 0 & -365 & -213 \\
0 & 0 & 0 & 0 & 0 & 252 & 0 & 0 & 0 & 0 & 0 & 0 & 0 & 0 & 0 \\
0 & 0 & 0 & 0 & 262 & 0 & 0 & 0 & 0 & 145 & -215 & 0 & 0 & 0 & -297 \\
0 & 0 & 0 & 0 & 0 & 0 & 0 & 0 & 0 & 0 & 0 & 0 & 0 & -387 & -352
\end{array}\right)$$

whose interaction pattern is again unconditionally identifiable.

The following tentative conclusions follow from the above analysis and can be compared with the conclusions given by [189]:

- The average level of hardness is about 784 ± 46.
- The gold alloys 6 and 7 are significantly better than the others. Gold alloy 8 is also better but not significantly so until it is taken into account that 6, 7 and 8 are the (Au Ca) alloys.
- Hand condensation C3 with -83 is significantly worse than the others.
- Dentists 3 and 4 with -36 are marginally worse than the others. Dentist 5 with -72 is significantly worse.
- 10 of the 14 interactions/outliers are associated with hand condensation C3.
- 10 of the 14 interactions/outliers are associated with the dentists 4 and 5.
- 7 of the 14 interactions/outliers are associated with the dentists 4 and 5 when using hand condensation C3. Consequently of these 16 cells 7 were identified as having interactions/outliers. In each case there was a sharp drop in hardness of, on average, -320.
- The gold alloys 1, 4 and 6 caused the fewest problems.

Chapter 7

Non-parametric Regression: Location

7.1 A Definition of Approximation

Nonparametric regression was treated briefly in Chapter 2.3. The ideas carry over to the general case as follows. The basic model is

$$X(t) = f(t) + \sigma Z(t), 0 \le t \le 1, \tag{7.1}$$

where $f : [0,1] \to \mathbb{R}$ is a function and $Z(t)$ is the standardized noise which is often taken to be Gaussian white noise. Given data $(\mathbf{t},\mathbf{x})_n = (t_j, x(t_j)), j = 1,\ldots,n$ the problem is to produce one or more functions f_n and values σ_n such that the model with $f = f_n$ and $\sigma = \sigma_n$ is an adequate approximation of the data.

Given data $(\mathbf{t},\mathbf{x})_n$ and a function $g : [0,1], \to \mathbb{R}$ put

$$w((\mathbf{t},\mathbf{x})_n, g, I) = \frac{1}{\sqrt{|I|}} \sum_{t_i \in I} (x(t_i) - g(t_i)) \tag{7.2}$$

where $I = [t_j, t_k]$ is an interval and $|I|$ denotes the number of points t_i in I. If σ is given an approximation region $\mathscr{A}_n$ for the regression function f is given by

$$\mathscr{A}_n = \mathscr{A}_n((\mathbf{t},\mathbf{x})_n, \mathscr{I}_n, \sigma, \alpha) = \left\{ g : \max_{I \in \mathscr{I}_n} |w((\mathbf{t},\mathbf{x})_n, g, I)| \le \sigma \sqrt{\tau_n(\alpha) \log n} \right\} \tag{7.3}$$

(see [45]) where $\mathscr{I}_n$ denotes a family of intervals.

As argued in Chapter 2.3 $\mathscr{I}_n$ should be a multiresolution scheme, that is, contain intervals of all sizes at all locations. One possible choice for $\mathscr{I}_n$ is family of all intervals. At first glance it would seem that the calculation of

$$\max_{I \in \mathscr{I}_n} |w((\mathbf{t},\mathbf{x})_n, g, I)|$$

for all intervals I is of complexity $O(n^2)$. However by making use of the special nature of the problem algorithms have been developed with a complexity of $O(n \log n)$ (see ([17]). Nevertheless there is an interest in reducing the complexity and this can be done by taking $\mathscr{I}_n$ to be a multiresolution scheme similar to the one defined by wavelets. The default scheme used here is of the form

$$\mathscr{I}_n(\lambda) = \Big\{ [t_{l(j,k)}, t_{u(j,k)}] : l(j,k) = (j-1)\lceil \lambda^k/8 \rceil + 1, u(j,k) = \min(l(j,k) + \lfloor \lambda^k \rfloor, n), \\ j = 1,\ldots,\lfloor (n-1)/\lceil \lambda^k/8 \rceil \rfloor, k = 1,\ldots,\lceil \log n / \log \lambda \rceil \Big\} \tag{7.4}$$

for some $\lambda > 1$. If the '8' in (7.4) is replaced by '1' then $\mathscr{I}_n(2)$ is a wavelet multiresolution scheme. Such a scheme contains $O(n)$ intervals with the corresponding decrease in complexity.

Under the model (7.1) with $Z(t)$ standard Gaussian white noise

$$\max_{I \in \mathscr{I}_n} |w((\mathbf{t}, \mathbf{X})_n, f, I)| = \max_{I \in \mathscr{I}_n} \frac{1}{\sqrt{|I|}} \Big| \sum_{t_i \in I} Z(t_i) \Big| .$$

Given $\alpha, 0 < \alpha < 1$, it is possible to choose $\tau_n(\alpha)$ such that

$$P\Big(\max_{I \in \mathscr{I}_n} \frac{1}{\sqrt{|I|}} \Big| \sum_{t_i \in I} Z(t_i) \Big| \leq \sqrt{\tau_n(\alpha) \log n} \Big) = \alpha. \tag{7.5}$$

With this choice of $\tau_n(\alpha)$ the approximation region (7.3) is an α-approximation region. As there are no restrictions on the regression function f it is in fact a universal, exact and non-asymptotic α–confidence region.

The values of $\tau_n(\alpha)$ for given n, $\mathscr{I}_n$ and α can be obtained by simulations. For $n = 1000$ and with $\mathscr{I}_n$ being the set of all intervals the values of $\tau_n(\alpha)$ are 2.771 ($\alpha = 0.9$) and 3.009 ($\alpha = 0.95$). These estimates are based on 5000 simulations. If $\mathscr{I}_n$ is the set of all possible intervals it follows from a result of [76] that $\lim_{n\to\infty} \tau_n(\alpha) = 2$ whatever the value of α. On the other hand for any $\mathscr{I}_n$ which contains all the degenerate intervals $[t_j, t_j]$ (as will always be the case) then again $\lim_{n\to\infty} \tau_n(\alpha) = 2$ whatever α. The exact asymptotic distribution of $\max_{1\leq i<j\leq n}(\sum_{l=i}^{j} Z_l)^2/(j-i+1)$ is given in [123]. The term containing α is of smaller order than $\sqrt{\log n}$. The consequence is that the values of $\tau_n(\alpha)$ are relatively insensitive to the values of α and the choice of $\mathscr{I}_n$. The fact that $\tau_n(\alpha)$ occurs inside the square-root reduces the sensitivity even further. Given all the other uncertainties this is of little consequence. It is even possible to take a fixed value of τ_n independent of n and α and to define the approximation region directly in terms of τ_n as $\mathscr{A}_n((\mathbf{t}, \mathbf{x})_n, \mathscr{I}_n, \sigma, \tau_n)$. For example with $\tau_n = 3$ simulations show that

$$P\big(f \in \mathscr{A}_n((\mathbf{t}, \mathbf{X})_n, \mathscr{I}_n, \sigma, 3)\big) \geq 0.95 \tag{7.6}$$

for all $n \geq 500$ and

$$\lim_{n\to\infty} \inf_f P\big(f \in \mathscr{A}_n((\mathbf{t}, \mathbf{X})_n, \mathscr{I}_n, \sigma, 3)\big) = 1. \tag{7.7}$$

In other words $\mathscr{A}_n$ is a universal, honest and non-asymptotic confidence region for f. This would be a conservative choice. A less conservative choice would be $\tau_n = 2$. The user is free to choose a value of τ_n which serves best for the data under consideration.

The approximation region (7.3) treats all intervals equally. An approximation region which down weights the importance of small intervals can be obtained as follows. The following extension Lèvy's uniform modulus of continuity of the Brownian motion $B(t)$

$$\sup_{0<s<t<1} \frac{\frac{(B(t)-B(s))^2}{t-s} - 2\log(1/(t-s))}{\log(\log(e^e/(t-s)))} < \infty \quad \text{a.s.} \tag{7.8}$$

was proved by [76]. The partial sums $\sum_{i\in I}^{j} Z(t_i)/\sqrt{|I|}$, $I \in \mathscr{I}_n$, can be embedded in n a standard Brownian motion to give

$$\sup_{I\in\mathscr{I}_n} \frac{(\sum_{t_j\in I} Z(t_j))^2/|I| - 2\log(n/|I|)}{\log(\log(e^e n/|I|)))} = \Gamma_n < \infty \quad \text{a.s.} \tag{7.9}$$

where the random variable Γ_n is bounded in probability as n tends to infinity. For any α there exist a $\gamma_n = \gamma_n(\alpha)$ such that

$$\begin{aligned}\mathscr{A}_n^*((\mathbf{t},\mathbf{x})_n, \mathscr{I}_n, \sigma, \gamma_n(\alpha)) = \{g : |w((\mathbf{t},\mathbf{x})_n, g, I)| \le \\ \sigma\sqrt{2\log(n/|I|) + \gamma_n(\alpha)\log(\log(e^e n/|I|))} \text{ for all } I \in \mathscr{I}_n)\}.\end{aligned} \tag{7.10}$$

is a universal, exact and non-asymptotic approximation region. The values of γ_n may be determined by simulation. For $\alpha = 0.95$ and with $\mathscr{I}_n = \mathscr{I}_n(2)$ a good approximation for $\gamma_n(\alpha)$ for $n \ge 100$ is given by

$$\gamma_n(0.95) \approx 5.77 - \exp(2.89 - 0.6\log(n)). \tag{7.11}$$

The approximation region (7.3) cannot be used as it stands as it requires a value for the noise level σ. The default value σ of which will be used is σ_n given by

$$\begin{aligned}\sigma_n &= \text{median}\{|x(t_2) - x(t_1)|, \ldots, |x(t_n) - x(t_{n-1})|\}/(\sqrt{2}\,\text{qnorm}(0.75)) \\ &= 1.048358\,\text{median}\{|x(t_2) - x(t_1)|, \ldots, |x(t_n) - x(t_{n-1})|\}.\end{aligned} \tag{7.12}$$

If applied to standard Gaussian white noise σ_n is a consistent estimate of σ and converges quickly. If the data have been generated under the model (7.1) it is seen that σ_n is positively biased. This of itself does not imply that replacing σ by σ_n in (7.3) will still result in an α-approximation region. One manner of obtaining an α-approximation region is the following. Put

$$W_i' = |X(t_{2i}) - X(t_{2i-1})|, i = 1, \ldots, \lfloor n+1 \rfloor/2 =: m$$

with order statistics $W_{(1)}' \le \ldots \le W_{(m)}'$. Then for moderate n

$$\mathbf{P}\big(1.4826|W'|_{(\lceil m/2+1.288\sqrt{m}\rceil)} \ge \sqrt{2}\sigma\big) \ge 0.995$$

which holds for any function f because of the positive bias (see [48]). On setting

$$\hat{\sigma}_n = \frac{1.4826}{\sqrt{2}}|W'|_{(\lceil m/2+1.288\sqrt{m}\rceil)} = 1.0484|W'|_{(\lceil m/2+1.288\sqrt{m}\rceil)} \tag{7.13}$$

it follows that $\mathbf{P}(\hat{\sigma}_n \ge \sigma) \ge 0.995$ and hence

$$\mathbf{P}\big(f \in \mathscr{A}_n((\mathbf{t},\mathbf{X})_n, \mathscr{I}_n, \hat{\sigma}_n, \tau_n(\alpha))\big) \ge \alpha - 0.005 \tag{7.14}$$

is an honest confidence region with $\alpha - 0.005$ in place of α. In practice it is not worth the extra effort so that the default σ_n of (7.12) will be used. The value of σ_n for the data of Figure 2.1 is 6.29 which gives rise to the approximation region of (2.10).

The approximation region $\mathscr{A}_n$ can be interpreted as the inversion of a multiscale test that the mean of the residuals is zero on all intervals $I \in \mathscr{I}_n$. Why one should test these hypotheses knowing them to be false is another question which can be avoided by not interpreting them as tests but as a definition of typical behaviour under the model. A similar idea phrased in terms of tests is to be found in [76] where tests are inverted to obtain confidence regions. The tests derive from kernel estimators with different locations and bandwidths and the kernels are chosen to be optimal for testing shape hypotheses.

Universal, exact and non-asymptotic approximation regions based on the signs of the residuals $\text{sign}(y(t_i) - g(t_i))$ rather than the residuals themselves are to be found in [36] and [45] based on the lengths of runs, and in [68, 69, 70, 72] based on the partial sums of the signs. These require only that the $Z(t)$ be independently distributed with median zero. As a consequence they do not require an auxiliary estimate of the noise such as (7.12).

The nonparametric regression problem described above is a location problem in a weak sense. If all the observations $x(t_i)$ are subject to the same one-dimensional affine transformations $x(t_i) \to ax(t_i) + b$ with $a \neq 0$, then in general any automatic procedure such lead to $ag + b$ being an adequate approximation for $a\mathbf{x}_n + b$ for any g which is an adequate approximation for the data $\mathbf{x}_n$.

7.2 Regularization

Not all functions in the approximation region $\mathscr{A}_n$ will be acceptable as an adequate approximation to the data, where the word 'adequate' is now interpreted in a wider sense. In particular any function which interpolates the data has residuals of zero and consequently belongs to $\mathscr{A}_n$. A less extreme example is the kernel estimator of Figure 2.3. Acceptable functions in $\mathscr{A}_n$ will in general be simple ones, where simple may be defined by shape or smoothness or a combination of both. Such functions can be seen as the result of a regularization within $\mathscr{A}_n$. This can be expressed more formally as

$$\text{minimize } J(g) \text{ subject to } g \in \mathscr{A}_n \tag{7.15}$$

where $J(g)$ is some measure of the complexity of the function g. Problems of this sort were first considered in [163]. In [164] the opposite direction is considered:

$$\text{minimize } \max_I |w((\mathbf{t},\mathbf{x})_n, g, I)| \text{ subject to } g \in S \tag{7.16}$$

where S is some specified class of functions.

The functional $J(g)$ considered in [45] was the number of local extreme values of g. In the one-dimensional situation it turns out that there is a close connection between minimizing the number of extreme values of g and minimizing the total variation $TV(g)$ of g subject to $g \in \mathscr{A}_n$. In the context of image analysis $J(g)$ is often taken to be the total variation of g with $\mathscr{A}_n$ being defined by wavelets, ridglets and curvelets as in [192], [25], [144], [143]. Other related articles are [28] and [79].

7.2.1 The Taut String

The software used in the following sections is available as an R-package [49].

The wriggles of the kernel estimator of Figure 2.3 can be interpreted as many local maxima. To remove as many as possible whilst still retaining an adequate approximation to the data is equivalent to minimizing the number of local extreme values of a function g subject to $g \in \mathscr{A}_n$. The problem is not an easy one but there are two efficacious methods of deriving a solution. The first is the so called 'taut string' algorithm. The algorithm goes back to one method of obtaining the monotonic least squares solution to a set of data points (see [11] where the term 'taut string' is first used). The extension from monotonic to unimodal in the context of densities is given in [105] where the term 'stretched string' is used. The further extension to minimizing the number of local extremes subject to the function being in $\mathscr{A}_n$ is treated in [45]. It is as follows. Let

$$S_j^\circ = S^\circ((\mathbf{t},\mathbf{x})_n)_j = \sum_{i=1}^{j} x(t_i),\ j = 1,\ldots,n, \tag{7.17}$$

denote the cumulative sums of the observations $x(t_i)$. Given $\varepsilon > 0$ a tube with upper bound U and lower bound L is defined by

$$L_j = S_j^\circ - \varepsilon, \quad U_j = S_j^\circ + \varepsilon,\ j = 1,\ldots,n.$$

The taut string is defined as follows. Imagine a string tied down at the midpoints of the two ends of the tube, then pulled to be taut but constrained to lie within the tube. The upper panel of Figure7.1 shows the situation for the data of Figure 2.1 with the width of the tube being $2\varepsilon = 2000$. The top panel shows the lower and upper bounds (dots) and the taut string in a neighbourhood of the peak. The taut string is spline of order one with automatically chosen knots. Its derivative (either the left or right hand side derivative), shown in the lower panel of Figure 7.1. It defines a piecewise constant function with just one peak as required but it is clearly not an adequate approximation to the data.

The idea is to decrease the width of the tube until an adequate approximation is found. This is shown in Figure 7.2. It has again has only one local extreme value and, as will be shown, has solved the problem of minimizing the number of local extreme values in the approximation region $\mathscr{A}_n$.

The taut string has many optimal properties [141] but the one of interest here is that its derivative has the smallest number of local extremes amongst all functions with the same end points and constrained to lie within the tube. Using this it is not difficult to prove that asymptotically the taut string will correctly estimate the number of local extremes of the function f for data generated under the model (7.1).

Although it is not guaranteed to produce the correct answer for any given data set it can often be proved post hoc that the solution given by the taut string is the correct one. For the data of Figure 2.1 $x(21.42) = 210$ and from (2.11) it follows that $g(21.42) \geq 210 - 27.31 = 182.69$ for any function g in $\mathscr{A}_n$. Similarly $g(21.35) \leq x(21.34) + 27.31 = 129 + 27.31 = 156.31$ and $g(21.48) \leq x(21.48) +$

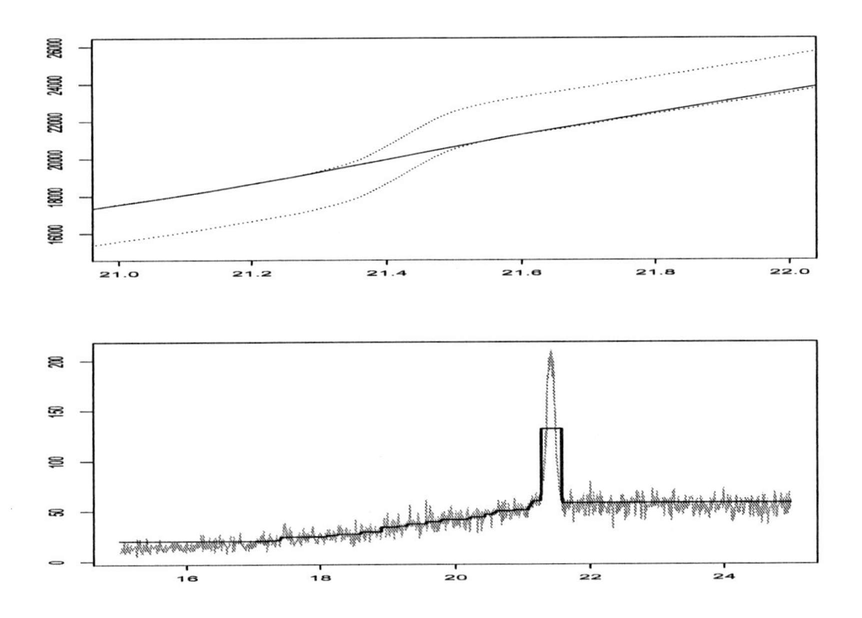

Figure 7.1 *Upper panel: Partial sums of data; lower and upper bounds (dots), the taut string (continuous line) ($\varepsilon = 1000$) together with the data. Lower panel: the derivative of the taut string.*

$27.31 = 153 + 27.31 = 180.31$. It follows that any function $g \in \mathscr{A}_n$ must have at least one local maximum. As the derivative of the taut string lies in $\mathscr{A}_n$ and has one local maximum it has indeed solved the minimization problem. More precise results can usually be obtained by extending the reasoning to cover intervals rather than single observations.

One can turn the argument around and ask how large the sample size must be to identify a given peak. Consider a peak of constant height Δ containing k_0 observations and suppose that to the left and right of the peak there are intervals each containing k_1 observations where the signal is zero. If the peak is not detected then there is a function g in the approximation region which is constant on the three intervals. Denoting the value of the constant by Δ the middle interval gives

$$\sqrt{k_0}(\Delta - m) \leq \sqrt{\tau_n \log n} + Z_0 \tag{7.18}$$

with Z_0 a standard normal random variable. The corresponding inequalities on the abutting intervals are

$$\sqrt{k_1}\, m \leq \sqrt{\tau_n \log n} + Z_i, \quad i = \pm 1 \tag{7.19}$$

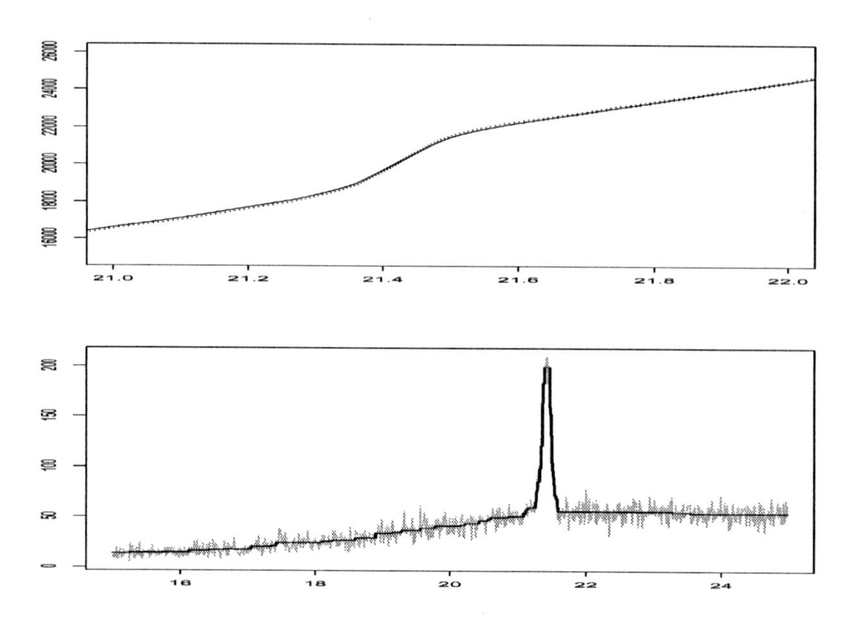

Figure 7.2 *Upper panel: Partial sums of data; lower and upper bounds (dots), the taut string (continuous line)* $(\varepsilon = 50)$ *together with the data.*

from which it follows that the probability of not detecting the peak is at most

$$\Phi\left(\left(\sqrt{\tau_n \log n\left(\frac{1}{k_0}+\frac{1}{k_1}\right)}-\Delta\right)\Big/\sqrt{\frac{1}{k_0}+\frac{1}{2k_1}}\right) \tag{7.20}$$

where Φ is the distribution function of a standard Gaussian random variable.

The taut string can be calculated with a complexity of $O(n)$ which when combined with the squeezing procedure leads to an overall complexity of $O(n \log n)$ (see [45]). The time required for the whole photon data set of 7001 observations is 0.3 seconds

The efficacy of the taut string methodology in controlling the number of peaks was discovered almost accidentally by Heiner Schwarte (see 'contributors()' in R) in the context of spectral densities. He applied it to a data set given in [93] and was surprised to see how all four peaks were recovered without any artifacts in spite of large differences in the orders of magnitude of the peaks. More details are given in [47].

As can be seen in the upper panel of Figure 7.1 the taut string switches from the upper boundary to the lower at a local maximum. At a local minimum it switches from the lower boundary to the upper. The effect is that the taut string underestimates the size of a local maximum and overestimates the size of a local minimum. This can be corrected by replacing the derivative of the taut string by the mean of the

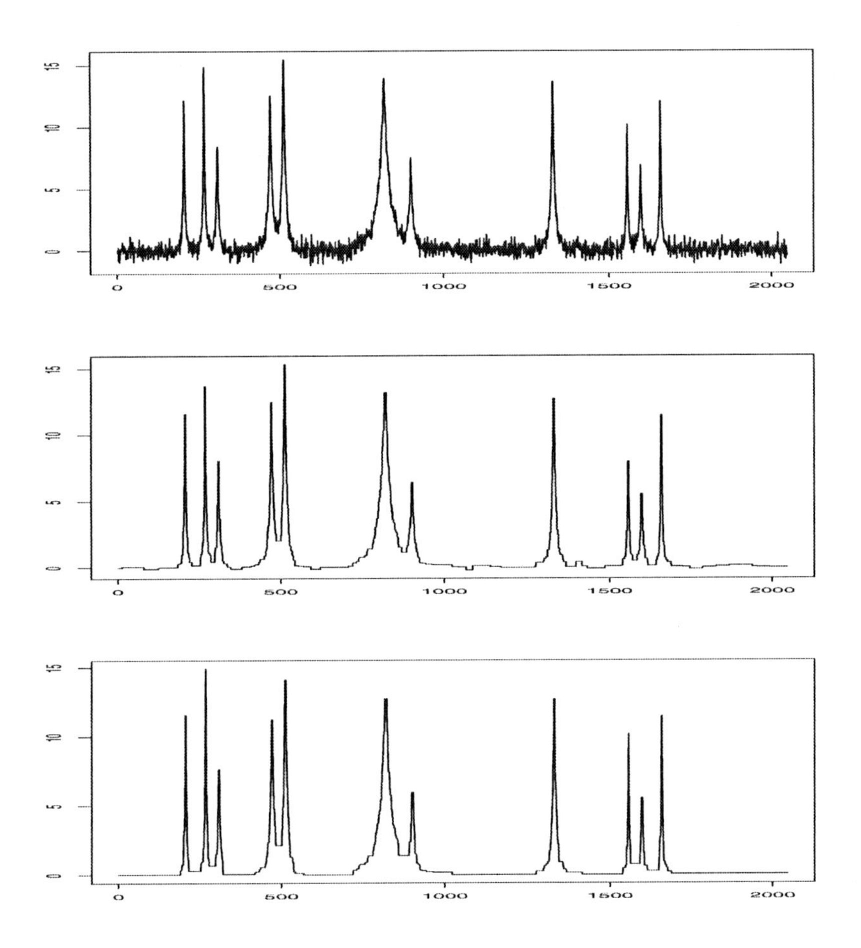

Figure 7.3 *Top panel: the bumps data of [62]. Centre panel: the result of the taut string procedure without local squeezing. Bottom panel: the taut string method with local squeezing.*

observations on the constant section which is a local extreme. This improves the fit without altering the modality. Even this can result in a value for a local maximum which seems to be too small. More elaborate methods such as a quadratic fit can be applied but these are minor details.

The taut string procedure worked well for the data of Figure 2.1 but this is not always the case.The top panel of Figure 7.3 shows the 'bumps' data taken from [62]. The function f used to generate the data has 21 local extreme values. The taut string procedure applied to this data set results in the function shown in the centre panel of Figure 7.3. It has twenty-nine local extreme values.

The reason for the additional local extreme values is the narrowness of the tube. In order to accommodate the peaks the tube must be narrow but then it is too narrow for those sections of the data with no peaks. The problem may be overcome by local squeezing, an idea due to Arne Kovac. Using the multiscale condition it is possible to determine those intervals I for which the inequality fails, that is

$$\frac{1}{\sqrt{|I|}}\left|\sum_{t_i \in I}(x(t_i) - g(t_i))\right| > \sigma_n \sqrt{\tau_n \log(n)}\,.$$

These are the intervals where the fit is unsatisfactory and the tube is squeezed only on these intervals. This is continued until all the multiscale conditions are fulfilled. The resulting tube is shown in the bottom panel of Figure 7.3. It has 21 local extreme values as required. Without local squeezing the taut string method could be classified as interesting but nothing more. Local squeezing turns it into a very effective method for controlling modality.

The taut string method with global squeezing is an asymptotically consistent method for estimating the number of local extreme values. Thus local squeezing is asymptotically of no advantage. That it is much superior for many data sets shows that local squeezing is a finite sample improvement. Although it is intuitively clear why this should be so, there is no proof.

Squeezing the tube with a constant tube width is equivalent to solving the problem

$$\text{argmin}_f\left(\sum_{i=1}^{n}(x(t_i) - f(t_i))^2 + \lambda\sum_{i=2}^{n}|f(t_i) - f(t_{i-1})|\right) \tag{7.21}$$

where λ is related to the width ε of the tube by $\varepsilon = 2\lambda$ ([145], [130]). This may be seen as a discrete version of

$$\text{argmin}_f\left(\sum_{i=1}^{n}(x(t_i) - f(t_i))^2 + \lambda\int_0^1 |f^{(1)}(t)|\,dt\right). \tag{7.22}$$

The total variation $TV(g)$ of a function on the grid $t_1,\ldots,t_n$ is defined by

$$TV(g) = \sum_{i=2}^{n}|g(t_i) - g(t_{i-1})|\,.$$

The continuous version for an absolutely continuous function is

$$TV(g) = \int_0^1 |g^{(1)}(t)|\,dt\,.$$

Thus (7.21) and (7.22) can be written as

$$\text{argmin}_f\left(\sum_{i=1}^{n}(x(t_i) - f(t_i))^2 + \lambda TV(f)\right). \tag{7.23}$$

The taut string with local squeezing is equivalent to

$$\text{argmin}_f \left(\sum_{i=1}^{n} (x(t_i) - f(t_i))^2 + \sum_{i=2}^{n} \lambda_i |f(t_i) - f(t_{i-1})| \right) \tag{7.24}$$

where the λ_i are data driven and related to the width of the tube at t_i ([73]). The continuous version is

$$\text{argmin}_f \left(\sum_{i=1}^{n} (x(t_i) - f(t_i))^2 + \int_0^1 \lambda(t) |f^{(1)}(t)| \, dt \right). \tag{7.25}$$

As in all cases the taut string minimizes the number of local extremes within the tube (see [141]).

7.2.1.1 *Generalizations*

In [73] a much more general version of the taut string idea is presented. The authors consider amongst other related minimization problems the following one:

$$\text{argmin}_{\mathbf{f}} \left(\sum_{i=1}^{n} \rho(X_i - f_i) + \sum_{i=1}^{n-1} \lambda_j |f_{j+1} - f_j| \right) \tag{7.26}$$

where ρ is some convex function. The standard taut string with local squeezing can be regarded as a special case with $\rho(x) = x^2$. The applications include quantile regression, Poisson regression, binary regression and, more generally, exponential families.

7.2.2 *Minimizing Total Variation*

It is seen from (7.24) that there is a close connection between the total variation and the number of local extreme values. This will be exploited below where the linear programming problem

$$\text{minimize } TV(g) \quad \text{subject to} \quad g \in \mathscr{A}_n \tag{7.27}$$

is considered. However there is no theorem which gives conditions under which the solution of (7.27) also solves the problem of minimizing the number of local extreme values. There are examples which show that the solution of (7.27) does not always minimize the number of local extreme values but they are hard to construct.

More generally linear programming problems of the form

$$\text{minimize } TV(g^{(k)}) \quad \text{subject to} \quad g \in \mathscr{A}_n \tag{7.28}$$

will be considered with additional shape constraints on g.

The first panel of Figure 7.4 shows the result of solving (7.27) for the photon data. The second panel and third panel show the results of solving (7.28) for $k = 1$

and $k = 2$ respectively. The case $k = 2$ exhibits additional local extremes caused by smoothing out the sharp corners at either side of the peak. This can be remedied by imposing the monotonicity conditions derived from the solution for $k = 0$. This shown in the third panel. The fourth panel shows the solution for $k = 2$ is in addition to the monotonicity constraints the convexity/concavity restraints obtained from $k = 1$ are imposed.

Higher derivatives are in principle possible but the limit seems to be the fourth. Minimizing $TV(g^{(5)})$ over $\mathscr{A}_n$ runs into numerical problems because of the different orders of magnitude of the of the fifth differences of g and the number of constraints specifying $\mathscr{A}_n$. Minimizing $TV(g^{(4)})$ for sample sizes of 1000 is about the limit in terms of the orders of magnitude and the number of constraints. Larger sets such as the complete photon data set with $n = 7000$ can be treated by decomposing the data into smaller, overlapping sets, solving the problem on each such set with careful splicing of the results. Figure 7.5 shows an excerpt of the photon data set analysed in the this manner. From top to bottom are the data with the smooth monotonicity and convexity/concavity constrained solution followed by the first four derivatives. The chaos just visible in the fourth derivative is due to numerical problems with the linear programming programme.

7.2.2.1 A Note on the Photon Data Set

The photon data are counts of X-ray photons. There are reasons for supposing that the number follows a Poisson distribution to a reasonable degree of accuracy. However this only holds for large data values. For small ones there is additional background noise which causes more variation than would be expected from the Poisson model. This leads to the model

$$X(t) = f(t) + \sigma(t)Z(t), 0 \le t \le 1, \tag{7.29}$$

where $\sigma(t) = \max\{\sigma, f(t)^{1/2}\}$ and σ represents the background noise. The model (7.29) uses the normal approximation for the Poisson distribution but it is sufficient for the photon data. In a first step an adequate function $\tilde{f}$ is calculated with $\sigma(t) = \sigma_n$ of (7.12). This function may have additional side peaks on the large peaks due to the noise level being underestimated. The final adequate function is then calculated with $\sigma(t) = \max\{\sigma, \tilde{f}(t)^{1/2}\}$. This done using the approximation region (7.3) modified in the obvious manner. A second problem is that the heights of the peaks are to be measured from a slowly varying base line which is apparent in the lower panels of Figure 7.4. Once the peaks have been identified they are to be expressed as mixtures of Gauss and other densities. A detailed analysis of such data sets is given in [43].

7.2.3 Piecewise Constant Regression

The following is based on [110] and [44].

There is sometimes interest in functions which are piecewise constant (see [92] and [21]). The obvious form of regularization is to minimize the number of intervals.

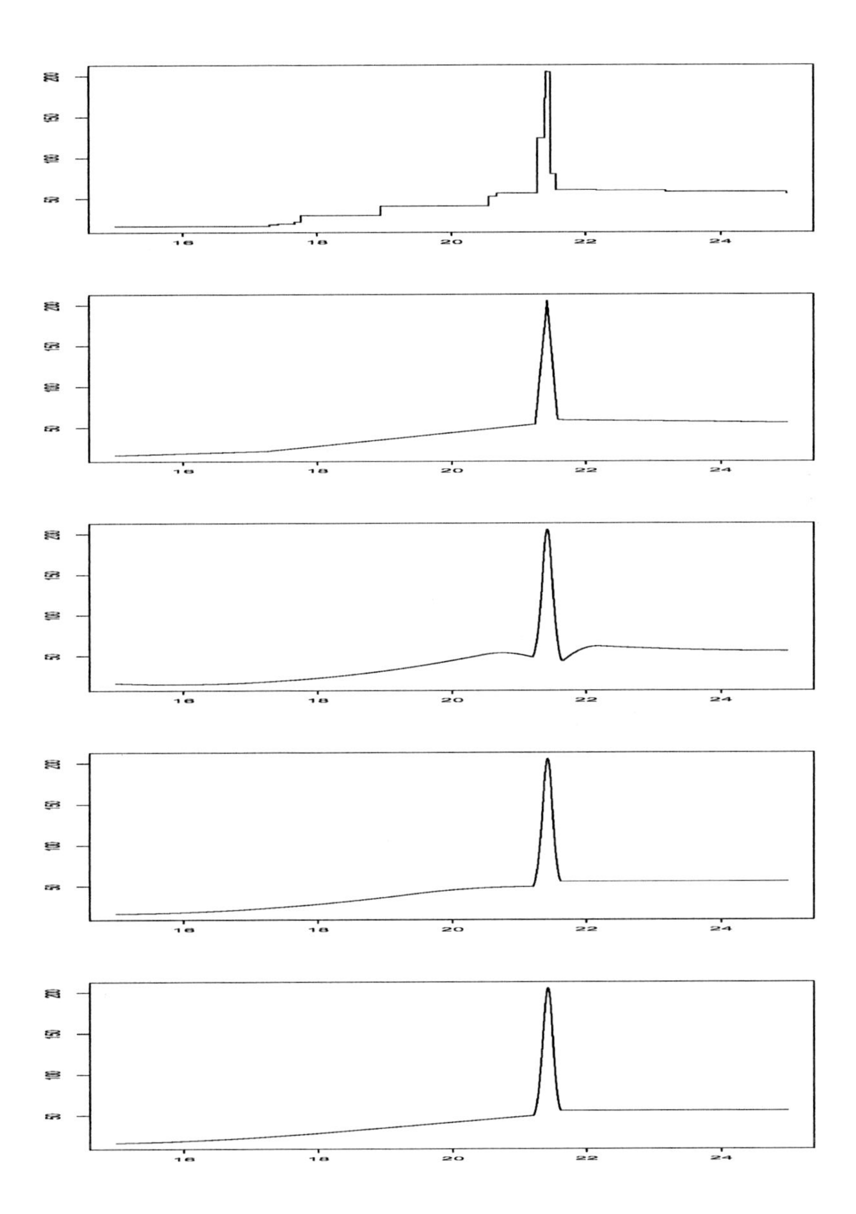

Figure 7.4 *The first 1000 observations of the photon data. From top to bottom: minimizing $TV(g)$; minimizing $TV(g^{(1)})$; minimizing $TV(g^{(2)})$; minimizing $TV(g^{(2)})$ under monotonicity constraints; minimizing $TV(g^{(2)})$ under monotonicity and convexity/concavity constraints.*

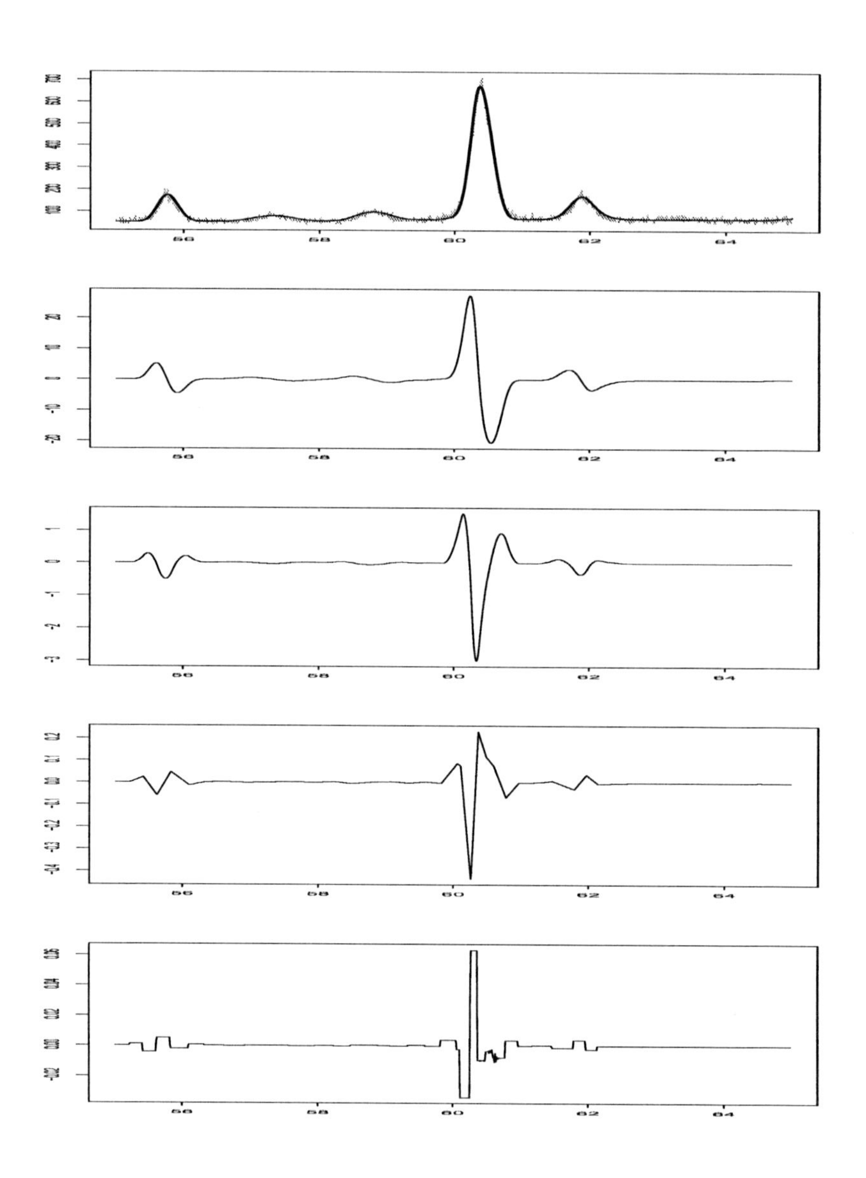

Figure 7.5 *First panel: A section of the photon data together with a smooth monotonicity and convexity/concavity constrained solution. Panels 2-5 give the first four derivatives.*

There is a simple algorithm which solves the problem if the definition of adequacy is changed from one of global approximation as in (7.3) to one of local approximation given by

$$\mathscr{A}_n^*((\mathbf{t},\mathbf{x})_n,\mathscr{I}_n,\sigma,\tau_n) = \left\{g : \max_{I' \in \mathscr{I}_n, I' \subset I} |w((\mathbf{t},\mathbf{x})_n,g,I) \le \sigma\sqrt{\tau_n \log n}\,,\ g \text{ constant on } I\right\}. \quad (7.30)$$

The difference is that the inequalities which define the approximation region are required only to hold on the sets where the function is constant.

Suppose for simplicity $\mathscr{I}_n$ is the set of all intervals and start with the first observation $x(t_1)$. Any function g which takes on the value m_1 at t_1 and lies in the approximation region must satisfy

$$lb(t_1) = x(t_1) - \sigma\sqrt{\tau_n \log n} \le m_1 \le x(t_1) + \sigma\sqrt{\tau_n \log n} = ub(t_1). \quad (7.31)$$

If the function is constant on the the interval $[t_1,t_2]$ then the following additional inequalities must hold:

$$x(t_2) - \sigma\sqrt{\tau_n \log n} \le m_1 \le x(t_2) + \sigma\sqrt{\tau_n \log n} \quad (7.32)$$

and

$$(x(t_1)+x(t_2))/2 - \sigma\sqrt{\frac{\tau_n \log n}{2}} \le m_1 \le (x(t_1)+x(t_2))/2 + \sigma\sqrt{\frac{\tau_n \log n}{2}}. \quad (7.33)$$

Taking the maximum of all the left hand inequalities and the minimum of all the right hand inequalities gives a lower bound $lb(t_2)$ and an upper bound $ub(t_2)$ for m_1. Continuing in the this manner yields an increasing sequence of lower bounds $lb(t_i)$ and a decreasing sequence of upper bounds $ub(t_i)$ with

$$lb(t_i) \le m_1 \le ub(t_i) \quad (7.34)$$

where $g(t) = m_1$ for t in $[t_1,t_i]$ and lies in the approximation region. If for some k, $lb(t_k) > ub(t_k)$ then no function g in the approximation region can be constant on $[t_1,t_k]$. At the latest a new interval of constancy must start at this point. This is a simple adaption of Algorithm 0 of [44]. Figure 7.6 shows the lower and upper bounds for a simulated data set. They cross at the observation $i = 344$.

The algorithm gives the minimum number of intervals for functions in the adequacy region. It does not however specify the values of any such function on its intervals of constancy. In the present context it would seem natural that this should be the mean. This requires

$$lb(t_i) \le \frac{1}{i}\sum_{j=1}^{i} x(t_j) \le ub(t_i) \quad (7.35)$$

and the problem is now how to minimize the number of intervals subject to this additional restriction. This problem can be solved by dynamic programming (see

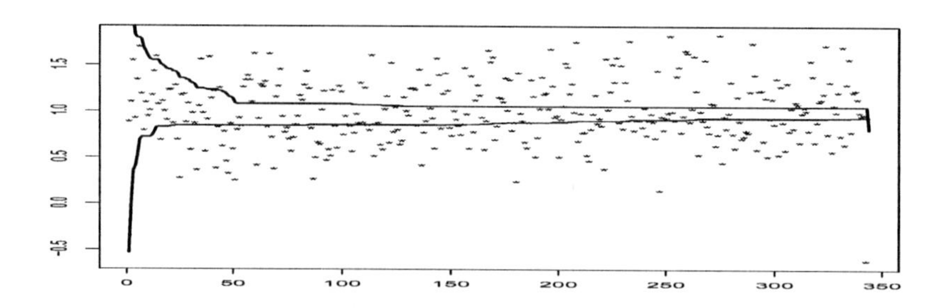

Figure 7.6 *Bounds for the value of a function on an interval of constancy.*

[92]). It was adapted to the present situation in [110]. Even if this is done there may be several solutions. In general a unique solution can be obtained by minimizing $\sum_{i=1}^{n}(x(t_i)-g(t_i))^2$ where g has the minimum number of intervals of constancy. Again this can be solved by dynamic programming.

Requiring (7.35) rather than just (7.34) is a stronger constraint and will therefore tend to increase the number of intervals required. For data generated under the model it will only rarely be the case that more intervals are required as the uniform bounds for the mean are essentially those for the approximation region. For real data however there may be a difference. The owner of the data will then have to decide what is more important, the smallest number of intervals or requiring (7.35). It is also possible to take the minimum number of intervals and then use the mean. The resulting function may then not lie in the approximation region but be sufficiently close for all practical purposes.

A similar approach to piecewise constant regression is given in [89] (written independently of [110] and [44]) where the authors consider change point inference in the context of exponential families using a likelihood based approach.

7.3 Rates of Convergence and Approximation Bands

This section is based on [50]. Rates of convergence are calculated on test beds where the model holds. Here are some typical results.

7.3.1 Rates of Convergence

Consider a function f which has a continuous second derivative and exactly one local maximum at the point t_0 in an open interval I_0 and for which

$$-c_2 \le f^{(2)}(t) \le -c_1 < 0, \quad t \in I_0\,. \tag{7.36}$$

Let f_n^* be a function in $\mathscr{A}_n$ which minimizes the number of local extremes. Let t_n^* be any point for which $f_n^*(t_n^*) \ge f_n^*(t), t \in I_0$. Then

$$|t_n^* - t_0| = \mathrm{O}_f\left(\left(\frac{\log n}{n}\right)^{1/5}\right), \tag{7.37}$$

$$f_n^*(t_n^*) \ge f(t_0) - \mathrm{O}_f\left(\left(\frac{\log n}{n}\right)^{2/5}\right), \tag{7.38}$$

$$f_n^*(t_n^*) \le f(t_0) + \sigma(\sqrt{3\log n} + 2.4). \tag{7.39}$$

An explicit upper bound for the constants in O_f in terms of c_1 and c_2 of (7.36) can be calculated.

If f is monotone on an open interval I_0 with a continuous first derivative then for any f_n^* in $\mathscr{A}_n$ which minimizes the number of local extremes

$$|f(t) - f_n^*(t)| \le 3^{4/3}\sigma^{2/3}|f^{(1)}(t)|^{1/3}\left(\frac{\log n}{n}\right)^{1/3}. \tag{7.40}$$

for any $t \in I_0$. Furthermore if $f^{(1)}(t) = 0$ on I_0, that is f is constant on I_0 then

$$|f(t) - f_n^*(t)| \le \frac{3^{1/2}\sigma}{\min\{\sqrt{t - t_l}, \sqrt{t_r - t}\}}\left(\frac{\log n}{n}\right)^{1/2}. \tag{7.41}$$

This shows that f_n^* adapts automatically to certain shape constraints on f.

The same argument shows that if

$$|f(t) - f(s)| \le L|t - s|^{\beta}$$

with $0 < \beta \le 1$ then

$$|f(t) - f_n^*(t)| \le cL^{1/(2\beta+1)}(\sigma/\beta)^{2\beta/(2\beta+1)}(\log n/n)^{\beta/(2\beta+1)} \tag{7.42}$$

where

$$c \le (2\beta + 1)3^{\beta/(2\beta+1)}\left(\frac{1}{\beta + 1}\right)^{1/(2\beta+1)} \le 4.327.$$

Apart from the value of c this corresponds to Theorem 2.2 of [76] and it is seen that an optimally rate of convergence with explicit constants has been obtained. There are corresponding results if f is convex or concave on an interval I_0.

Rates of convergence for pointwise constant regression can also be obtained. The situation is much simpler, the arguments are correspondingly simpler and the results more precise. Assume the design points are equidistant on the interval $[0,1]$ and consider the accuracy with which the location of a jump of size Δ can be detected. Suppose that there are two intervals of length γ to the left and to the right of the jump and that the jump is not detected. Adopting the reasoning which leads to (7.20) gives

$$\sqrt{n\gamma}\Delta \le 2\sqrt{\tau_n \log n} + Z_1 + Z_2 \tag{7.43}$$

where Z_1 and Z_2 are i.i.d. standard Gaussian random variables. The probability of detecting the jump with an accuracy of γ is therefore at least

$$\Phi\left(\Delta\sqrt{\frac{n\gamma}{2}} - \sqrt{2\tau_n \log n}\right). \tag{7.44}$$

This gives a rate of $O(\log n/(\Delta^2 n))$ which can be compared with (7.37) for the precision with which smooth peaks can be localized.

7.3.2 Approximation Bands

7.3.2.1 Shape Restriction

Confidence or approximations bands based on the signs of residuals are given in [36] and [45] and more sophisticated ones in [68, 69, 70, 72] for shaped restricted functions.

Suppose for simplicity that the generating function f is monotone on $[0,1]$. An approximation band for monotone functions in $\mathscr{A}_n$ is given by $[lb_n(t_i), ub_n(t_i)]$ where

$$lb_n(t_i) = \min\{g(t_i) : g \in \mathscr{M}^+ \cap \mathscr{A}_n\}, \tag{7.45}$$

$$ub_n(t_i) = \max\{g(t_i) : g \in \mathscr{M}^+ \cap \mathscr{A}_n\}. \tag{7.46}$$

and $\mathscr{M}^+$ denotes the set of monotone functions on $[0,1]$.

The calculation of $lb_n(t_i)$ and $ub_n(t_i)$ requires solving a linear programming problem and, although this can be done, it is practically impossible for larger sample sizes because of exorbitantly long calculation times. If the family of intervals $\mathscr{I}_n$ is restricted to a wavelet multiresolution scheme ($\lambda = 2$ in (7.4)) then samples of size $n = 1000$ can be handled. Fast honest bounds can be attained as follows. If $g \in \mathscr{M}^+ \cap \mathscr{A}_n$ then for any i and k with $i + k \le n$

$$\sqrt{k+1}\, g(t_i) \ge \frac{1}{\sqrt{k+1}} \sum_{j=0}^{k} x_n(t_{i-j}) - \sigma\sqrt{\tau_n \log n}.$$

From this the lower bound

$$lb_n(t_i) = \max_{0 \le k \le i-1} \left(\frac{1}{k+1}\sum_{j=0}^{k} x_n(t_{i-j}) - \sigma\sqrt{\frac{\tau_n \log n}{k+1}}\right) \tag{7.47}$$

and the corresponding upper bound

$$ub_n(t_i) = \min_{0 \le k \le n-i} \left(\frac{1}{k+1}\sum_{j=0}^{k} x_n(t_{i+j}) + \sigma\sqrt{\frac{\tau_n \log n}{k+1}}\right). \tag{7.48}$$

follow. These bounds are of algorithmic complexity $O(n^2)$. Even faster bounds can be obtained by putting

$$lb_n(t_i) = \max_{0 \le \theta(k) \le i-1} \left(\frac{1}{\theta(k)+1} \sum_{j=0}^{\theta(k)} x_n(t_{i-j}) - \sigma \sqrt{\frac{\tau_n \log n}{\theta(k)+1}} \right) \quad (7.49)$$

$$ub_n(t_i) = \min_{0 \le \theta(k) \le n-i} \left(\frac{1}{\theta(k)+1} \sum_{j=0}^{\theta(k)} x_n(t_{i+j}) + \sigma \sqrt{\frac{\tau_n \log n}{\theta(k)+1}} \right) \quad (7.50)$$

where $\theta(k) = \lfloor \theta^k - 1 \rfloor$ for some $\theta > 1$. These latter bounds are of algorithmic complexity $O(n \log n)$. The fast bounds are not necessarily non–decreasing but can be made so by putting

$$\begin{aligned} ub_n(t_i) &= \min(ub_n(t_i), ub_n(t_{i+1})), \quad i = n-1, \ldots, 1, \\ lb_n(t_i) &= \max(lb_n(t_i), lb_n(t_{i-1})), \quad i = 2, \ldots, n. \end{aligned}$$

The upper panel of Figure 7.7 shows data generated by

$$X(t) = \exp(5t) + 5Z(t) \quad (7.51)$$

evaluated on the grid $t_i = i/1000, i = 1, \ldots, 100$ together with the three lower and three upper bounds with σ replaced by σ_n of (7.12) and $\tau_n = 3$. The lower bounds are those given by (7.45) with $\mathscr{I}_n$ a dyadic multiresolution scheme, (7.47) and finally (7.49) with $\theta = 2$. The times required for were about 12 hours, 19 seconds and less than one second respectively with corresponding times for the upper bounds (7.46), (7.48) and (7.50). The differences between the bounds are small. The methods of Section 7.3 can be applied to show that all the uniform bounds are optimal in terms of rates of convergence.

Convexity and concavity can be treated similarly but with somewhat more effort especially for the lower bound for a convex function (see also [68, 69, 70, 72]). The fastest bands for the upper band corresponding to (7.50) has an algorithmic complexity of $O(n \log n)$, the lower bound corresponding to (7.49) one of $O(n(\log n)^2)$.

The lower panel of Figure 7.7 shows the same data as in the upper panel but with the fast convexity bounds. The calculation of the lower bound took about five seconds. As the function is both increasing and convex the minimum of the upper and the maximum of the lower bounds gives an honest band for non-decreasing convex functions.

The ideas can be extended to functions which are piecewise monotone and piecewise concave/convex. In the case of a piecewise monotone function possible positions of the local extremes can in theory be determined by solving the appropriate linear programming problems. The taut string methodology is however extremely good and very fast so that in practice it can be used to identify possible positions of the local extremes. The approximation bounds depend on the exact location of the local extreme values. Given the intervals of constancy of the taut string solution which correspond to the local extreme values it is possible to calculate the approximation bounds for any function which has its local extreme values at points in these intervals. The result for the function $f(t) = \sin(\pi t)$ is shown in the top panel of Figure 7.8. The

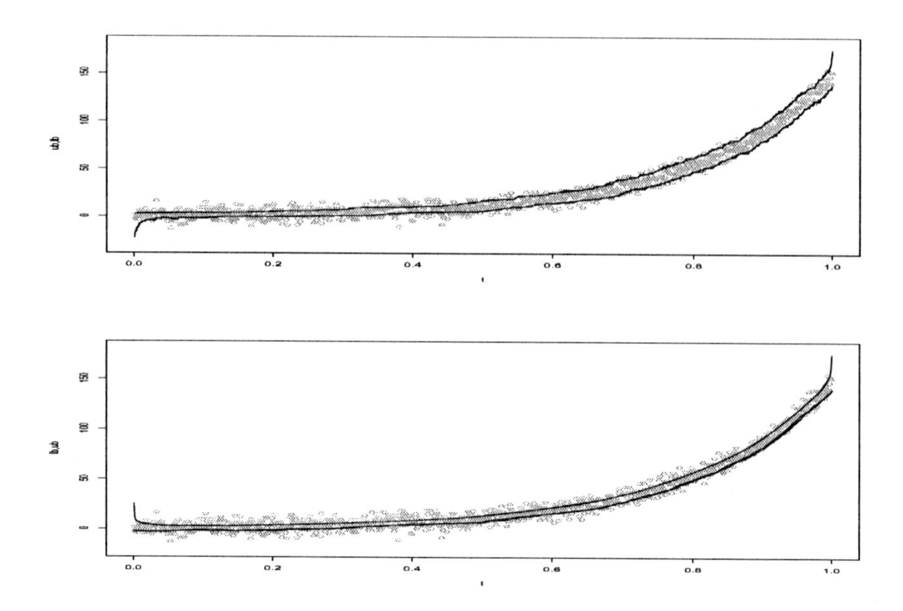

Figure 7.7 *Upper panel: monotonic bounds. Lower panel: convex bounds.*

bounds are those given by (7.49) and (7.50) with $\theta = 1.5$. If the mid-points of the taut string intervals are used as default choices for the positions of the local extreme values the calculations are much simplified. Figure 7.8 shows the result for the function $f(t) = \sin(4\pi t)$. The user can of course specify the locations of the local extreme values. The programme will check if these are consistent with the linear constraints which define the approximation region $\mathscr{A}_n$ and, if so, calculate the bounds.

For functions which are piecewise concave/convex the intervals of concavity and convexity can be determined by minimizing the total variation of the first derivative to give a piecewise linear function $\hat{f}_n$. The default choice of the switch is the midpoint of the interval belongs to a concave and a convex section of $\hat{f}_n$. The upper panel of Figure 7.9 shows the result for convexity/concavity which corresponds to Figure 7.8. The lower panel of Figure 7.9 shows the result of imposing both monotonicity and convexity/concavity constraints.

7.3.2.2 *Bands Based on Smoothness*

Approximation bands based on smoothness are possible but are much more difficult to justify. Honest bands are attainable only if f is restricted to a set $\mathscr{F}_n = \{g : \|g^{(2)}\|_\infty \le K\}$ with a specified K. A lower bound is given by

$$lb(i/n) \le \min_k \left(\frac{1}{2k+1} \sum_{j=-k}^{k} x((i+j)/n) + \left(\frac{k}{n}\right)^2 K + \sigma\sqrt{\frac{3\log n}{2k+1}} \right) \tag{7.52}$$

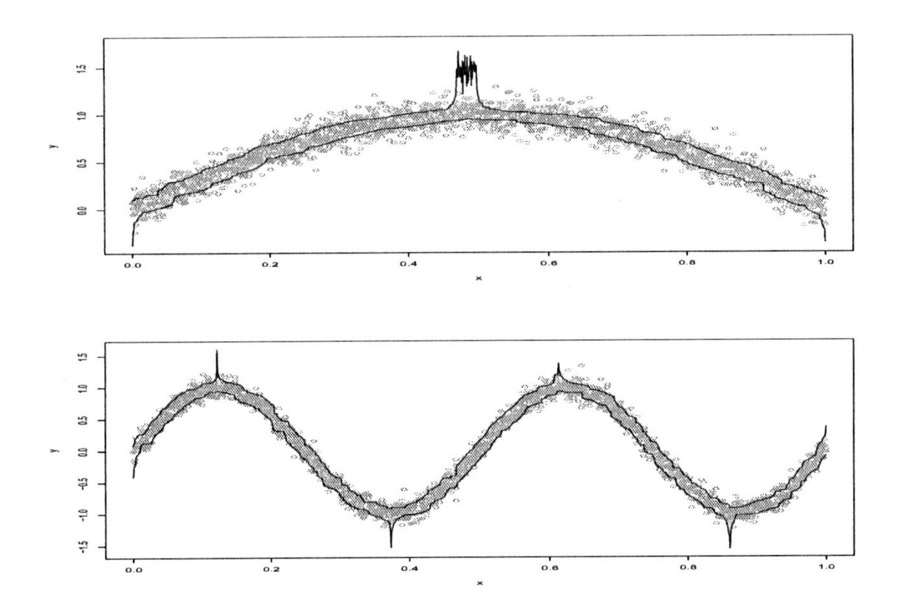

Figure 7.8 *Approximation bounds without (upper panel) and with (lower panel) the specification of the precise positions of the local extreme values. The positions in the lower panel are the default choices obtained from the taut string reconstruction ([131]). The bounds are the fast bounds (7.49) and (7.50) with* $\theta = 1.5$.

and an upper bound by

$$ub(i/n) \geq \max_k \left(\frac{1}{2k+1} \sum_{j=-k}^{k} x((i+j)/n) - \left(\frac{k}{n}\right)^2 K - \sigma \sqrt{\frac{3 \log n}{2k+1}} \right). \qquad (7.53)$$

Minimizing $\|g^{(2)}\|_\infty$ over $\mathscr{A}_n$ results in a unique function $\tilde{f}_n$ with $\|\tilde{f}_n^{(2)}\|_\infty = K_{\min}$. Bands based on this value of K are then degenerate as is shown in the upper panel of Figure 7.10. The function f used to generate the data is $f(t) = \sin(4\pi t)$ with $\|f^{(2)}\|_\infty = 16\pi^2 = 157.9$. The value of K_0 is 117.7. Non-degenerate bands require specifying a value of $K > K_0$ and it is not clear how this can be done except for rare cases where there are scientific reasons which give a numerical bound. The lower panel of Figure 7.10 shows the bands based on $\|g^{(2)}\|_\infty \leq K$ for

$$K = 137.8(= (117.7 + 157.9)/2),\ 157.9 \quad \text{and} \quad 315.8(= 2 \times 157.9).$$

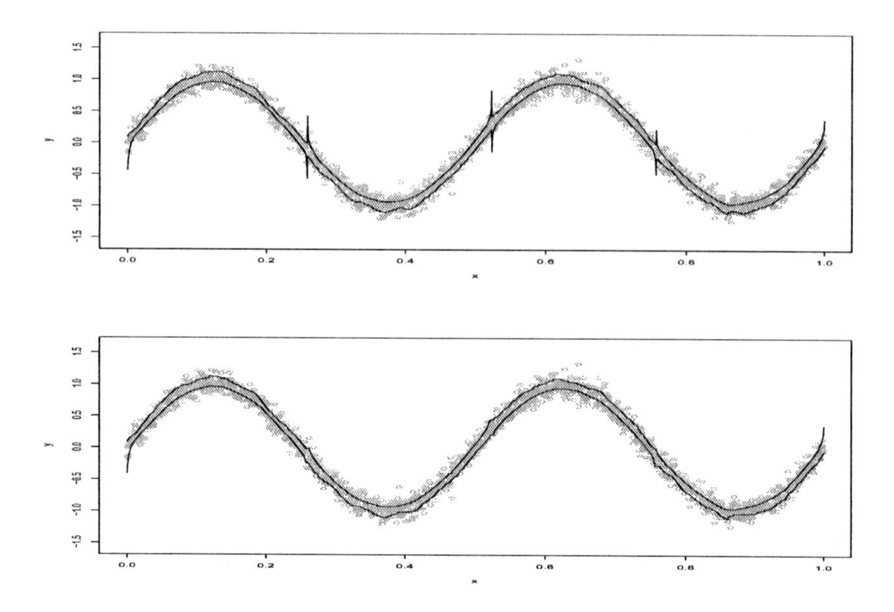

Figure 7.9 *Upper panel: Approximation bounds with default choices for the intervals of convexity/concavity. Lower panel: combined approximation bounds for monotonicity and concavity/convexity.*

7.3.2.3 Bands For Piecewise Constant Regression

Bands for the regression function can be obtained from the bands shown in Figure 7.6. These can be calculated going from left to right for the whole intervals and the break points give upper bounds to the positions of the jumps. Going from right to left gives lower bounds. Given now any point in the interval it is a simple matter to deduce bounds for the function at this point. It is only necessary to distinguish points in intervals where there are no jumps from points in intervals where there is a jump. Figure 7.11 shows a simulated data set with bounds (dashed) and the function used to generate the data (solid).

7.4 Choosing Smoothing Parameters

Solutions of the optimization problem (7.15) may prove too difficult to obtain or be too expensive in terms of computing time. Often however it is possible to give a sequence of functions $g_j, j = 1, 2, \ldots$ can be calculated quickly and are of increasing complexity. The idea is to choose the first g_j which lies in the approximation region $\mathscr{A}_n$. An example is the taut string procedure: the tube is squeezed giving a sequence

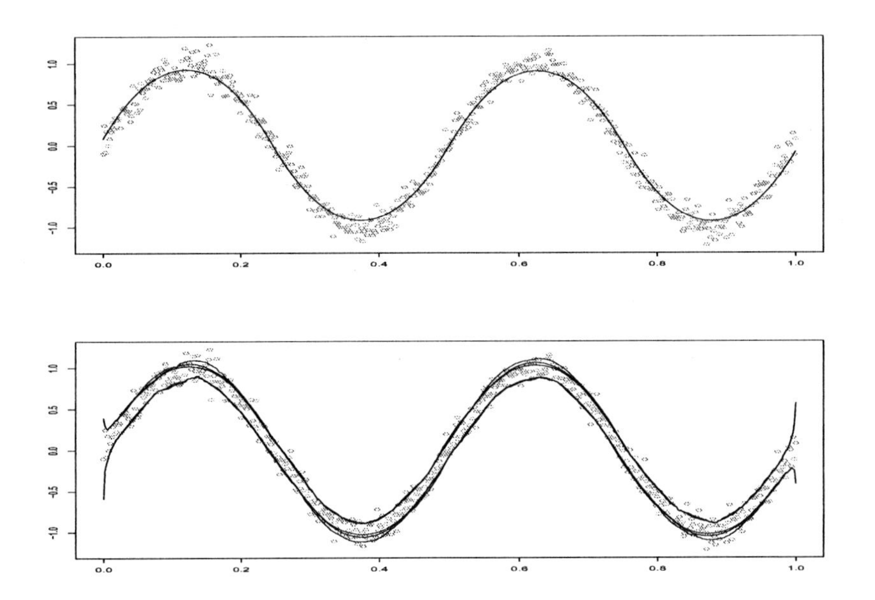

Figure 7.10 *Smoothness approximation bounds for* $g \in \mathscr{A}_n$ *with* $\|g_n^{(2)}\|_\infty \leq K$ *for data generated according to (7.1) with* $f(t) = \sin(4\pi t)$, $\sigma = 0.2$ *and* $n = 500$. *Upper panel: degenerate band for* $K = K_{min} = 117.7$. *Lower panel: bands for* $K = 137.8$, 157.9 *and* 315.8.

of functions and the process is terminated when the function lies in the approximation region.

The idea can be extended to many methods in non-parametric regression which involve smoothing or penalty parameters whose purpose is to regularize the problem. Examples are the bandwidth h in kernel methods

$$f_h(t) = \frac{\sum_{i=1}^n x(t_i) K\left(\frac{t-t_i}{h}\right)}{\sum_{i=1}^n K\left(\frac{t-t_i}{h}\right)}$$

and the penalty parameter λ in least penalized squares

$$f_\lambda = \operatorname{argmin}_f \left(\sum_{i=1}^n (x(t_i) - f(t_i))^2 + \lambda \int_0^1 f^{(2)}(t)^2 \, dt \right). \tag{7.54}$$

One starts with parameter values which yield smooth solutions and then alters the values either globally, or on intervals where the multiscale conditions are violated, until the multiscale conditions are fulfilled.

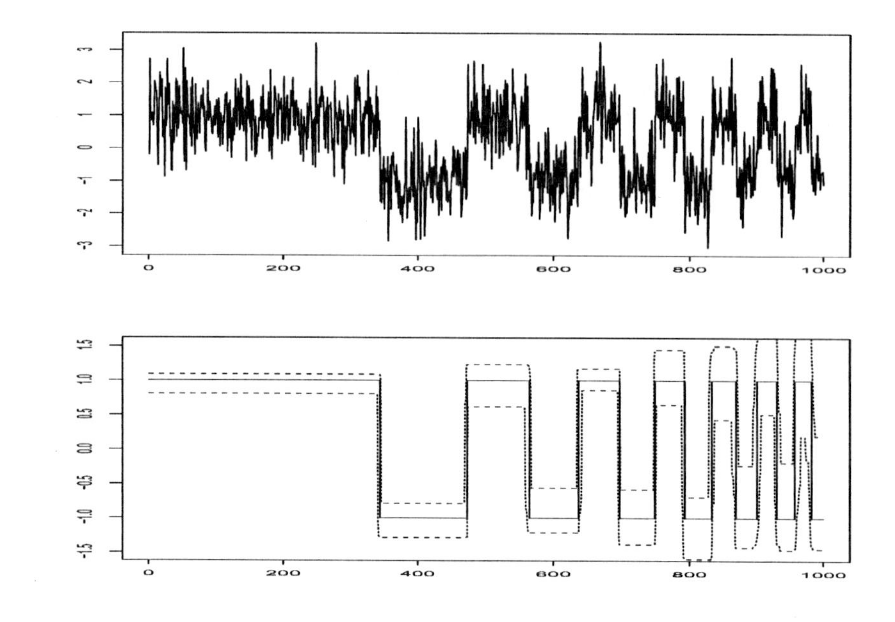

Figure 7.11 *Upper panel: the data. Lower panel: lower and upper bounds (dashed) and the generating function (solid).*

7.4.1 *Global Bandwidth*

This is the simplest situation. Initially a large global bandwith is used and then gradually decreases until the resulting function lies in the adequacy region. For the first thousand observations of the photon data the resulting bandwidth is $h = 0.062$. The corresponding function is shown in Figure 2.3.

7.4.2 *Local Bandwidth*

This section is based on [153]. Given a bandwidth function h the approximating function $\hat{f}_h$ is given by

$$\hat{f}_h(t) = \frac{\sum_{i=1}^{n} x(t_i) K\left(\frac{t-t_i}{h(t)}\right)}{\sum_{i=1}^{n} K\left(\frac{t-t_i}{h(t)}\right)}.$$

In practice the function will only be evaluated at the design points t_i and hence it is only necessary to consider the bandwidths $h(t_i) := h_i$ at these points. In all there are now n possible values instead of the single value for the global bandwidth. The strategy of local squeezing for the taut string can be adapted to this case: reduce the bandwidths h_i at all points t_i which lie in intervals I where the multiresolution condition is not fulfilled. Initially $h_1 = \ldots = h_n$ all large and then h_i is reduced by

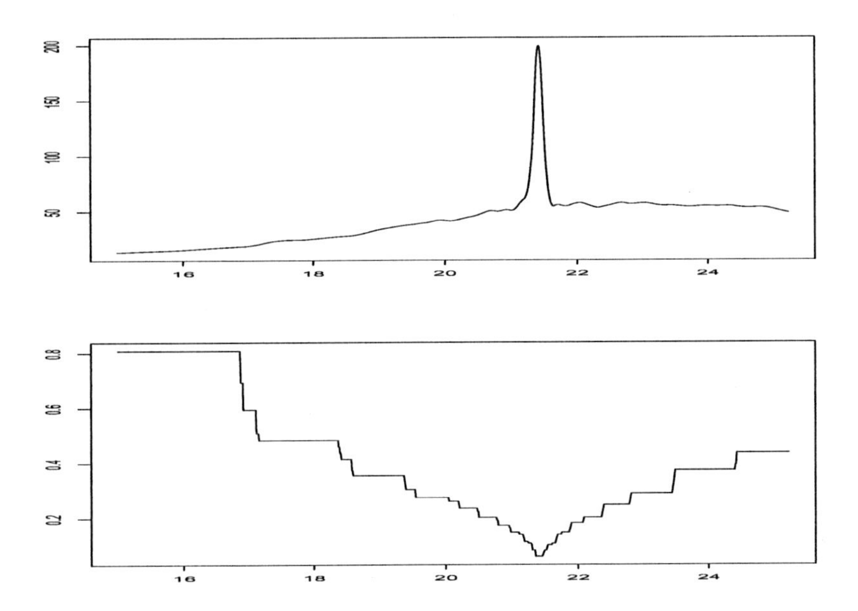

Figure 7.12 *Upper panel: approximation for the data of Figure 2.1 using local bandwidths and the taut string strategy for choosing them. Lower panel: the resulting bandwidths.*

a constant factor $q = 0.95$ if t_i lies in an interval where the multiresolution condition is not fulfilled. The upper panel of Figure 7.12 shows the result for the data of Figure 2.1. The lower panel shows the bandwidths. One problem with the taut string strategy for choosing bandwidths is that a large peak in an interval can cause the multiresolution condition to fail. The strategy then reduces the bandwidths at all points in this interval although this may not be necessary. The result is that the function oscillates more than is necessary. This can be avoided to some extent by first reducing the bandwidths in intervals I for which the absolute value of

$$w((\mathbf{t},\mathbf{x})_n, \hat{f}_h, I) = \frac{1}{\sqrt{|I|}} \sum_{t_i \in I} (x(t_i) - \hat{f}_h(t_i))$$

is large. This strategy results in Figure 7.13. The bandwidths obtained using the second strategy are much larger than those using the taut string strategy as can be seen by comparing the lower panels of Figures 7.12 and 7.13. The function in the upper panel of Figure 7.13 oscillates less than the function of the upper panel of Figure 7.12. It does however exhibit discontinuities which are caused by the discontinuities of the bandwidth function h. These can be reduced somewhat by smoothing the bandwidth function but the result remains unsatisfactory. Moreover the computational cost is quite high.

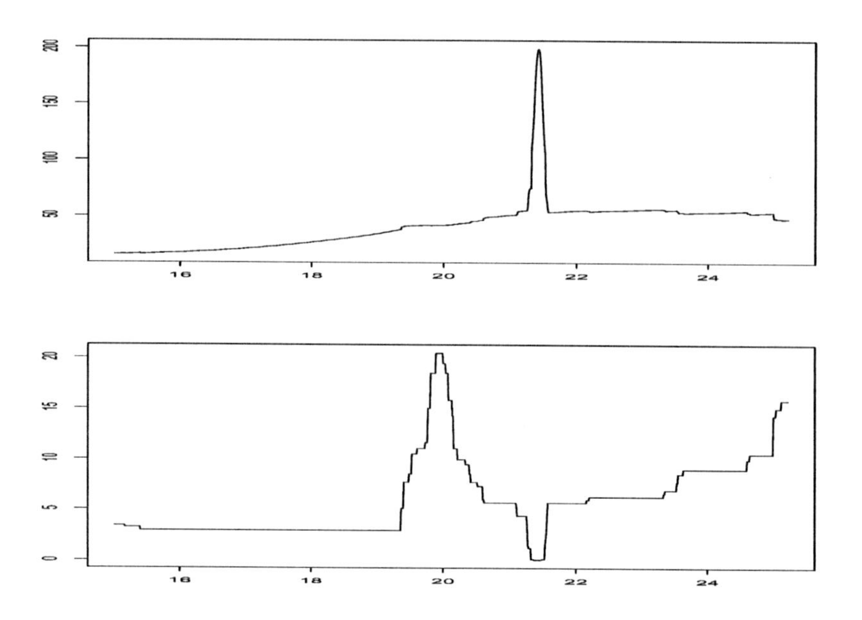

Figure 7.13 *Upper panel: approximation for the data of Figure 2.1 using local bandwidths and the second strategy for choosing them. Lower panel: the resulting bandwidths.*

7.4.3 Smoothing Splines 1

The following is based on [154] and [51]. The standard smoothing spline problem of (7.54) is replaced by the weighted spline problem

$$f_{\mathbf{w}} = \operatorname{argmin}_f \left(\sum_{i=1}^{n} w(i)(x(t_i) - f(t_i))^2 + \int_0^1 f^{(2)}(t)^2\, dt \right) \tag{7.55}$$

where $\mathbf{w} = (w_1, \ldots, w_n)$ is now the smoothing parameter. For equally spaced design points. The finite sample version may be written as

$$f_{\mathbf{w}} = \operatorname{argmin}_f \left((\mathbf{x} - \mathbf{f})^t \mathbf{W} (\mathbf{x} - \mathbf{f}) + \mathbf{f}^t \mathbf{D}^t \mathbf{D} \mathbf{f} \right) \tag{7.56}$$

where $\mathbf{f} = (f(t_1, \ldots, f(t_n))$, $\mathbf{W}$ is the diagonal matrix with diagonal elements w_i and $\mathbf{D}$ is the difference matrix

$$\mathbf{D} = \begin{pmatrix} -1 & 1 & 0 & 0 & 0 & \cdot & \cdot & \cdot & 0 \\ 1 & -2 & 1 & 0 & 0 & \cdot & \cdot & \cdot & 0 \\ 0 & 1 & -2 & 1 & 0 & \cdot & \cdot & \cdot & 0 \\ 0 & \cdot & \cdot & \cdot & \cdot & \cdot & \cdot & \cdot & 0 \\ 0 & \cdot & \cdot & \cdot & \cdot & \cdot & \cdot & \cdot & 0 \\ 0 & \cdot & \cdot & \cdot & \cdot & \cdot & \cdot & \cdot & 0 \\ 0 & \cdot & \cdot & \cdot & 0 & 1 & -2 & 1 & 0 \\ 0 & \cdot & \cdot & \cdot & 0 & 0 & 1 & -2 & 1 \\ 0 & \cdot & \cdot & \cdot & 0 & 0 & 0 & 1 & -1 \end{pmatrix}. \tag{7.57}$$

The solution is

$$\mathbf{f_w} = (\mathbf{W} + \mathbf{D}^t\mathbf{D})^{-1}\mathbf{W}\mathbf{x}. \tag{7.58}$$

Starting with very small weights w_i they can be adjusted using either of the two strategies outlined above but this time the weights w_i are increased by a fixed factor if t_i lies in an interval where the multiresolution criterion does not hold. If n is large the solution cannot be calculated directly from (7.58) but it can be computed iteratively using the conjugate gradient method and making use of the fact that the matrix $\mathbf{W} + \mathbf{D}^t\mathbf{D}$ is sparse. The time for the whole data set was 18.2 seconds.

Figure 7.14 shows the results for the same section of the photon data as Figure 7.5.

7.4.4 *Smoothing Splines 2*

The second method for smoothing splines is based on the formulation (7.25) of the local squeezing procedure for the taut string. The problem is to solve

$$f_\lambda = \operatorname{argmin}_f \left(\sum_{i=1}^{n} (x(t_i) - f(t_i))^2 + \int_0^1 \lambda(t) f^{(2)}(t)^2 \, dt \right) \tag{7.59}$$

for a given weight function λ. The finite sample version may be written as

$$f_\lambda = \operatorname{argmin}_f \left(\|x - f\|_2^2 + f^t D^t \Lambda D f \right) \tag{7.60}$$

and the solution is given by

$$f_\lambda = (I + D^t \Lambda D)^{-1} x. \tag{7.61}$$

The starting weights λ_i are taken to be very large. The weight λ_i is decreased by a fixed factor if t_i lies in an interval where the multiresolution criterion does not hold. As the term λ_i appears also with the terms $f(t_{i-2}), f(t_{i-1}), f(t_i), f(t_{i+1})$ and $f(t_{i+2})$ all the weights λ_j for $j = i-2, i-1, i, i+1, i+2$ should be decreased. The matrix $I + D^t \Lambda D$ is a banded matrix with bandwidth $m = 5$ so the solution (7.60) can be calculated very quickly. Figure 7.14 shows the results for the same section of the photon data as Figures 7.5 and 7.14. For a completely different approach on the smoothing spline method see [171].

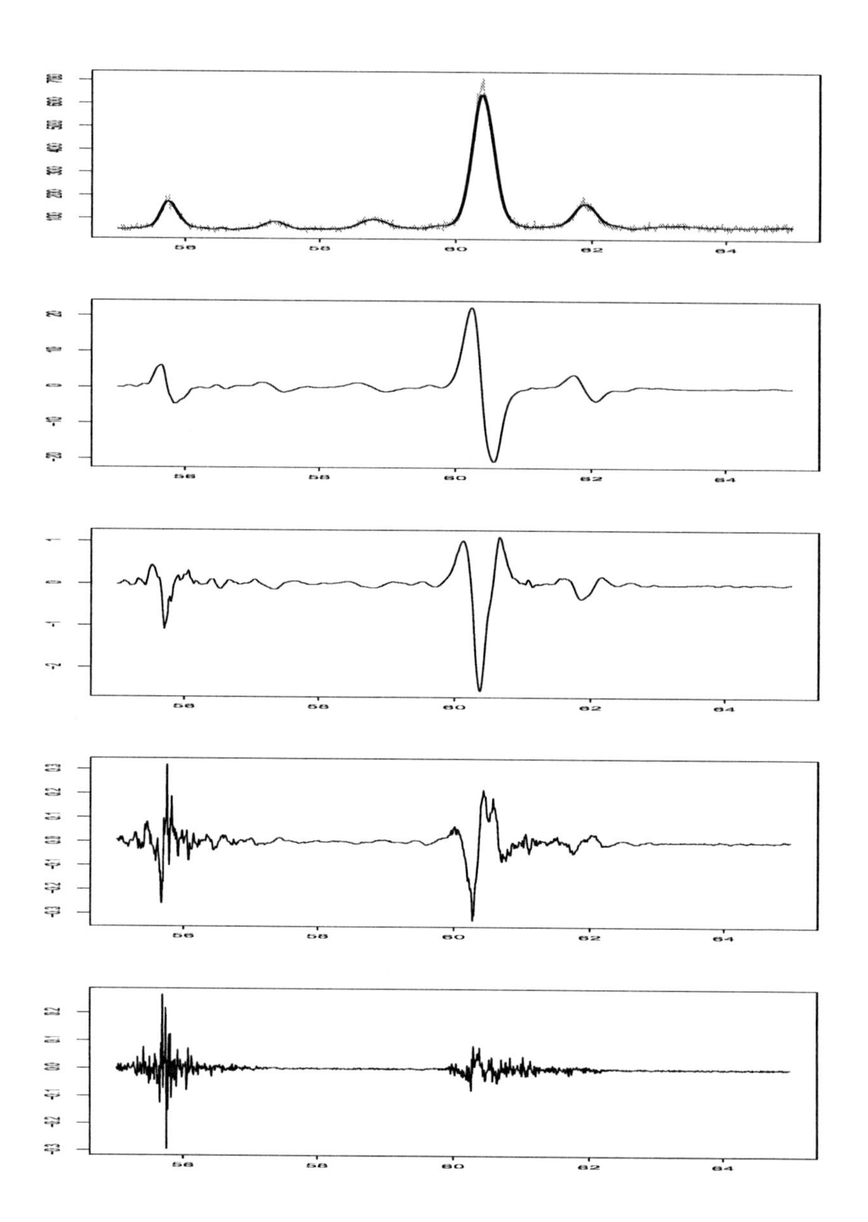

Figure 7.14 *Top panel: excerpt 55-65 degrees of the reconstruction of the photon data. The remaining panels from top to bottom show the first four derivatives.*

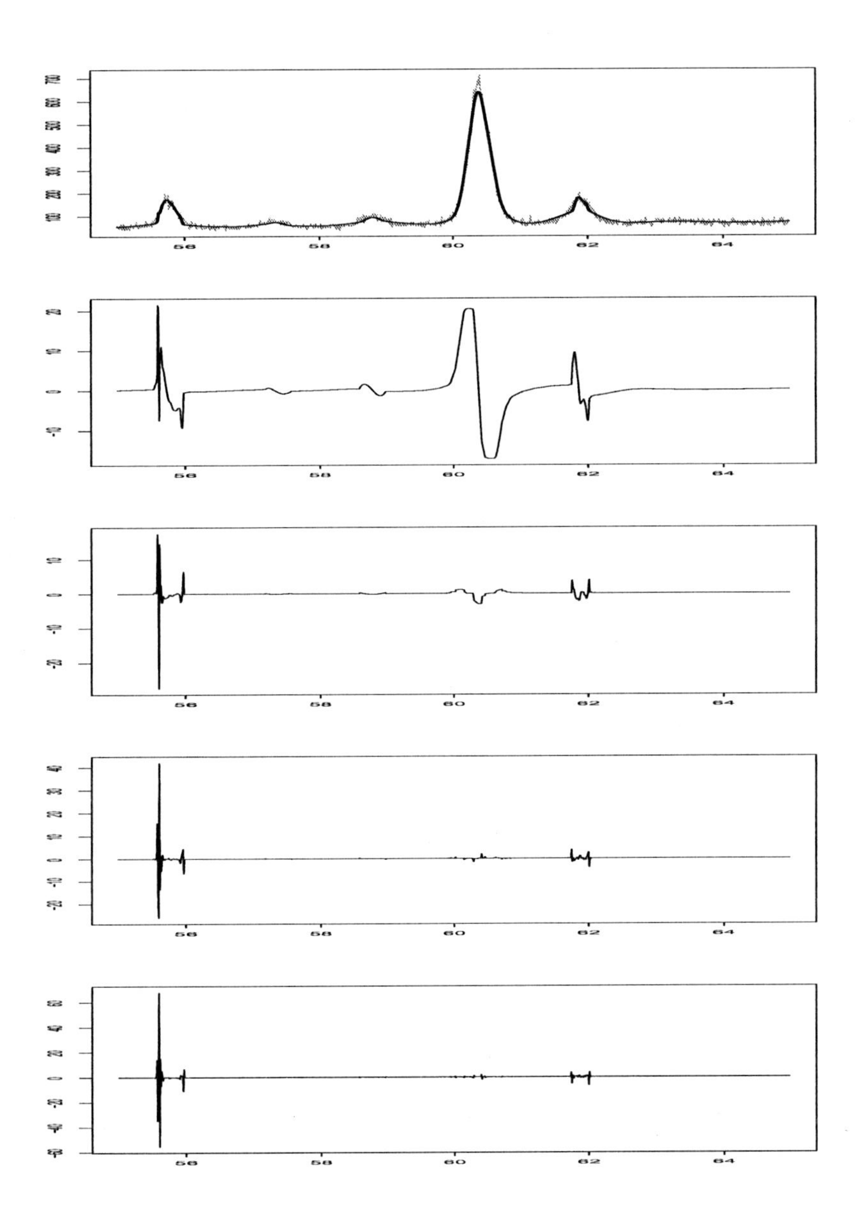

Figure 7.15 *Top panel: excerpt 55-65 degrees of the reconstruction of the photon data. The remaining panels from top to bottom show the first four derivatives.*

7.4.5 Diffusion

The following us based on [195]. A kernel approximation $\hat{f}_h$ with global bandwidth h and Gaussian kernel

$$K(x) = \frac{1}{\sqrt{2\pi}} \exp\left(-\frac{1}{2}x^2\right)$$

is given by

$$\hat{f}_h(t) = \frac{\sum_{i=1}^n x(t_i)\exp\left(-\frac{1}{2}\left(\frac{t-t_i}{h}\right)^2\right)}{\sum_{i=1}^n \exp\left(-\frac{1}{2}\left(\frac{t-t_i}{h}\right)^2\right)}.$$

On letting n tend to infinity with $t_i = i/n$ the expression becomes

$$\hat{f}_h(t) = \frac{\int_{-\infty}^{\infty} x(s)\exp\left(-\frac{1}{2}\left(\frac{t-s}{h}\right)^2\right) ds}{\int_{-\infty}^{\infty} \exp\left(-\frac{1}{2}\left(\frac{t-s}{h}\right)^2\right) ds} = \frac{1}{\sqrt{2\pi h^2}} \int_{-\infty}^{\infty} x(s)\exp\left(-\frac{1}{2}\left(\frac{t-s}{h}\right)^2\right) ds.$$

On replacing h by $\tau = 2h^2$ this becomes

$$\hat{f}_\tau(t) = \frac{1}{\sqrt{\pi\tau}} \int_{-\infty}^{\infty} x(s)\exp\left(-\frac{(t-s)^2}{\tau}\right) ds.$$

At this point a change of notation is necessary to make it compatible with the standard notation for the heat equation. The function $x(t)$ will be replaced by $y(x)$ so that $y(x)$ now represents the data at the point $x \in [0,1]$. The variable τ is replaced by t to represent time. The result is

$$\hat{f}(x,t) = \hat{f}_t(x) = \frac{1}{\sqrt{\pi t}} \int_{-\infty}^{\infty} y(u)\exp\left(-\frac{(x-u)^2}{t}\right) du.$$

The function $\hat{f}$ satisfies the heat equation

$$\frac{\partial \hat{f}(x,t)}{\partial t} = \frac{\partial^2 \hat{f}(x,t)}{\partial^2 x} \tag{7.62}$$

and the initial condition

$$\hat{f}(x,0) = y(x).$$

The interpretation is the following. The temperature at time $t = 0$ at the point x is $y(x)$; at time t the temperature is $\hat{f}(x,t)$. As $t \to \infty$ the temperature tends to zero. This assumes that the thermal conductivity or diffusivity is constant over $(-\infty,\infty)$. In nonparametric regression the solution is to be constrained to the interval $[0,1]$. This can be done by replacing (7.62) by

$$\frac{\partial \hat{f}(x,t)}{\partial t} = \alpha(x)\frac{\partial^2 \hat{f}(x,t)}{\partial^2 x} \tag{7.63}$$

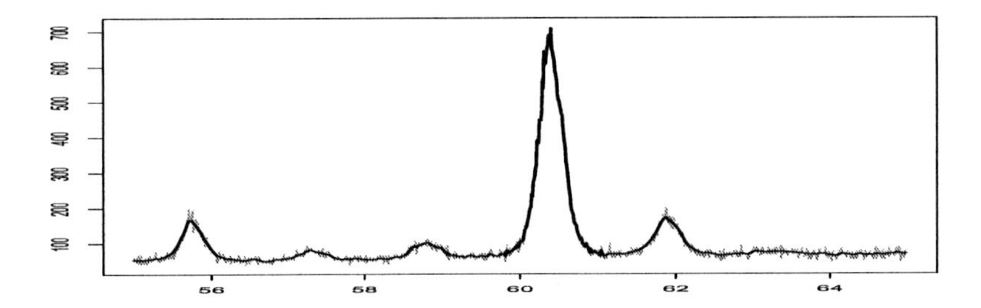

Figure 7.16 *Diffusion approximation to the photon data.*

where the thermal conductivity $\alpha(x)$ is now explicitly included. If $\alpha(x) = 0$ for $x \leq 0$ and $x \geq 1$ no heat escapes from the interval $[0,1]$ and consequently, assuming $\alpha(x) > 0$ for $x \in (0,1)$,

$$\lim_{t\to\infty} \hat{f}(x,t) = \begin{cases} \int_0^1 y(u)\,du, & 0 \leq x \leq 1, \\ 0, & \text{otherwise.} \end{cases}$$

The adaption to nonparametric regression is as follows. The diffusion is stopped at time $t = 1$ so that $\hat{f}(x,1)$ represents the approximating function. Just as for the smoothing spline of Section 7.4.4 the calculation of $\hat{f}(x,1)$ can be done very quickly exploiting the banded nature of the iteration scheme. The time required for the complete diffraction data set was 0.157 seconds. Initially the thermal conductivity $\alpha(x)$ is constant and large so that at $t = 1$ the equilibrium distribution has almost been attained. The multiresolution conditions are then used to detect those points interval I in which the approximation is not satisfactory. For points x in such intervals the thermal conductivity $\alpha(x)$ is reduced by a fixed proportion $q\alpha(x)$ for some $q, 0 < q < 1$. At such points the the temperature changes more slowly. This is continued until a satisfactory approximation is attained.

Figure 7.16 shows the diffusion approximation to the same section of the photon data as the top panels of Figures 7.14 and 7.15.

7.4.6 *Penalizing the Number of Intervals*

In [21] a different approach to the problem of piecewise constant regression. The following problem

$$\min_{g\in\mathscr{C}} \left(\sum_i^n (x(t_i) - g(t_i))^2 + \lambda \ell_0(g) \right) \tag{7.64}$$

was considered where $\mathscr{C}$ is the space of piecewise constant functions, $\ell_0(g)$ is the number of intervals of constancy of g and λ is the smoothing parameter. It was shown that if λ is taken to be the largest value such that the corresponding $g_{n,\lambda} \in \mathscr{A}_n$ then $g_{n,\lambda}$ is a consistent estimate of f for data generated as $X(t) = f(t) + \sigma Z(t)$ with

$f \in \mathscr{C}$ and $\ell_0(f)$ finite. In this case the authors used the global approximation region (7.3) rather than the local one of (7.30).

7.5 Joint Approximation of Two or More Samples

The following discussion is based on [48]. It will be mainly restricted to the case of two samples as the ideas can be carried over to three or more samples with no difficulty. Although the topic is somewhat neglected it does have its own special points of interest.

Given two samples $(\mathbf{t},\mathbf{x})_{1n_1}$ and $(\mathbf{t},\mathbf{x})_{2n_2}$ the problem is to decide whether both samples can be adequately approximated by some common function g. The approximation regions for the two data sets are

$$\mathscr{A}_{1n_1}((\mathbf{t},\mathbf{x}_1)_{n_1},\mathscr{I}_{n_1},\sigma_1,\tau_{n_1}) = \left\{g : \max_{I\in\mathscr{I}_{n_1}} |w((\mathbf{t},\mathbf{x}_1)_{n_1},g,I)| \leq \sigma_1\sqrt{\tau_{n_1}\log n_1}\right\}$$

and

$$\mathscr{A}_{2n_2}((\mathbf{t},\mathbf{x}_2)_{n_2},\mathscr{I}_{n_2},\sigma_2,\tau_{n_2}) = \left\{g : \max_{I\in\mathscr{I}_{n_2}} |w((\mathbf{t},\mathbf{x}_2)_{n_2},g,I)| \leq \sigma_2\sqrt{\tau_{n_2}\log n_2}\right\}$$

where $w((\mathbf{t},\mathbf{x})_n,g,I)$ is as given in (7.2). The values of $\tau_{n_i}, i = 1,2$ or, equivalently, that of α may be adjusted to take into account that there are two samples. The approximation regions $\mathscr{A}_{1n_1}$ and $\mathscr{A}_{1n_2}$ can be replaced by $\mathscr{A}^*_{1n_1}$ and $\mathscr{A}^*_{1n_2}$ corresponding to the approximation region $\mathscr{A}^*_n$ of (7.10).

A joint approximating function exists if and only if $\mathscr{A}_{1n_1} \cap \mathscr{A}_{2n_2} \neq \emptyset$. In principle this can be decided by linear programming but for large data sets it may not be practically possible. A simple method is the following, although it is not guaranteed to provide the correct answer. It applies to k samples independently of the supports. For a point $t_i \in \mathbf{t}_n = \cup_{j=1}^k \mathbf{t}_{jn_j}$ define the function g as the linear interpolation of the function defined at t_i by

$$g(t_i) = \frac{\sum_{j=1}^k \sum_{t\in\mathbf{t}_{jn_j},t=t_i} x_{jn_j}(t)/\sigma_j^2}{\sum_{j=1}^k \sum_{t\in\mathbf{t}_{jn_j},t=t_i} 1/\sigma_j^2}. \tag{7.65}$$

If the supports are disjoint the function g simply linearly interpolates the data and all residuals are zero. If $g \in \mathscr{A}_{jn_j}, j = 1,\ldots,k$ then there is at least one joint approximating function, namely g. If $g \notin \mathscr{A}_{jn_j}$ for some j the conclusion is that no joint approximation exists although this has not actually been shown.

Given that that $\mathscr{A}_{1n_1} \cap \mathscr{A}_{2n_2}$ is not empty it contains all the functions which are adequate approximations for the two data sets. All questions concerning joint approximating functions can be answered by reference to it. This is a functional version of the approach to the one-way table given in Chapter 6.1. In particular it is possible to regularize within $\mathscr{A}_{1n_1} \cap \mathscr{A}_{2n_2}$ to exhibit, for example, a joint approximation with the smallest number of local extremes. The taut string methodology can often be adapted to this case as will be shown below.

The discussion below will be restricted to the two extreme cases where the supports $\mathbf{t}_{1n_1}$ and $\mathbf{t}_{2n_2}$ are either equal or disjoint.

7.5.1 Equal Supports

If the supports of the two data sets are the same $n = n_1 = n_2$ and $\mathbf{t}_{1n_1} = \mathbf{t}_{2n_2}$, then under the model the function $f = f_1 = f_2$ disappears on taking the differences

$$X_1(t_j) - X_2(t_j) = \sigma_1 Z_1(t_j) - \sigma_2 Z_2(t_j) \stackrel{D}{=} \mathfrak{N}(0, \sigma_1^2 + \sigma_2^2).$$

One possibility is simply to check the multiscale conditions for the differences using the default values σ_{1n} and σ_{2n} given by (7.12):

$$\frac{1}{\sqrt{I}} \left| \sum_{i \in I} (x_1(t_i) - x_2(t_i)) \right| \le \sqrt{\sigma_{1n}^2 + \sigma_{2n}^2} \sqrt{\tau_n \log n}, \; I \in \mathscr{I}_n. \tag{7.66}$$

As it stands this is not satisfactory as no such approximating function has been exhibited. However if (7.66) holds there always is a joint approximating function $\tilde{f}_n$, namely that given by (7.65)

$$\tilde{f}_n(t_i) = \left(\frac{x_1(t_i)}{\sigma_{2n}^2} + \frac{x_2(t_i)}{\sigma_{1n}^2} \right) \Big/ \left(\frac{1}{\sigma_{1n}^2} + \frac{1}{\sigma_{2n}^2} \right). \tag{7.67}$$

It can be checked that if (7.66) holds then $\tilde{f}_n \in \mathscr{A}_{1n} \cap \mathscr{A}_{2n}$. Nevertheless given that there is a joint approximating function there will in general be an interest in regularizing, for example minimizing the number of local extreme values. This can be done by treating $\tilde{f}_n(t_i)$ as a third sample and applying the taut string methodology to it checking the multiscale constraints for each sample individually. The top panel of Figure 7.17 shows two data sets generating according to according to (7.1) with

$$f(t) = \sin(10\pi t), \; \sigma_1 = 0.1, \; \sigma_2 = 1, \; n = 500$$

and the design points uniform on $[0, 1]$. The centre panel of Figure 7.17 shows the data set $\tilde{f}_n(t_i)$: the bottom panel gives the taut string approximation with the minimum number of local extreme values.

Just as it is possible to quantify the sample size required to detect a local extreme as in (7.19) it is possible to quantify the sample size required to detect a given difference between the functions f_1 and f_2. The theorem is formulated in terms of σ_1 and σ_2 rather than the default version for data.

Theorem 7.5.1 *Let $\mathbf{X}_{1n}$ and $\mathbf{X}_{2n}$ be generated by (7.1) with $\mathbf{t}_{1n_1} = \mathbf{t}_{2n_2} = \mathbf{t}_n$ and with noise levels σ_1 and σ_2 respectively. Further let $\mathscr{I}_n$ be the set of all intervals. Suppose $f_1(t) > f_2(t) + \eta_n$ on an interval I_n containing $|I_n|$ support points of t and set*

$$\zeta_n = \left(\max \left\{ 0, \eta_n \sqrt{|I_n|} - (\sigma_1 + \sigma_2) \sqrt{\tau_n \log n} \right\} \right) / \sqrt{\sigma_1^2 + \sigma_2^2}. \tag{7.68}$$

Then with probability at least $2\Phi(\zeta_n) - 1$ there will be no joint approximation for the data sets $\mathbf{X}_{1n}$ and $\mathbf{X}_{2n}$.

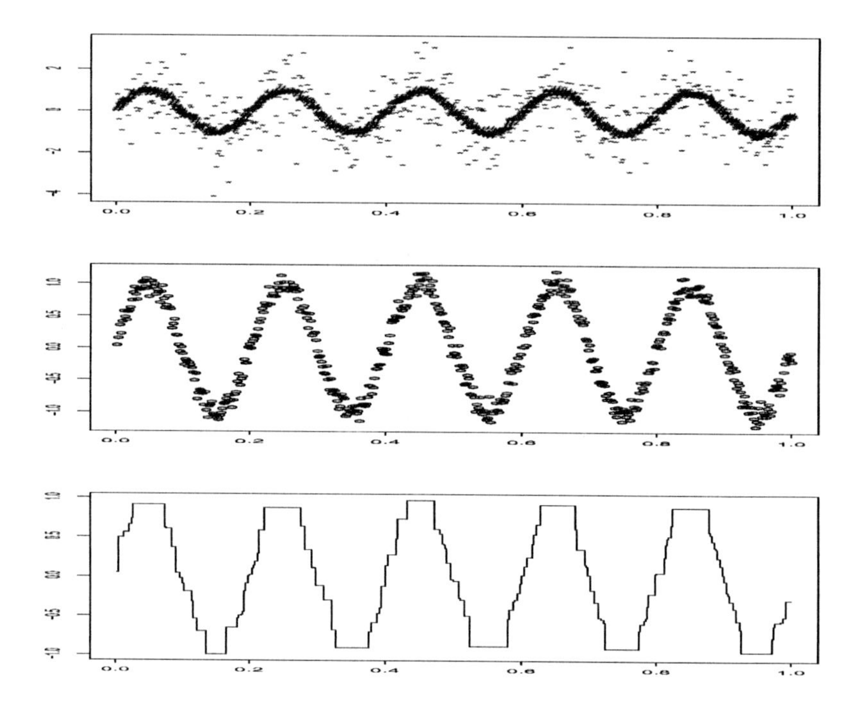

Figure 7.17 *Top panel: the two data sets. Centre panel: the constructed data set of (7.67). Bottom panel: a joint approximating function.*

The proof is given in [48].

The same analysis can be carried through for the approximation region $\mathscr{A}_n^*$ of (7.10). In the theorem ζ_n is replaced by

$$\zeta_n^* = \frac{\max\left\{0, \eta_n\sqrt{|I_n|} - (\sigma_1+\sigma_2)\sqrt{2\log(n/|I_n|) + \gamma_n(\alpha)\log\log(e^e n/|I_n|)}\right\}}{\sqrt{\sigma_1^2+\sigma_2^2}}. \tag{7.69}$$

If $\limsup |I_n|/n > 0$ then Theorem 7.5.1 gives a rate of convergence of $O(\sqrt{\log n/n})$ whereas (7.69) implies a rate of $O(1/\sqrt{n})$. The extra $\sqrt{\log n}$ term reflects the cost of detecting differences on small intervals.

Finally if $\mathscr{I}_n(\lambda)$ of (7.4) is used, then (7.68) and (7.69) are to be replaced by

$$\zeta_n(\lambda) = \left(\max\left\{0, \eta_n\sqrt{|I_n|/\lambda} - (\sigma_1+\sigma_2)\sqrt{\tau_n\log(n)}\right\}\right)/\sqrt{\sigma_1^2+\sigma_2^2} \tag{7.70}$$

and

$$\zeta_n^*(\lambda) = \frac{\max\left\{0, \eta_n\sqrt{|I_n|/\lambda} - (\sigma_1+\sigma_2)\sqrt{2\log(n\lambda/|I_n|) + \gamma_n(\alpha)\log\log(e^e n\lambda/|I_n|)}\right\}}{\sqrt{\sigma_1^2+\sigma_2^2}} \tag{7.71}$$

respectively.

7.5.1.1 A Comparison with Two Other Procedures

Three other procedures for testing the equality of regression functions are [53], [86] and [165]. In the case of equal supports $\mathbf{t}_{1n_1} = \mathbf{t}_{2n_2}$ which are moreover uniform over [0, 1], that is $\mathbf{t}_n = \{1/n, 2/n, \ldots, 1\}$, the tests proposed by [53] and [165] are the same. The test statistic is

$$T_n = \sqrt{n}\max_{1\le j\le n}|R(j)|/s_n^* = \max_{1\le j\le n}\left|\sum_{i=1}^{j}(Y_1(t_i) - Y_2(t_i))\right|/(\sigma_n\sqrt{n}) \tag{7.72}$$

where σ_n is some quantifier of the noise. Under the null hypothesis $f_1 = f_2 = f$ the distribution of T_n does not depend on f. Under H_0 the distribution of T_n converges to that of $\max_{0\le t\le 1}|B(t)|$ where B is a standard Brownian motion. The 0.95-quantile is approximately 2.24 which leads to rejection of H_0 if

$$T_n \ge 2.24. \tag{7.73}$$

To analyse the behaviour of the test consider the situation of Theorem 7.5.1 and suppose that $f_1(t) = f_2(t)$ apart from t in an interval I_n of length δ_n where $f_1(t) \ge f_2(t) + \eta_n$. It follows from (7.73) that H_0 will be rejected with high probability if

$$\delta_n\eta_n \ge 4.48\sigma/\sqrt{n} \tag{7.74}$$

where $\sigma^2 = \sigma_1^2 + \sigma_2^2$. As already noted if $\delta_n = 1$ deviations of the order of $O(\sigma/\sqrt{n})$ can be picked up. This contrasts contrasts with the $O(\sigma\sqrt{\log(n)/n})$ obtainable from Theorem 7.5.1. If $\mathscr{A}_n^*$ is used instead of $\mathscr{A}_n$ then it follows from (7.69) that deviations of order $O(\sigma/\sqrt{n})$ will eventually be detected. If however $\delta_n = 1/\sqrt{n}$ then it follows from (7.74) that the test statistic T_n will pick up deviations of the order of σ. In this situation the methods based on $\mathscr{A}_n$ and $\mathscr{A}_n^*$ will both pick up deviations of the order of $O(\sigma\sqrt{\log(n)/\sqrt{n}})$.

The test due to [86] is also applicable in this situation. It is based on the Fourier transform of the data sets denoted by by $\tilde{\mathbf{X}}_{1n}$ and $\tilde{\mathbf{X}}_{2n}$. If these are ordered as described in [86] the test statistic reduces to

$$T_n^* = \max_{1\le m\le n}\left|\frac{1}{\sqrt{m}}\sum_{i=1}^{m}((\tilde{X}_{2n}(i) - \tilde{X}_{1n}(i))^2/\tilde{\sigma}_n^2 - 1)\right| \tag{7.75}$$

where $\tilde{\sigma}_n$ is some estimate of the standard deviation of $\tilde{\mathbf{X}}_{2n} - \tilde{\mathbf{X}}_{1n}$. For data generated under the model (7.1) the critical value of T_n^* can be obtained by simulations. It is not as simple to determine the size of the deviations which can be detected by the test (7.75) as the test statistic is a function of the Fourier transforms. Differences in the functions must consequently be translated into differences in the Fourier transforms. The first member of the sum in (7.75) is the difference of the means and this is given the largest weight.

In the following simulation study $n = 500$ and samples are generated of the form

$$X_{1n_1}(i/n) = Z_1(i/n), \quad X_{2n_2}(i/n) = g(i/n) + Z_2(i/n) \quad i = 1, \ldots, n = 500 \quad (7.76)$$

where the $Z_j(i/n)$ are i.i.d $N(0,1)$ random variables and the function g is one of the following

$$\left.\begin{array}{lcl} g_1(t) & = & \eta\{0 \le t \le 1\}, \\ g_2(t) & = & \eta\{0.432 \le t \le 0.570\}, \\ g_3(t) & = & \eta\{0.492 \le t \le 0.510\}, \\ g_4(t) & = & \eta(\{0.462 \le t \le 0.540\} - \{0.662 \le t \le 0.740\}) \end{array}\right\}. \quad (7.77)$$

The four procedures, Delgado–Neumeyer–Dette, Fan–Lin and those based on $\mathscr{A}_n$ and $\mathscr{A}_n^*$ were calibrated to give tests of size 0.05 for testing $g \equiv 0$. The critical values for Delgado–Neumeyer–Dette and Fan–Lin tests are 2.22 and 6.97 respectively. The value of τ_n for the test based on $\mathscr{A}_n$ is 1.46 and the corresponding value of γ_n for that based on $\mathscr{A}_n^*$ is 0.66.

Figure 7.18 shows the power of the tests for different values of η. The panels give from top to bottom the results for the four functions $g_1, ..., g_4$ of (7.77). The lines are as follows: Delgado–Neumeyer–Dette dotted, the Fan–Lin dotdash, $\mathscr{A}_n$ solid and $\mathscr{A}_n^*$ dashed. The results confirm the analysis given above. The Delgado–Neumeyer–Dette and Fan–Lin tests are better for large intervals of difference represented by g_1. For smaller intervals represented by g_2 and g_3 the procedures based on $\mathscr{A}_n^*$ and $\mathscr{A}_n$ are better, especially if the mean difference is zero as represented by g_4.

7.5.1.2 An Application

An example with some real data uses two samples of the thin-film diffraction an example of which is shown in Figure 2.1. Two extracts from such data sets are shown in the top panel of Figure 7.19; the differences $x_{1n_1}(t_i) - x_{2n_2}(t_i)$ are shown in the bottom panel. Each data set is composed of 4806 measurements and the design points are the same. The samples differ in the manner in which the thin film was prepared. One of the questions to be answered is whether the results of the two methods are substantially different. The noise levels for the data sets are the same, namely 8.317, which is explainable by the fact that the data are integer valued. The differences between the two data sets are mainly concentrated on two intervals each containing about 40 observations.

The following is expressed in the language of test theory rather than approximation to enable a comparison with the tests of Delgado–Neumeyer–Dette and Fan–Lin. The estimate (7.74) suggests that the differences will have to be of the order of 92

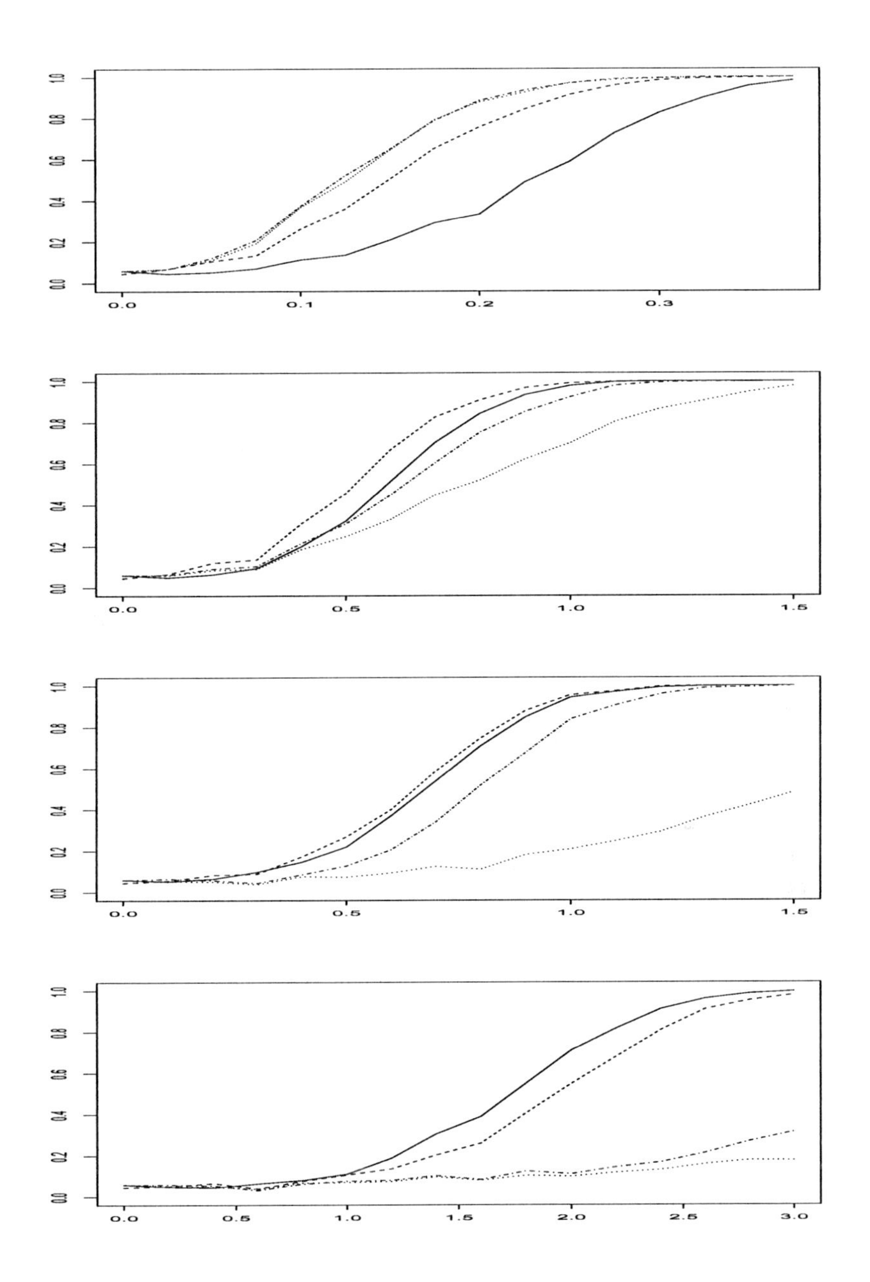

Figure 7.18 *From top to bottom the power functions of the four tests with* $g = g_1$ *to* $g = g_4$ *of (7.77) in order with Delgado–Neumeyer–Dette dotted, Fan–Lin dotdash,* $\mathscr{A}_n$ *solid and* $\mathscr{A}_n^*$ *dashed.*

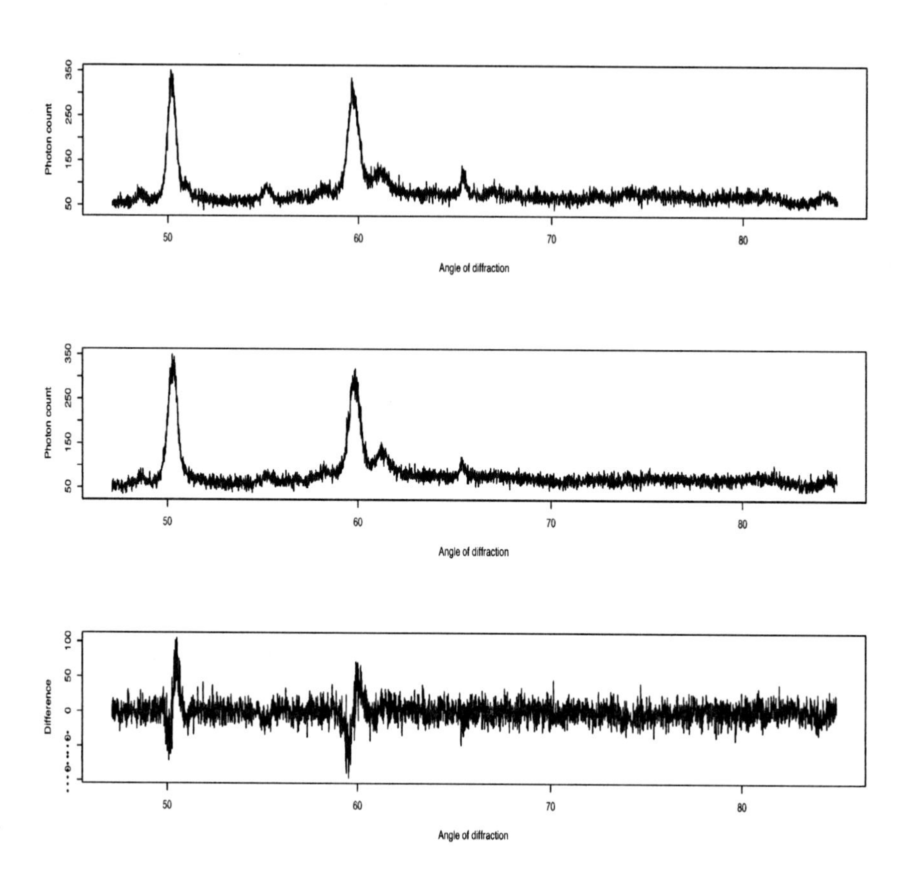

Figure 7.19 *Top and centre panels: two extracts from thin-film data sets each of 4806 observations with the same design points. Bottom panel: the differences of the two samples.*

to be detected with a degree of certainty by the Delgado–Neumeyer–Dette test. The actual differences are of about this order as can be seen from the bottom panel of Figure 7.19. In fact the test just fails to reject the null hypothesis at the 0.1 level. The realized value of the test statistic is 1.734 as against the critical value of 1.90 given in (7.73). The Fan–Lin test (7.75) rejects the null hypothesis at the 0.01 level. The realized value of the test statistic is 111.66 as against the critical value of 12.44 for a test of size $\alpha = 0.01$. Finally the tests based on $\mathscr{A}_n$ and $\mathscr{A}_n^*$ both reject the null hypothesis at the 0.01 level. The realized value of τ_n is 43.15 as against the critical value of 1.50. The realized value of γ_n is 53.27 as against the critical value of 0.733.

7.5.2 *Disjoint Supports*

If the supports are disjoint $\mathbf{t}_{1n_1} \cap \mathbf{t}_{2n_2} = \emptyset$ then there always is an adequate joint approximating function for the two data sets. Two data sets of size $n_1 = n_2 = 500$ were generated according to (7.1) with $f_1(t) = \exp(1.5t), f_2(t) = \exp(1.5t) + 4$ and $\sigma_i = 0.25, i = 1,2$. The support of the first sample is $j/500, j = 0,\ldots,499$ and of the second sample $(2j+1)/1000, j = 0,\ldots,499$. The top panel of Figure 7.20 shows the two data sets, the centre panel a joint approximating function and the bottom panel data generated using this function.

This may seem rather extreme and would seem to deny the the possibility of rejecting the claim of no joint approximation. There are at least two ways of avoiding this conclusion. The first is to place constraints on the joint approximating function, for example a first derivative $f^{(1)}$ with $\|f^{(1)}\|_\infty \le 1$ or that it should be non-decreasing. The second possibility is to make assumptions about the support points, for example that they be interchangeable.

7.5.2.1 *Regularization of a Joint Approximating Function*

Many articles in the literature propose tests as to whether a joint approximating function exists or not. In order not to be vacuous such a test must place constraints on the allowable functions. Typically the constraints are smoothness ones: [98], a bounded first derivative; [104], Hölder continuity; [129], at least uniform continuity; [134], [135], a continuous second derivative; [161], Hölder continuity of order $\beta > 1/2$; [54], a continuous rth derivative: [137], a second derivative which is uniformly Lipschitz of order $\beta, 0 \le \beta < 1$; [165], continuous derivatives of order $d \ge 2$. All these conditions are of a qualitative nature and cannot therefore solve the problem. If the function which linearly interpolates the data points is convoluted with a Gaussian kernel with a very small bandwidth the result is an infinitely differentiable function which approximates both data sets. In order to exclude such a joint approximating functions it is necessary to place quantitative conditions on the smoothness such as $\|f^{(1)}\|_\infty \le 1$. Such a condition would clearly eliminate the joint approximating function of Figure 7.20. There is an understandable reluctance to do this as there may be little justification for any chosen qualitative restriction. Shape restrictions offer an alternative and these may be more acceptable. Requiring the joint approximation to be monotonic is also sufficient to exclude the joint approximating function of Figure 7.20.

If the owner of the data is reluctant to place constraints on the joint approximating function the best the statistician can do is to provide a simplest such function and ask whether it is acceptable or not. If the two data sets of Figure 7.20 are moved closer together with $f_2 = f_1 + 0.079$ then a monotone joint approximating function exists. One such is shown in the upper panel of Figure 7.21 together with the data sets. If the functions are slightly further apart $f_2 = f_1 + 0.080$ then no monotonic joint approximating function exists. The smoothest monotone joint approximating function in the sense of minimizing $TV(g^{(2)})$ is shown in the lower panel. The provider of the data may judge this function to be acceptable.

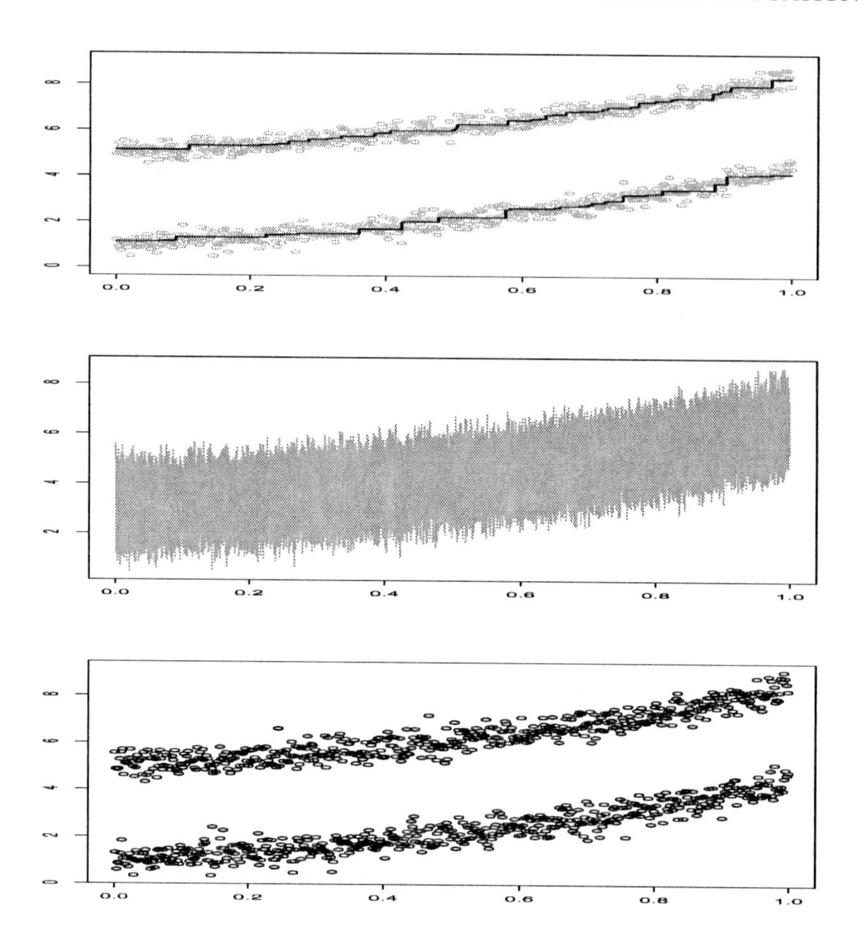

Figure 7.20 *Top panel: two data sets with disjoint supports and individual approximating monotone functions. Centre panel: a joint approximating function with the fewest number of local extreme values. Bottom panel: data generated using the function in the centre panel.*

7.5.2.2 *Exchangeable Supports*

The joint approximating function shown in the centre panel of Figure 7.20 may be regarded as unacceptable. It is however possible to think of situations where the joint approximating function is of this complexity. The function is characterized by many very thin peaks so that a random choice of design points will often miss them all (needles in a haystack). On the basis of some additional information (metal detectors) the design points of the second sample are chosen intentionally and the peaks (needles) are discovered. If however both sets of design points were chosen at random, then it seems clear that there is no joint approximating function. More generally it may be

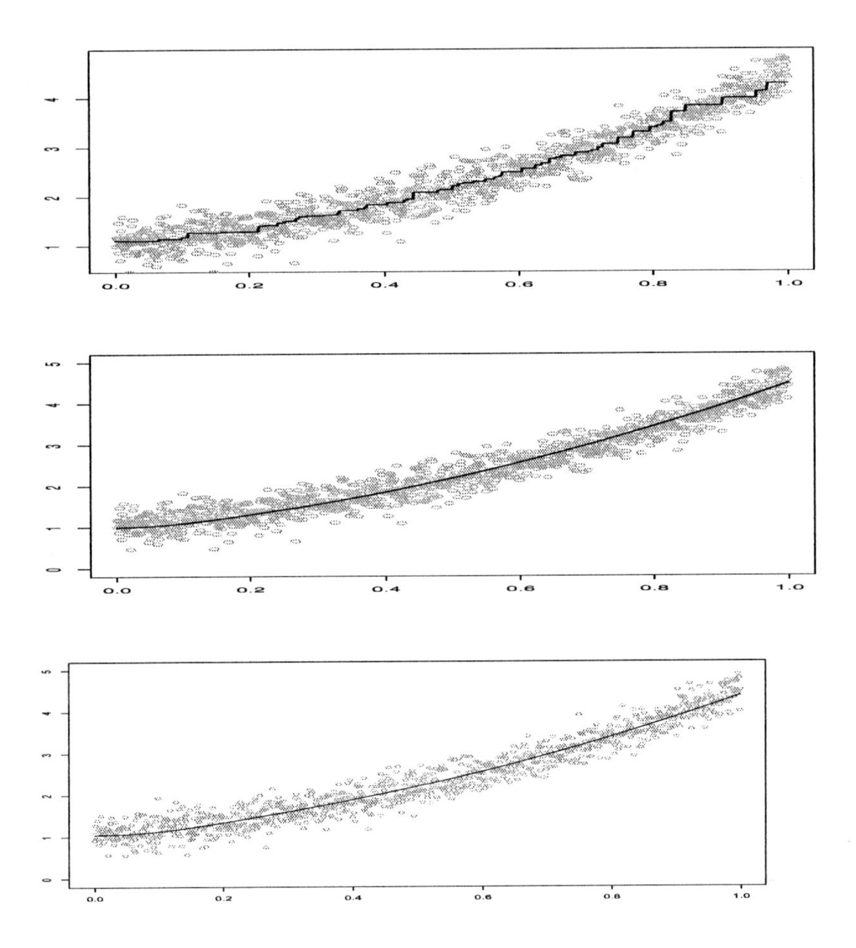

Figure 7.21 *Upper panel: the two data sets of Figure 7.20 moved closer together* $f_2 = f_1 + 0.079$ *with a joint approximating monotone function. Lower panel: the smoothest* $\min(TV(g^{(2)}))$ *joint approximating function with the same monotonic behaviour.*

the case that the design points may be considered to be interchangeable. If one set of design points were chosen at random and the second on the basis of metal detectors they would not be regarded as interchangeable.

The following is an improvement on the argument in [48]. Consider two samples $(\mathbf{t}_{1n_1}, \mathbf{x}_{1n_1})$ and $(\mathbf{t}_{2n_2}, \mathbf{x}_{2n_2})$ generated under the model (7.1) with $f_1 = f_2$, $\sigma_1 = \sigma_2$ and interchangeable supports. Of the $n = n_1 + n_2$ observations allocate at random n_1 to a new first sample $(\tilde{\mathbf{t}}_{1n_1}, \tilde{\mathbf{X}}_{1n_1})$ and the remaining n_2 to a new second sample $(\tilde{\mathbf{t}}_{2n_2}, \tilde{\mathbf{X}}_{2n_2})$. Given a multiscale family of intervals $\mathscr{I}_1$ of $[1, \ldots, n_1]$ calculate the

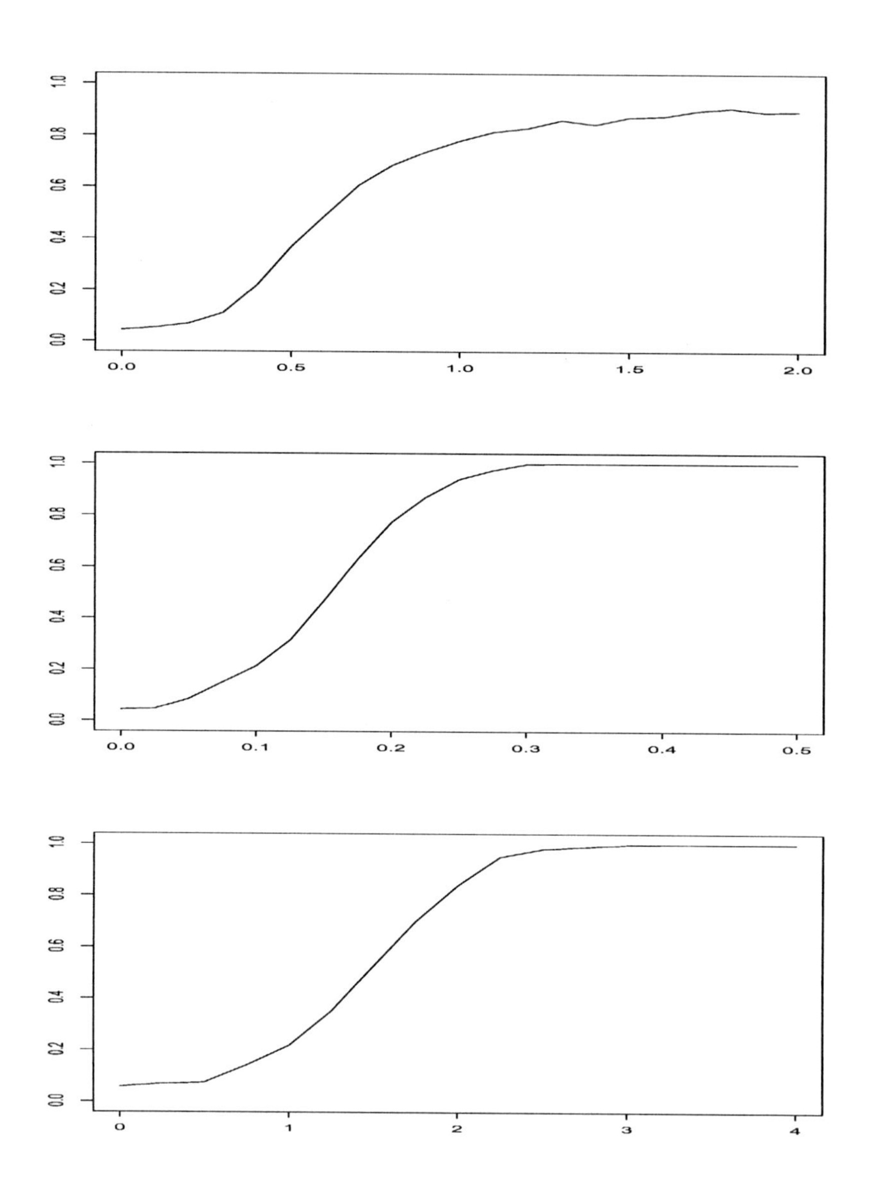

Figure 7.22 *Probability of rejecting a joint approximation as a function of* δ *for the rank method for data generated as in (7.1) with parameters given by (7.78) from top to bottom.*

means

$$\tilde{A}_1(I) = \frac{1}{|I|}\sum_{i\in I}\tilde{X}_{1n_1}(t_{1i}), I \in \mathscr{I}_1$$

and similarly for the second sample $(\tilde{\mathbf{t}}_{2n_2}, \tilde{\mathbf{X}}_{2n_2})$. Repeat this $m-1$ times and include the actual data as the mth 'simulation'. For each I in $\mathscr{I}_1$ the ranks for each simulation $r_1(I,j), j = 1,\ldots,m$ are then calculated to give the smallest $r_1^-(j) = \min_I r_1(I,j)$ and largest $r_2^+(j) = \max_I r_2(I,j)$ ranks. The procedure is repeated for the second sample. Given α denote the $(1-\alpha)/4$ quantiles of $r_1^-(j), j = 1,\ldots,m$ and $r_2^-(j), j = 1,\ldots,m$ by q_1^- and q_2^- respectively. Similarly the $(1-\alpha/4)$ quantiles of $r_1^+(j), j = 1,\ldots,m$ and $r_2^+(j), j = 1,\ldots,m$ are denoted by q_1^+ and q_2^+ respectively. If one of

$$r_1^-(m) \le q_1^-, \quad r_2^-(m) \le q_2^-, \quad r_1^+(m) \ge q_1^+, \quad r_1^+(m) \ge q_1^+$$

holds then it is concluded that no joint approximating function exists.

The procedure requires more or less equal noise levels. If this is not the case then the noise level of the less noisy data can be artificially increased so that the levels are the same. More precisely using the noise quantification of (7.12), if $\sigma_{n1} < \sigma_{n2}$ then Gaussian white noise with variance $\sigma_{n2}^2 - \sigma_{n1}^2$ is added to the observations of the first sample. This seems rather odd but it seems to work.

Figure 7.22 shows the results of a small simulation study to demonstrate the method. Data were generated according to (7.1) with $n_1 = 200, n_2 = 300, \sigma_1 = 0.25, f_1(t) = \exp(1.5t)$ and from top to bottom

$$\begin{array}{lcl} \sigma_2 = 0.25 & , & f_2(t) = f_1(t) + \delta\{0.5 \le t \le 0.55\} \\ \sigma_2 = 0.25 & , & f_2(t) = f_1(t) + \delta \\ \sigma_2 = 5 & , & f_2(t) = f_1(t) + \delta\,. \end{array} \tag{7.78}$$

The value of α was set to 0.95 and the support points were independently and uniformly distributed over $[0,1]$ in all cases. The family $\mathscr{I}_n$ of subsets of $[0,1]$ was taken to be $\mathscr{I}_n(1.25)$. The figure shows the probability of rejecting a joint approximation as a function of δ.

7.6 Inverse Problems

A linear inverse problem is of the form

$$X(t) = K \circ f(t) + \sigma Z(t), 0 \le t \le 1, \tag{7.79}$$

where K is a known linear operator. If K has a bounded inverse K^{-1} then (7.79) can be transformed into the standard model (7.1) by taking inverses

$$K^{-1} \circ X(t) = f(t) + \sigma K^{-1} \circ Z(t) \tag{7.80}$$

with the one difference that the noise $K^{-1} \circ Z(t)$ is no longer white. As K is known the latter is not a principle problem as the bounds required for the approximation region can be obtained from simulations. If K is invertible but the inverse K^{-1} not bounded then the transformation from (7.79) to (7.79) can still be performed but is of little use. The unboundedness of K^{-1} means that small changes in the noise can lead to large changes in $K^{-1} \circ X(t) - f(t)$ which makes the consistent estimation of f impossible.

The standard method of solving inverse problems is to introduce a penalty term for f. One such method based on a penalty term is to minimize

$$I(f,\lambda) := \|X(t) - K \circ f(t)\|_2^2 + \lambda \int_0^1 f^{(2)}(t)^2 \, dt. \tag{7.81}$$

with respect to f. This problem is well defined with a unique solution, but the choice of λ remains.

Conceptually the easiest method is to minimize the total variation of some derivative of f as in Sections 7.2.2. The approximation region corresponding to (7.82) is

$$\mathscr{A}_n = \mathscr{A}_n((\mathbf{t},\mathbf{x})_n, \mathscr{I}_n, \sigma, \tau_n) = \left\{g : \max_{I \in \mathscr{I}_n} |w((\mathbf{t},\mathbf{x})_n, g, I)| \le \sigma\sqrt{\tau_n \log n}\right\} \tag{7.82}$$

where

$$w((\mathbf{t},\mathbf{x})_n, g, I) = \frac{1}{\sqrt{|I|}} \sum_{t_i \in I} (x(t_i) - K \circ g(t_i)). \tag{7.83}$$

As K is linear, minimizing the total variation of some derivative g subject to $g \in \mathscr{A}_n$ results in a linear programming problem.

The upper panel of Figure 7.23 shows a data set of size $n = 1000$ with the support points equidistant on $[10^{-3}, 1]$, $t_i = i/1000, i = 1, \ldots, 1000$. The lower panel shows the first 200 data points. The data were generated as follows. The linear operator K is a convolution defined by

$$K \circ g(t) = \int_0^t h(t-u) g(u) \, du \tag{7.84}$$

with h given by

$$h(t) = 1 + \sin(2\pi t)^2.$$

The function f is taken to be

$$f(t) = 1 + \exp(2t^2) + \cos(2\pi t) + 100 \exp(-1000(t-0.4)^2)$$

and the noise level in (7.79) is set to $\sigma = 100$. The top panel of Figure 7.24 shows the result of minimizing the total variation of the second derivative $TV(g^{(2)})$ with function values in the interval $[0, 10]$. The function f is solid and the reconstructions dashed. The oscillation of the reconstruction on the left of the peak is an artifact due to the use of the second derivative. The centre panel shows the result of minimizing $TV(g^{(2)})$ but subject to the monotonicity behaviour obtained from minimizing $TV(g)$. The bottom panel shows the result of minimizing $TV(g^{(2)})$ but now subject to the monotonicity behaviour of the centre panel and the convexity behaviour obtained from minimizing $TV(g^{(1)})$ with the latter also subject to the monotonicity behaviour of the centre panel.

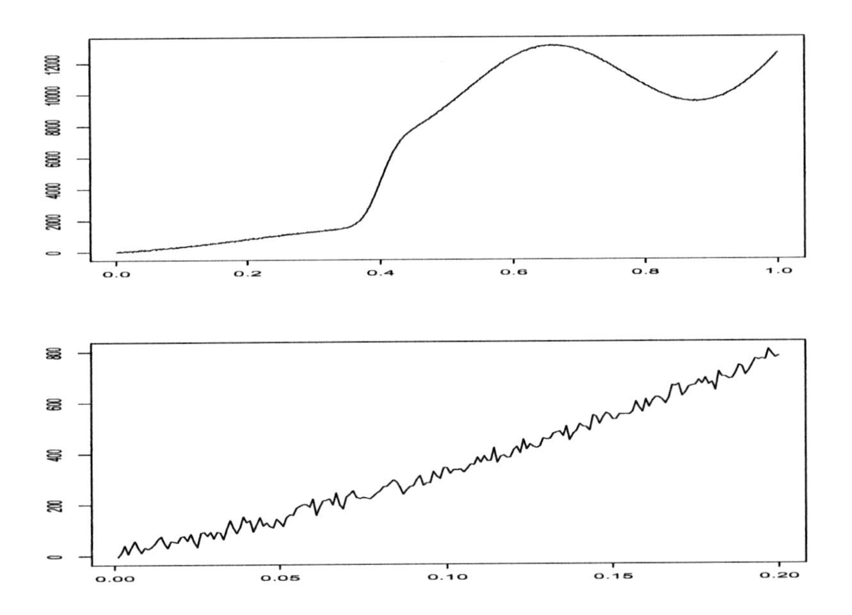

Figure 7.23 *Upper panel: the data. Lower panel: the first 200 data points.*

7.7 Heterogeneous Noise

The model (7.1) assumes homogeneous noise. It is a simple matter to extend it to heterogeneous noise

$$X(t) = f(t) + \sigma(t)Z(t), 0 \le t \le 1. \tag{7.85}$$

This was done already for the Poisson model (7.29) with $\sigma(t) = \max\{\sigma, f(t)^{1/2}\}$. As mentioned in Chapter 7.2.2.1 the approximation regions can be modified in the obvious manner assuming that $\sigma(t)$ is known. In general this will not be the case. This is the topic of the following chapter.

The procedures based on minimzing total variation (Chapter 7.2.2), on piecewise constant regression (Chapter 7.2.3) and on smoothing splines (Chapters 7.4.3 and 7.4.4) can be adopted to heterogeneous noise in a straight forward manner.

The taut string method can also be adapted as follows. The partial sums (7.17) are replaced by

$$S_j^\circ = S^\circ((\mathbf{t},\mathbf{x})_n)_j = \sum_{i=1}^{j} x(t_i)/\sigma(t_i)^2, \; j = 1,\ldots,n, \tag{7.86}$$

and plotted against

$$SC_j^\circ = SC^\circ((\sigma,\mathbf{x})_n)_j = \sum_{i=1}^{j} 1/\sigma(t_i)^2, \; j = 1,\ldots,n. \tag{7.87}$$

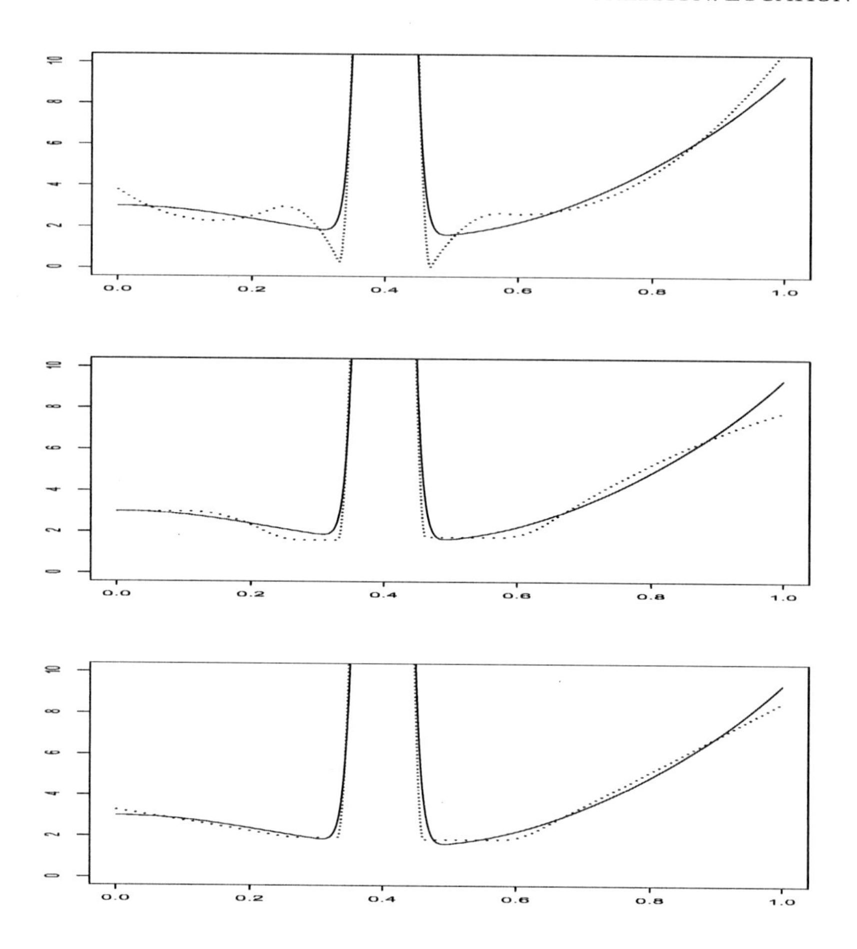

Figure 7.24 *The true function is the solid line. Top panel: minimizing the total variation of the second derivative. Centre panel: minimizing the total variation of the second derivative subject to the monotonicity constraints. Bottom panel: minimizing the total variation of the second derivative subject to the monotonicity and convexity constraints.*

The derivative is calculated as

$$\frac{S(t_j) - S(t_i)}{SC(t_j) - SC(t_i)}$$

and the local squeezing procedure is carried out as before. This is somewhat ad hoc as there are many advantages in altering the local squeezing procedure but attempts to do this did not result in any improvements. Nevertheless the resulting procedure works well in the presence of heterogeneous noise.

Chapter 8

Non-parametric Regression: Scale

8.1 The Standard Model and a Concept of Approximation

The standard model for the non-parametric scale problem is

$$R(t) = \Sigma(t)Z(t), 0 \le t \le 1. \tag{8.1}$$

where $R(t)$ represents the observations, $\Sigma(t)$ the scale and $Z(t)$ the random noise. In the context of financial data $R(t)$ is the rate of return and $\Sigma(t)$ is the volatility. The complete specification of the model requires the specification of the noise. For the most part this will be taken to be standard Gaussian white noise but t distributions and non-symmetric noise will also be considered.

If $Z(t)$ is taken to be standard Gaussian white noise then $R(t)^2/\Sigma(t)^2 = Z(t)^2$ follows a χ^2-distribution with one degree of freedom. For any interval $I = [t_j, t_k]$ this implies

$$\sum_{t_i \in I} \frac{R(t_i)^2}{\Sigma(t_i)^2} \stackrel{D}{=} \chi^2_{|I|} \tag{8.2}$$

where χ^2_ν denotes a χ^2 random variable with ν degrees of freedom and $|I|$ denotes the number of points $t_i \in I$. It follows that for any $\alpha, 0 < \alpha < 1$,

$$\mathbf{P}\left(\text{qchisq}((1-\alpha)/2,|I|) \le \sum_{t_i \in I} \frac{R(t_i)^2}{\Sigma(t_i)^2} \le \text{qchisq}((1+\alpha)/2,|I|)\right) = \alpha \qquad (8.3)$$

where qchisq(α,ν) denotes the α quantile of the χ^2 distribution with ν degrees of freedom. Extending this to a family $\mathscr{I}_n$ of intervals implies for any α there exists an α_n such that

$$\mathbf{P}\left(\text{qchisq}((1-\alpha_n)/2,|I|) \le \sum_{t_i \in I} \frac{R(t_i)^2}{\Sigma(t_i)^2} \le \text{qchisq}((1+\alpha_n)/2,|I|)\, \forall I \in \mathscr{I}_n\right) = \alpha. \qquad (8.4)$$

For data $\mathbf{r}_n = (r(t_1),\ldots,r(t_n))$ an α-approximation region for the scale function σ is given by

$$\begin{aligned} \mathscr{A}_n(\mathbf{r}_n,\mathscr{I}_n,\alpha) = \Bigg\{\sigma : &\text{qchisq}((1-\alpha_n)/2,|I|) \le \sum_{t_i \in I} \frac{r(t_i)^2}{\sigma(t_i)^2} \\ &\le \text{qchisq}((1+\alpha_n)/2,|I|) \text{ for all } I \in \mathscr{I}_n\Bigg\} \end{aligned} \qquad (8.5)$$

with the only restriction that $r(t_i) \neq 0$ for all t_i. For data generated under the model (8.1) with $Z(t)$ standard Gaussian white noise

$$\mathbf{P}(\Sigma \in \mathscr{A}_n(\mathbf{R}_n,\mathscr{I}_n,\alpha)) = \alpha \qquad (8.6)$$

for all Σ for which $\Sigma(t) \neq 0$ for all t.

For finite n the values of α_n can be obtained by simulation. The following approximations reasonable for $100 \le n \le 20000$

$$\alpha_n = 1 - 0.020\exp(-1.14\log(n)), \quad \alpha = 0.90, \qquad (8.7)$$
$$\alpha_n = 1 - 0.0116\exp(-1.18\log(n)), \quad \alpha = 0.95. \qquad (8.8)$$

Using the results of [194] (see also [110]) it can be shown that

$$\lim_{n\to\infty} \frac{1}{n}\log(1-\alpha_n) = -1$$

so that asymptotically the approximations are conservative.

Given the approximation region $\mathscr{A}_n = \mathscr{A}_n(\mathbf{r}_n,\mathscr{I}_n,\alpha)$ the choice of σ requires some form of regularization. One possibility is to minimize the total variation of σ subject to $\sigma \in \mathscr{A}_n$ just as in the location case. Unfortunately this and similar regularizations lead to non-linear programming problems which are more difficult to solve. However the other methods for the location case can be adopted to scale and this will be the path taken in this chapter.

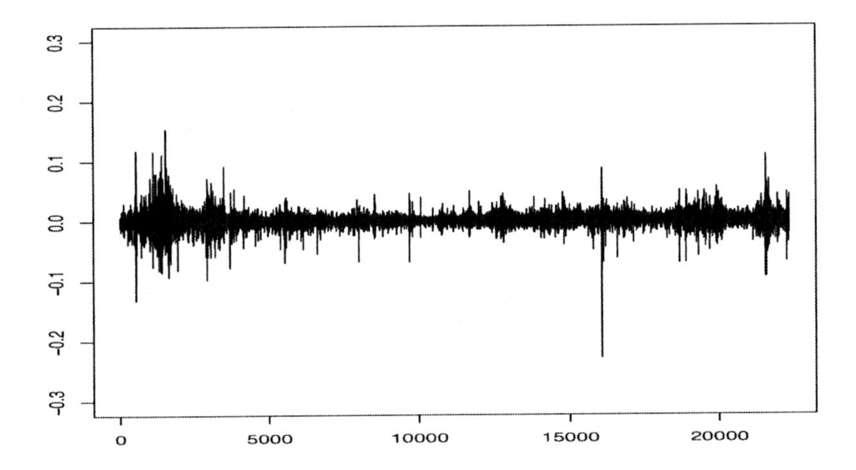

Figure 8.1 *The daily rates of return of the Standard and Poor's index. There are 22378 observations of which 21717 are non-zero.*

8.2 Piecewise Constant Scale and Local Approximation

This section is mainly based on [110] and [44] and follows the location case described in Chapter 7.2.3.

Piecewise constant functions volatility functions are sometimes of interest in their own right. Financial data often exhibit long periods of constant volatility followed by abrupt changes. This can be seen to some extent in Figure 8.1 which shows the daily rate of returns of the Standard and Poor's index over a period of about eighty years. It will be a standard example.

Just as in the location case (7.30) a local concept of approximation will be used. It derives from [110]. The approximation region is defined as

$$\tilde{\mathscr{A}}_n(\mathbf{r}_n, \mathscr{I}_n, \alpha) = \Big\{ \sigma : \text{qchisq}((1-\alpha_n)/2, |I'|) \le \sum_{t_i \in I'} \frac{r(t_i)^2}{\sigma(t_i)^2} \le \text{qchisq}((1+\alpha_n)/2, |I'|) \text{ for all } I' \subset I \in \mathscr{I}_n \;\; \sigma \text{ constant on } I \Big\}. \tag{8.9}$$

Clearly $\mathscr{A}_n \subset \tilde{\mathscr{A}}_n$ so that if the data are generated according to the model (8.1) with $Z(t)$ standard Gaussian white noise and Σ is piecewise constant then

$$\mathbf{P}\left(\Sigma \in \tilde{\mathscr{A}}_n(\mathbf{R}_n, \mathscr{I}_n, \alpha)\right) \ge \alpha. \tag{8.10}$$

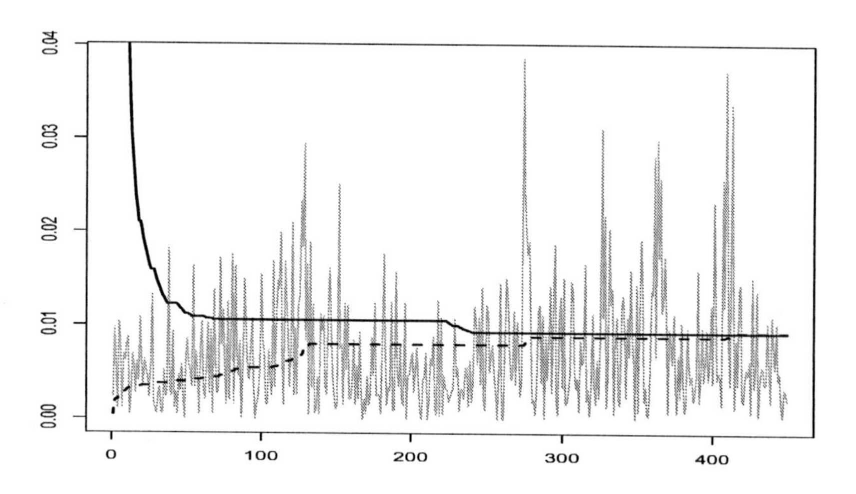

Figure 8.2 *Absolute values of the first 450 daily rates of return of the Standard and Poor's index with the upper and lower bounds. The lower bound first exceeds the upper at observation 414.*

The algorithm for producing a solution to the problem of minimizing $\ell_0(\sigma)$ subject to $\sigma \in \tilde{\mathscr{A}}_n$ follows the argument of (7.31)-(7.33) with the obvious changes (Algorithm 0 of [44]). Figure 8.2 shows the absolute values of the first 450 values of the Standard and Poor's data in black together with the lower and upper bounds. The bounds cross at observation 414.

Just as in the location case of Chapter 7.2.3 the algorithm gives the minimum number of intervals but no values for the volatility on the intervals. Under the model it seems reasonable to require σ_I^2 to be the mean of the $r(t_i)^2$ on an interval of constancy I:

$$\sigma_I^2 = \left(\sum_{t_j \in I} r_j^2 \right) / |I|. \tag{8.11}$$

Indeed, many practitioners would not accept a solution which did not have this property. Again as in the location case this can be achieved using dynamic programming as in [110] and [44]. The algorithm is based on [92].

This does not produce a unique solution but in this can be achieved by minimizing

$$\sum_{i=1}^{n} (r(t_i)^2 - \sigma(t_i)^2)^2. \tag{8.12}$$

Again, this problem can be solved using dynamic programming. The relevant algorithm is due to Höhenrieder [110]. Although it is possible to defined data sets for which even minimizing (8.12) does not yield a unique solution it may be supposed

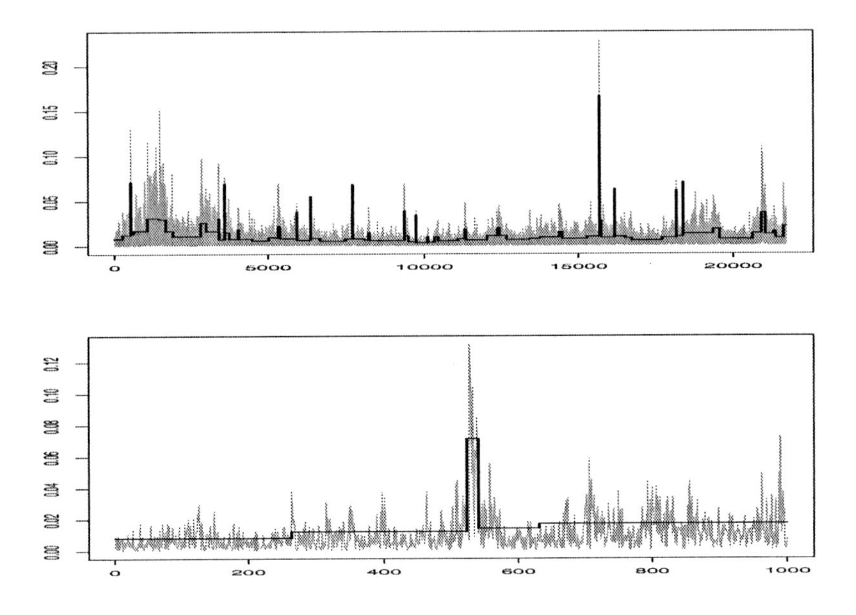

Figure 8.3 *Upper panel:Absolute values of the daily rates of return of the non-zero Standard and Poor's index together with the piecewise constant volatility function with 77 intervals of constancy. Lower panel: The same for the first 1000 observations.*

that in practice the solution is unique. The results for the Standard and Poor's data with $\alpha = 0.9$ are shown in Figure 8.3. The upper panel shows the complete data set, the lower panel shows the first 1000 observations. The volatility function has 77 intervals of constancy, the shortest being one day (there are four such days) and the longest being the interval [8242,9355] of 1114 days.

According to the model the 'residuals', $R(t)/\Sigma(t)$ are i.i.d. $N(0,1)$. The upper panel of Figure 8.5 shows the q-q-plot of the empirical residuals $r(t_j)/\sigma(t_j)$ together with the 45-degree line. There is a clear divergence showing that the empirical residuals are heavier than is consistent with the model. This is not an artifact of the procedure as can be shown by simulations. Figure 8.4 shows an artificial data set generated under the model using the empirical volatility function of the Standard and Poor's data. The lower panel of Figure 8.5 shows the resulting q-q-plot. The results are essentially the same for all simulations.

The above results indicate that the model (8.1) with Gaussian white noise captures many of the features of the Standard and Poor's data but there remain significant divergences. Whether these are of practical relevance can only be judged when such models are used for practical purposes. Nevertheless it is of statistical interest to see if the model can be improved to reduce the divergences to an acceptable level. The obvious starting point is to replace the Gaussian distribution by one with a heavier

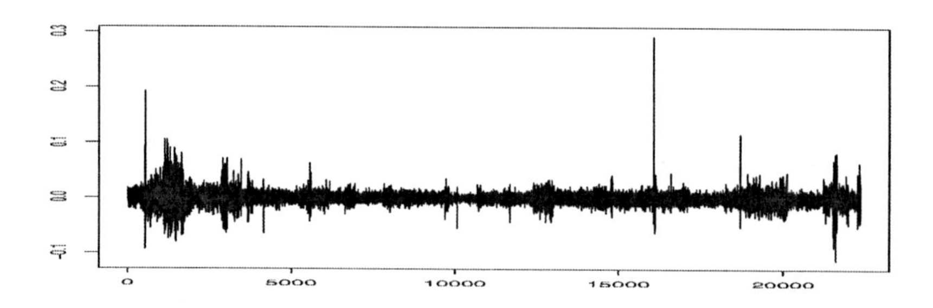

Figure 8.4 *Data simulated according to the model using the volatility function of the Standard and Poor's data.*

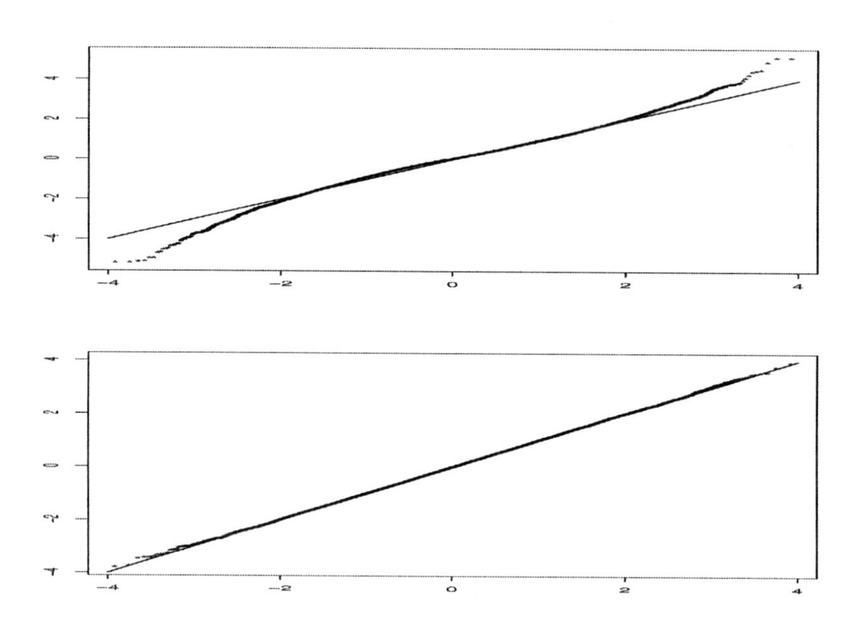

Figure 8.5 *Upper panel: the q-q-plot of the empirical residuals* $r(t_j)/\sigma(t_j)$ *together with the 45-degree line. Lower panel: the same for the simulated data of Figure 8.4.*

tail, for example a t distribution with a given number of degrees of freedom. The model is now

$$R(t) = \Sigma(t) T_\nu(t), 0 \le t \le 1, \tag{8.13}$$

where the $T_\nu(t)$ are i.i.d. t random variables with ν degrees of freedom. In principle the method goes through as before, it is only necessary to replace the quantiles of sums of squares of the normal distribution by the corresponding quantiles of sums of squares of T_ν random variables. In practice this fails as there seem to be no methods for determining these with sufficient accuracy. A method of overcoming this is required. The following is partially based on [10].

A t random variable T_ν can be transformed into a χ^2 random variable $h(T_\nu)$ with one degree of freedom by means of the function

$$h(x) = \text{qchisq}(2\text{pt}(x, \nu) - 1, 1) \tag{8.14}$$

where qchisq$(\cdot, 1)$ denotes the quantile function of a χ_1^2 random variables and pt$(\cdot, \nu)$ the distribution function of a T_ν random variable. It follows that

$$h\left(\frac{R(t)}{\Sigma(t)}\right) \stackrel{D}{=} \chi_1^2$$

and hence, corresponding to (8.2),

$$\mathbf{P}\Big(\text{qchisq}((1-\alpha_n)/2, |I|) \le \sum_{t_i \in I} h\left(\frac{R(t_i)}{\Sigma(t_i)}\right) \\ \le \text{qchisq}((1+\alpha_n)/2, |I|) \text{ for all } I \in \mathscr{I}_n\Big) = \alpha. \tag{8.15}$$

The resulting approximation region is

$$\mathscr{A}(\mathbf{r}_n, \mathscr{I}_n, \alpha) = \Big\{\sigma : \text{qchisq}(\alpha_n/2, |I|) \le \sum_{t_i \in I} h\left(\frac{r(t_i)}{\sigma(t_i)}\right) \\ \le \text{qchisq}((1+\alpha_n)/2, |I|) \text{ for all } I \in \mathscr{I}_n\Big\}. \tag{8.16}$$

If, as above, interest centres on piecewise constant volatility functions the algorithms of [110] require only a minor modification. It is only necessary to determine $\sigma_{\ell,|I|}$ and $\sigma_{u,|I|}$ which satisfy

$$\sum_{t_i \in I} h\left(\frac{r_i}{\sigma_{\ell,|I|}}\right) = \text{qchisq}((1+\alpha_n)/2, |I|) \tag{8.17}$$

and

$$\sum_{t_i \in I} h\left(\frac{r_i}{\sigma_{u,|I|}}\right) = \text{qchisq}((1-\alpha_n)/2, |I|) \tag{8.18}$$

respectively. As the function h of (8.14) is strictly monotone increasing and continuous with $h(0) = 0$ and $h(\infty) = \infty$ it is straight forward to solve (8.17) and (8.18). In practice however the numerical costs are considerable for large data sets. They can be reduced by approximating $h(x)$ over a finite interval by a sum of powers of x

$$h(x) \approx q(x) = \sum_{i=1}^{k} \delta_i x^{\gamma_i}, \; 0 \le x \le c,$$

where q is strictly monotone increasing on $[0,c]$. Using q instead of h in (8.17) and (8.18) leads to

$$\sum_{j=1}^{k} \delta_j \sigma_{\ell,|I|}^{-\gamma_j} \left(\sum_{t_i \in I} |r_i|^{\gamma_j} \right) = \text{qchisq}((1+\alpha_n)/2, |I|) \tag{8.19}$$

and

$$\sum_{j=1}^{k} \delta_j \sigma_{u,|I|}^{-\gamma_j} \left(\sum_{t_i \in I} |r_i|^{\gamma_j} \right) = \text{qchisq}((1-\alpha_n)/2, |I|). \tag{8.20}$$

The numerical complexity of solving (8.19) and (8.20) is now independent of I once the partial sums of the $|r_i|^{\gamma_j}$ have been calculated for each j. Occasionally values of the $|r_i|$ are such that the approximation of $h(x)$ by $q(x)$ fails and it is not possible to solve (8.19) and (8.20). In such cases recourse is taken to solving (8.17) and (8.18).

The use of the approximation makes it possible to analyse large data sets such as the Standard and Poor's data. The additional overload compared with the use of the Gaussian model is considerable with a running time of 1370 seconds compared with 1.3 seconds for the Gaussian model.

The results are as follows using a t_9-distribution in (8.13). There are 50 intervals of constancy shown in upper panel of Figure 8.6: the lower panel shows the first 100 observations corresponding to the lower panel of Figure 8.3. Figure 8.7 shows the q-q-plot corresponding Figure (8.5). It is clear that the fit is much better than for the Gaussian model but it is also clear that the negative returns have a heavier tail than the positive ones. This suggest non-symmetric noise with different degrees of freedom for the negative and positive returns. The upper panel of Figure 8.8 shows the q-q-plot of the positive returns against the quantiles of a $|T_7|$ random variable: the lower panel shows the same for the negative returns and a $|T_5|$ random variable.

One stylized fact about long range finance data is the so called 'long range memory' which is formalized as the slow decay of the autocorrelation function of the absolute returns. Figure 8.9 shows the autocorrelation function of the absolute daily returns $|r(t_i)|$ (upper panel), the lower panel shows the same for the standardized returns $|r(t_i)/\sigma(t_i)|$. The long range memory effect for the real data has been eliminated for the standardized residuals but there remains a short-term dependency over about 40 days. The same holds even if the Gaussian model is replaced by a model based on a t distribution. It may be decided that the remaining short-term dependency of the volatility is of no consequence but nevertheless it is of interest to remove it. This can be done by generating dependent $|T_\nu|$ random variables.

Let $Z(i)$ be i.i.d. $N(0,1)$ random variables and define a moving average process $X(i)$ by

$$X(i) = Z(i) + \alpha \sum_{j=1}^{\infty} \exp(-j\lambda) Z(i-j). \tag{8.21}$$

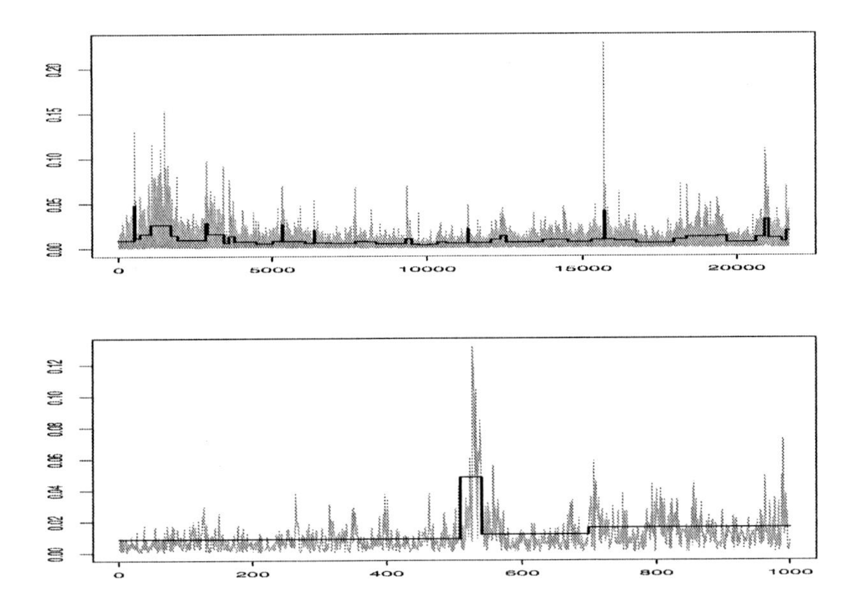

Figure 8.6 *Upper panel: the Standard and Poor's data together with the piecewise constant volatility using the t_9-model. Lower panel: the first 1000 observations.*

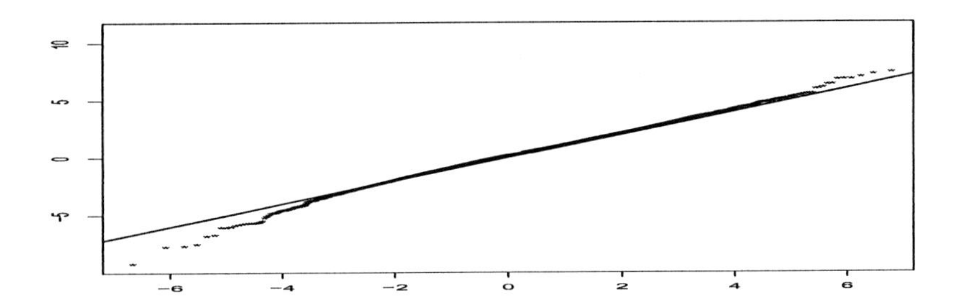

Figure 8.7 *The q-q-plot for the t_9-model.*

with $\lambda > 0$. The $X(i)$ are identically $N(0, \sigma^2)$ distributed with

$$\sigma^2 = 1 + \frac{\alpha^2 \exp(-2\lambda)}{1 - \exp(-2\lambda)}$$

but they are dependent if $\alpha \neq 0$. On putting

$$|T_\nu(i)| = \text{qt}(2\text{pnorm}(X(i)/\sigma) - 1, \nu) \tag{8.22}$$

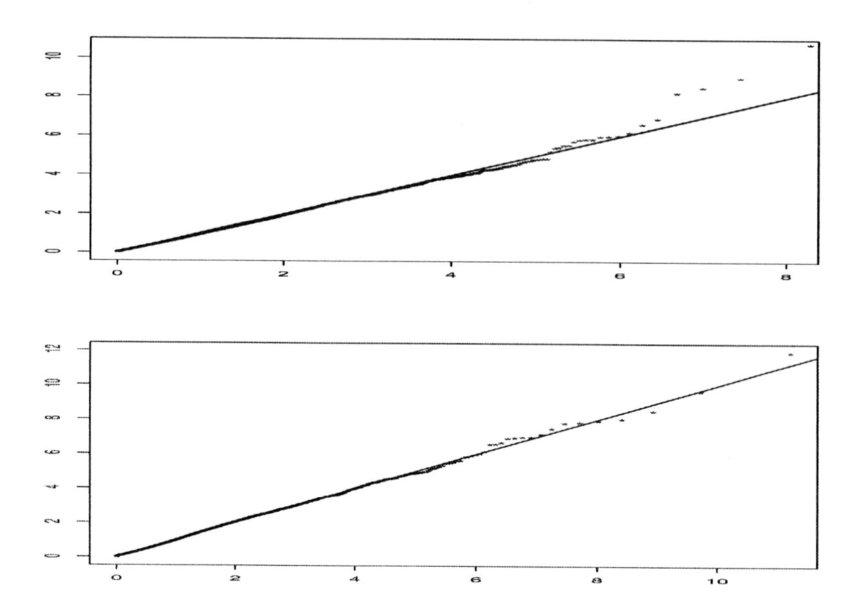

Figure 8.8 *Upper panel: the q-q-plot of the positive returns against the quantiles of a $|T_7|$ random variable. Lower panel: same for the negative returns and a $|T_5|$ random variable.*

it is seen that the $|T_\nu(i)|$ are identically distributed as the absolute value of a t_ν-random variables but they are dependent. Figure 8.10 shows a simulated autocorrelation function of the $X(i)$ (solid) and the $|T_\nu(i)|$ (dashed) for $\nu = 9$, $\alpha = 0.1$ and $\lambda = 0.1$. For empirical data the problem is to reverse the order, start with the empirical $|t(i)|$ and choose values of α and λ to give uncorrelated $z(i)$. The following naïve approach seems to work well. Put

$$x(1) = \text{qnorm}(2 * \text{pt}(|t(1)|, \nu) - 1), \quad z(1) = x(1)$$

and then recursively

$$x(i) = \text{qnorm}(2 * \text{pt}(|t(i)|, \nu) - 1),$$

$$z(i) = x(i) + (1 - \alpha)\exp(-\lambda)z(i-1) - \exp(-\lambda)x(i-1).$$

The results for the Standard and Poor's data are shown in Figure 8.11 for $\alpha = \lambda = 0.05$. These values were chosen by trial and error but this can be replaced by a search over a grid of values whereby that pair is chosen which minimizes the maximum of the absolute autocorrelations up to a lag of 100.

As mentioned in Chapter 7 multiscale change point inference for exponential families, which covers the model (8.1) with Gaussian noise but not with t noise, was considered in a similar manner in [89].

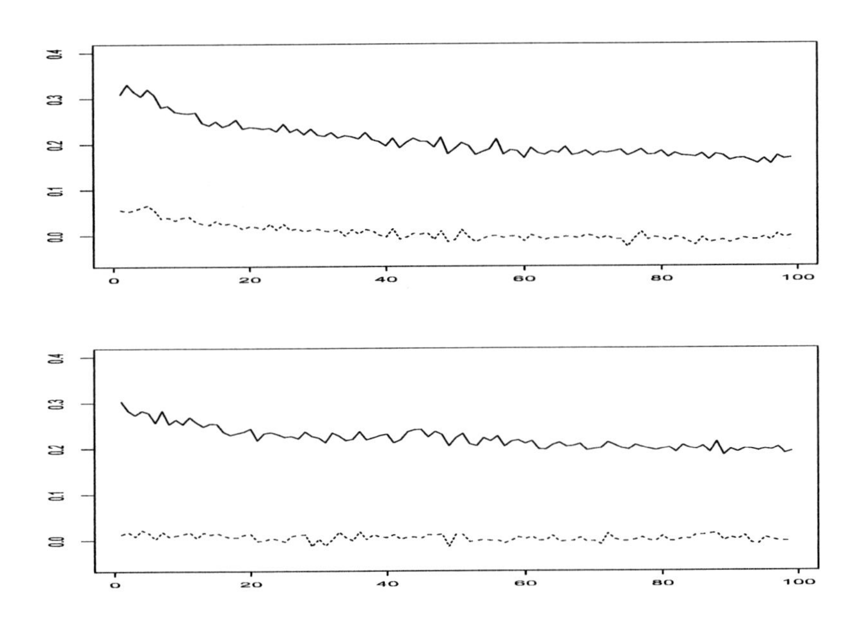

Figure 8.9 *Upper panel: the autocorrelation function of the absolute returns (continuous) and of the standardized returns (dashed). Lower panel: the same for the simulated data of Figure 8.4.*

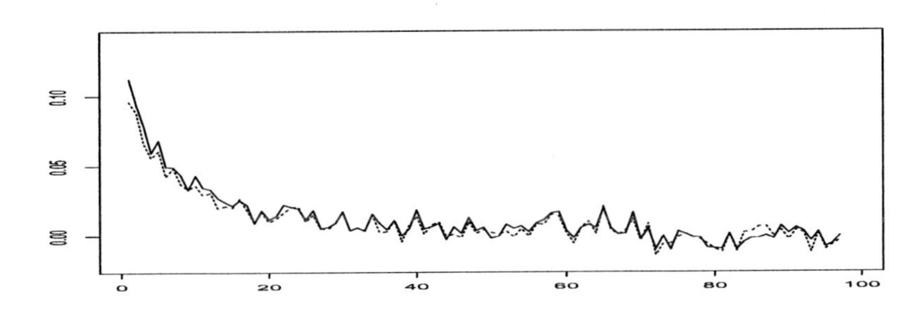

Figure 8.10 *The autocorrelation function of the* $X(i)$ *of (8.21) (continuous) and the associated one of the* $|T_\nu(i)|$ *of (8.22) with* $\nu = 9, \alpha = 0.1, \lambda = 0.1$ *and based on a sample of size 20000.*

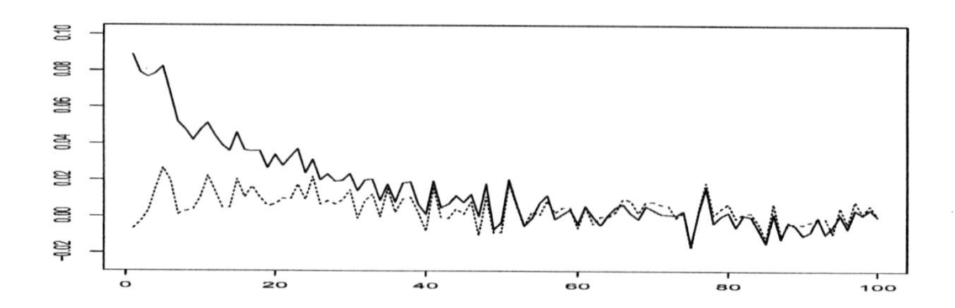

Figure 8.11 *The autocorrelation function of the absolute residuals of the Standard and Poor's data with* $\nu = 9$ *(solid) and the autocorrelation function of the corresponding* z_t *(dashed) with* $\alpha = 0.05$ *and* $\lambda = 0.05$.

8.3 GARCH Segmentation

This section is mainly based on [110] and [44]. One popular class of models for analysing financial data is the class of (G)ARCH models ([83], [20]). A GARCH(p,q) model as defined by

$$\Sigma(t)^2 = \alpha(0) + \sum_{i=1}^{p} \alpha(i)\Sigma(t-i)^2 + \sum_{j=1}^{q} \beta(j)R(t-j)^2 \tag{8.23}$$

where

$$\alpha(0) \geq 0, \quad 0 \leq \alpha(i) < 1,\ i = 1,\ldots,p, \quad 0 \leq \beta(j) < 1,\ j = 1,\ldots,q.$$

It is known that the GARCH(p,q) process is weakly stationary if and only if

$$\sum_{i=1}^{p} \alpha_i + \sum_{j=1}^{q} \beta_j < 1\,. \tag{8.24}$$

The restriction (8.24) may or may not be imposed. In the following it will be imposed so that interest centres on weakly stationary models.

Simple GARCH models are too restrictive to be satisfactory for long data sets such as Standard and Poor's (see [200]). This often leads to the conclusion that such data sets cannot be modelled by stationary processes in their entirety, but that shorter sections can be so modelled. The breaks between the sections are interpreted as a change of structure ([156], [158], [96]). Alternatively the whole data set can be modelled using a non-parametric approach although the two different approaches are not clearly separated ([157], [155]).

The segmentation approach can be carried out using the concept of local approximation. Given a simple GARCH(p,q) model (p and q small) the data is to be segmented in such a manner that a GARCH(p,q) model is adequate on each segment

(p,q)	(0,1)	(1,0)	(0,2)	(1,1)	(2,0)	(0,3)	(1,2)
k	49	43	43	12	29	41	12
(p,q)	(2,1)	(3,0)	(0,4)	(1,3)	(2,2)	(3,1)	(4,0)
k	12	21	42	10	10	11	15

Table 8.1 *Minimum number of intervals for the GARCH segmentation of the Standard and Poor's data as a function of* (p,q).

although with different parameter values. Given (p,q) a GARCH segmentation consists of k disjoint intervals $I_1,\ldots,I_k$ whose union is $\{1,\ldots,n\}$ and such that on each interval I_ν there exist parameters $\alpha_\nu(i), i=0,\ldots,p$ and $\beta_\nu(j), j=1,\ldots,q$ such that

$$\chi^2_{|J|,\frac{1-\alpha_n}{2}} \leq \sum_{t\in J}\frac{r(t)^2}{\sigma_{I_\nu}(t)^2} \leq \chi^2_{|J|,\frac{1+\alpha_n}{2}}, \ \forall J\subset I_\nu, \nu=1,\ldots,k \tag{8.25}$$

where for $I_\nu = \{t_{\nu,1},\ldots,t_{\nu,\ell_\nu}\}$

$$\sigma_{I_\nu}(t)^2 = \alpha_\nu(0)+\sum_{i=1}^{p}\alpha_\nu(i)\sigma_{I_\nu}(t-i)^2+\sum_{j=1}^{q}\beta_\nu(j)r(t-j)^2,$$
$$t_{\nu,\max(p,q)}+1\leq t\leq t_{\nu,\ell_\nu}.$$

The problem is to minimize the number of intervals subject to (8.25). The idea of the first algorithm of Chapter 8.2 can be used by performing a search over the possible values of the parameters. For small p and q such a search is possible as long as the grid used is not too fine. For $p+q\leq 3$ a grid width of 0.01 is possible for $\alpha(i), 1\leq i\leq p$ and $\beta(j), 1\leq j\leq q$: for $p+q=4$ one of 0.02 is possible. These should be sufficiently fine for all practical purposes. In [44] the parameter $\alpha(0)$ was bounded using the results for the piecewise constant volatility. In the following an improved algorithm is used based on the simple observation that if the upper bound in (8.25) is violated then $\alpha(0)$ can be increased. Analogously if the lower bound in (8.25) is violated then $\alpha(0)$ can be decreased. This is continued until the lower and upper bounds for $\alpha(0)$ are sufficiently close. The improvement leads to smaller numbers of breaks.

If more than one parameter constellation minimizes the number of intervals the one with minimum $\sum_i((r(t)^2-\sigma(t)^2)^2$ is chosen. This is not the same as minimizing the sum of the quadratic deviations given the minimum number of intervals as this would require a dynamic programming approach with a considerable additional computational overhead. Table 8.1 gives the results for the Standard and Poor's data for $p+q\leq 4$.

On the basis of these results the GARCH(1,1) model would seem to be best. It is noticeable that by far the worst results are with $p=0$. The results may be compared with the piecewise constant approach which has 77 breaks.

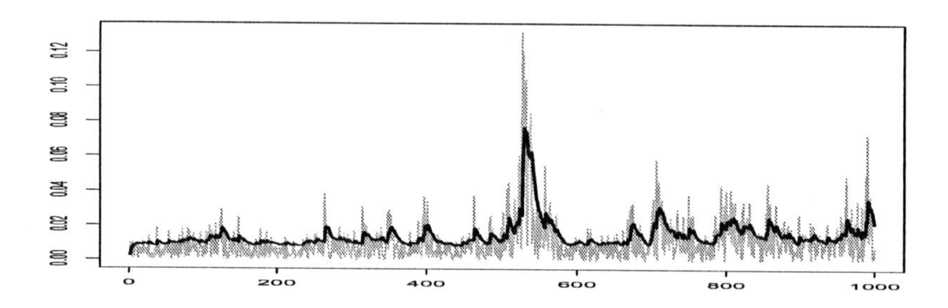

Figure 8.12 *The first 1000 of the Standard and Poor's data and the volatility as derived from the GARCH(1,1) model.*

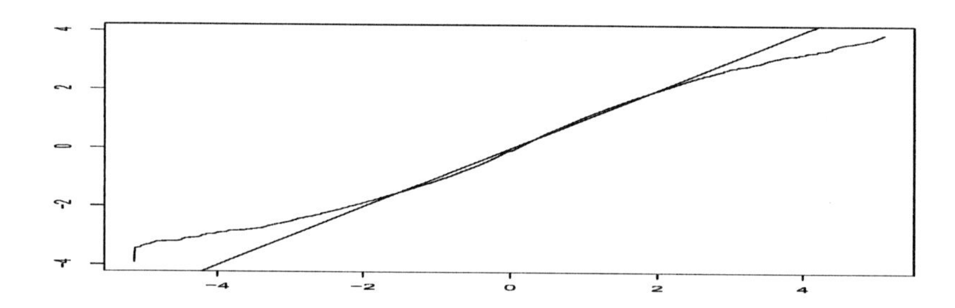

Figure 8.13 *The q-q-plot for the GARCH(1,1)-segmentation-model with $Z(t)$ standard Gaussian white noise.*

Figure 8.12 shows the first 1000 observations of the Standard and Poor's data together with the volatility as derived from the GARCH(1,1) model with parameters for this section given by

$$\alpha_0 = 0.0020,\ \alpha_1 = 0.16,\ \beta_1 = 0.81\,.$$

The GARCH segmentation completely removes the long range dependencies of the absolute values without any remaining short term dependencies. However just as for the piecewise constant Gaussian model the residuals differ significantly from the model assumption as is shown in the q-q plot of Figure 8.13.

8.4 The Taut String and Scale

The taut string method of Section 7.2.1 can be adapted to the case of scale: the observations x_i are replaced by the $|r_i|$ and the 'residuals' are $\sqrt{2/\pi}\, r_i/s_{n,i}$ where s_n

is the scale derived from the taut string. The factor $\sqrt{2/\pi}$ is due to

$$\mathbf{E}(|X|) = \sqrt{2/\pi} \tag{8.26}$$

for a $N(0,1)$ random variable X. The approximation region will be taken to be (8.5) where, for reasons of fast calculation, the family $\mathscr{I}_n$ will be a multiresolution scheme of the form (7.4). For given α and λ the values of $\alpha_n = \alpha_n(\lambda)$ can be determined by simulations. For example for $\alpha = 0.9$ and $\lambda = 1.5$

$$\alpha_n = 1 - 2\exp(-4.0 - 1.06 * \log(n)).$$

On intervals $I \in \mathscr{I}_n$ where the conditions of the approximation region are violated, the tube is squeezed locally and the s_n recalculated. This is continued until $\sqrt{\pi/2}s_n \in \mathscr{A}_n(\mathbf{r}_n, \mathscr{I}_n, \alpha)$. The local squeezing factor has a default value of 0.95, that is at each point where the tube is squeezed its width is reduced by a factor of 0.95.

The method can be adopted for other than the Gaussian model by transforming the residuals as in Chapter 8.2. For the t distribution with ν degrees of freedom the factor $\sqrt{2/\pi}$ of (8.26) must be replaced by

$$\mathbf{E}(|X_\nu|) = \frac{2\nu\Gamma((\nu+1)/2)}{\sqrt{\pi\nu}(\nu-1)\Gamma(\nu/2)}. \tag{8.27}$$

The algorithm is sufficiently fast for (8.17) and (8.18) to solved exactly rather than using the polynomial approximation to the function h.

The upper panel of Figure 8.14 shows the results for the first 1000 Standard and Poor's observations based on the Gaussian model. The lower panel shows the same for the t_9 model. Under the Gaussian model the volatility function for the complete data set has 161 internal local extreme values which contrasts with the 48 of Figure 8.3. The reason is the sensitivity of the taut string methodology to heavy tailed data. If it is applied to the simulated Gaussian data of (8.4) where the underlying volatility has 48 local extreme values it results in 44 local extreme values. If data is simulated by multiplying the piecewise constant volatility function of (8.3) with t_9 noise, the taut string methodology results in 190 local extreme values and a correspondingly poor reconstruction of the underlying volatility.

The t_9-model results in 40 local extreme values as against the 31 for the piecewise constant approach. Again the difference is largely due to the fact that the empirical tails are heavier than those corresponding to a t_9 distribution. If the volatility function of the simulated data of Figure 8.3 with 55 local extreme values is multiplied by t_9 noise the taut string method based on the t_9 model yields 30 local extreme values.

The q-q-plots for both models based on the taut string methodology are similar to the corresponding ones for the piecewise constant volatility approach although the deviations from the 45-degree straight line are not quite as pronounced.

8.5 Smooth Scale Functions

As in Chapter 7.2.2 it is theoretically possible to minimize the total variation of the scale function s_n subject to $s_n \in \mathscr{A}_n$ and possibly additional monotonicity constraints.

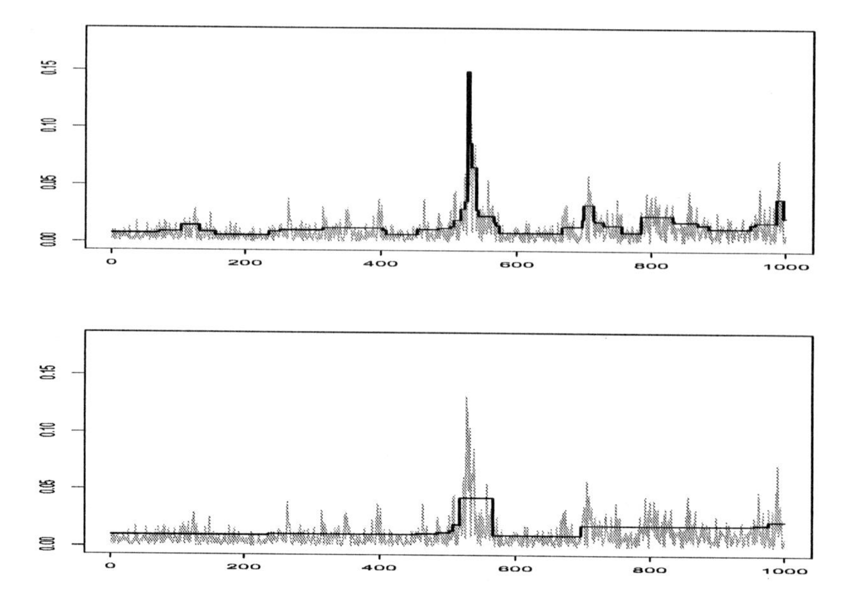

Figure 8.14 *Upper panel: the first 1000 observations of the Standard and Poor's data with the volatility calculated using the taut string method based on the Gaussian model. Lower panel: the same but for the t model with $\nu = 9$ degrees of freedom.*

In the location case this becomes a linear programming problem. For scale the problem is no longer linear. It can be made linear by minimizing the total variation of $1/s_n$ but this is not what is wanted. Smooth scale functions can be obtained by adapting the methods of Chapter 7.4.3 and Chapter 7.4.4. There is no guarantee that they are the smoothest functions in any sense, nor is it possible to take monotonicity constraints into account.

The weighted spline approach of Chapter 7.4.3 leads to the problem of minimizing

$$\sum_{i=1}^{n} w_i(|r_n(t_i)| - \sigma_n(t_i))^2 + \int_0^1 \sigma_n^{(2)}(t)^2 \, dt \tag{8.28}$$

where the w_i are the weights. The model is (8.1) with the $Z(t)$ standard Gaussian white noise. The solution s_n of the minimization of (8.28) is calculated: the residuals for checking the inequalities of the approximation region are $\sqrt{2/\pi}\, r_n(t_i)/s_n(t_i)$. Initially all the weights w_i are small: 10^{-8} is the default value. For points t_i in intervals where the inequalities are violated the weights w_i are increased by a fixed factor fct with default value $fct = 2$. The new solution s_n is calculated and the process continued until $\sqrt{\pi/2}\, s_n \in \mathscr{A}_n$. The upper panel of Figure 8.15 shows the result for the first 1000 observations of the Standard and Poor's data.

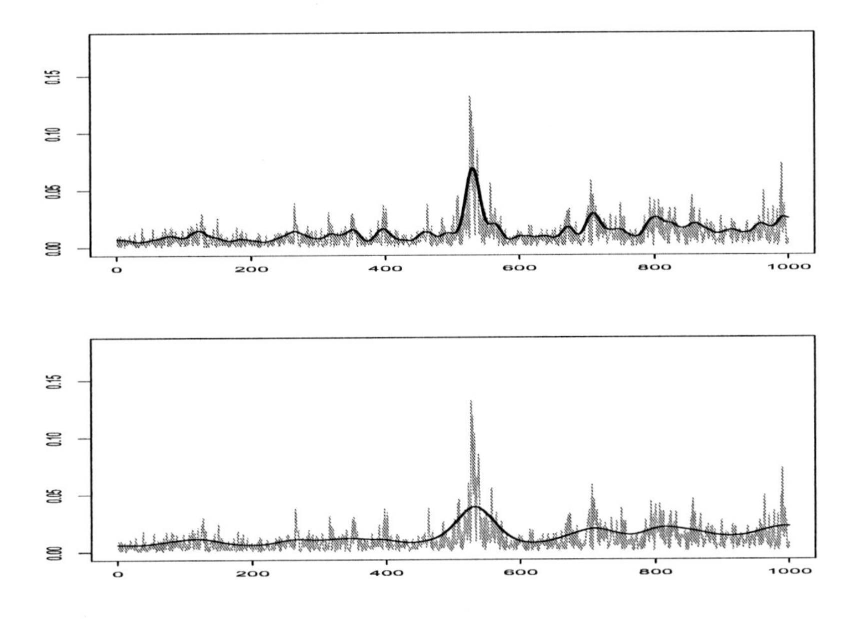

Figure 8.15 *Upper panel: the first 1000 observations of the Standard and Poor's data and the smoothing spline volatility. Lower panel: the same for the weighted penalized method.*

The second method of Chapter 7.4.3 is based on minimizing

$$\sum_{i=1}^{n}(|r_n(t_i)| - \sigma_n(t_i))^2 + \int_0^1 w(t)\sigma_n^{(2)}(t)^2\,dt \tag{8.29}$$

where the function w now weights the penalty term as in the local squeezing variant of the taut string methodology. The model is again (8.1) with the $Z(t)$ standard Gaussian white noise. The default starting values of w are large 10^8. The solution s_n is calculated and the inequalities defining $\mathscr{A}_n$ are checked for the residuals $\sqrt{2/\pi}\,r_n(t_i)/s_n(t_i)$. On intervals I where the inequalities are violated the values of $w(t)$ are reduced by a factor fct with default values 0.95. The lower panel of Figure 8.15 shows the results for the first 1000 observations of the Standard and Poor's data. Both methods can be extended to cover t distributions just as for the taut string method.

8.6 Comparison of the Four Methods

The scale problem is much more difficult than the location problem. If two sets of data are generated in the regression case as $X_i = 10 + Z_i$ and $Y_i = 1 + Z_i$ with the Z_i standard Gaussian white noise then it will require a huge sample before the smallest X_i is smaller than the largest Y_i. In other words the two samples are clearly separated.

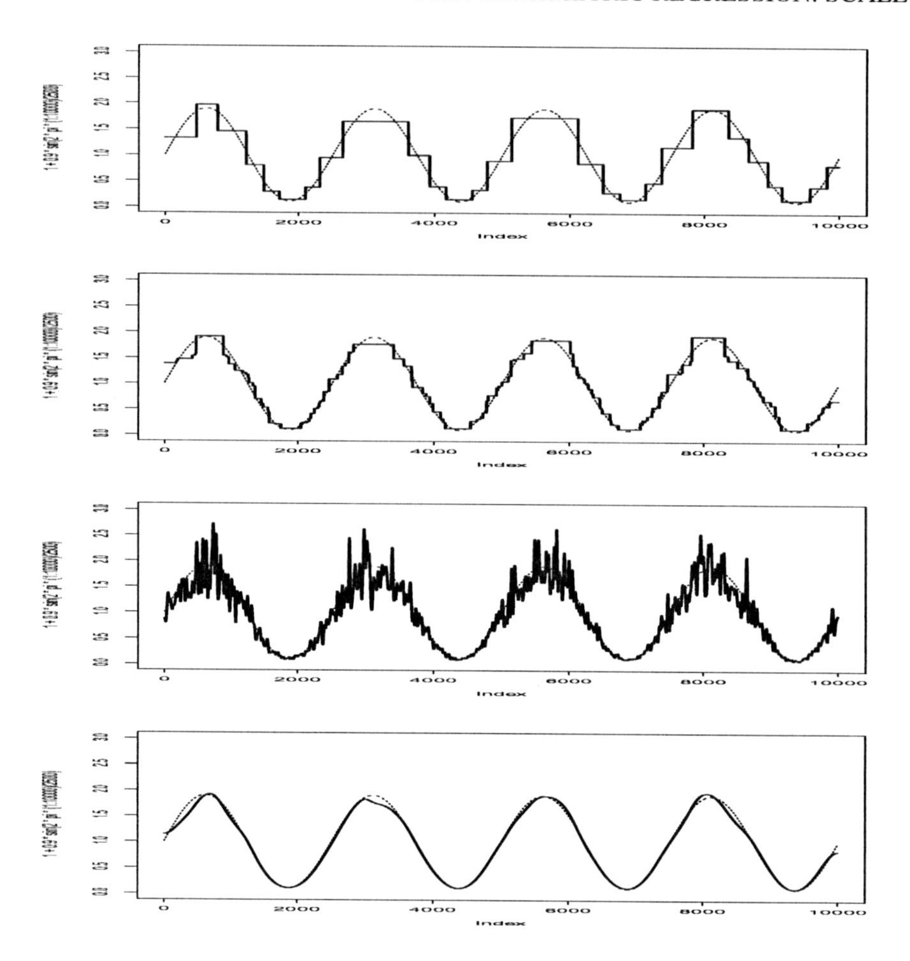

Figure 8.16 *The function $\Sigma(t)$ for the model (8.30) together with the reconstructions from top to bottom: (1) the piecewise constant scale method, (2) the taut string method, (3) the weighted spline method and (4) the weighted penalty spline method. The sample size is $n = 10000$.*

In the scale case $X_i = 10Z_i$ and $Y_i = Z_i$ then the separation is poor. Simulations show that $\min(|X_i|) < \max(|Y_i|)$ with a probability of about 0.78 for samples of size $n = 10$. As a consequence much larger samples are required to get reasonable estimates of the scale when compared with those required for location.

Figure 8.16 shows the results of applying the four methods for a data set of size $n = 10000$ data generated according to the model

$$R(t) = \Sigma(t)Z(t) = (1 + 0.9\sin(8\pi t))Z(t) \tag{8.30}$$

The weighted spline method performs the worst by far, not only in this simulation but in others. For this reason it will not be included in the further comparisons.

8.7 Location and Scale

Location and scale can be incorporated into the same model to give

$$X(t) = f(t) + \Sigma(t)Z(t), 0 \le t \le 1, \tag{8.31}$$

where mostly, but not always, $f(t)$ represents the signal and $\Sigma(t)$ is a nuisance parameter. As they stand the two functions f and Σ in (8.31) are confounded. They can only be separated on the basis of further information about relationship between the two. Typically this will be that the signal f is much smaller than the scale Σ, that is $|f(t)| << \Sigma(t)$. In this case the noise can be quantified using the model (8.1). Once this has been done the signal $f(t)$ can be quantified with the noise level taken as known. One example is the Standard and Poor's data, another is data from diffusion process where the incremental scale is of order $\sqrt{dt}$ and the drift of order dt.

If the signal is not small compared to the noise the level of noise can still be estimated by taking the differences of the observations. This will work if the signal is smooth. Define $\Delta_h(g(t)) = g(t+h) - g(t)$. Then

$$\begin{aligned} \Delta_h(X(t)) &= \Delta_h(f(t)) + \Delta_h(\Sigma(t)Z(t)) \\ &= \Delta_h(f(t)) + \Delta_h(\Sigma(t))Z(t+h) + \Sigma(t+h)\Delta_h Z(t) \\ &\approx \Delta_h(f(t)) + \Sigma(t)\Delta_h(Z(t)) \\ &\approx \Sigma(t)\Delta_h(Z(t)), 0 \le t \le 1. \end{aligned} \tag{8.32}$$

if h is small and f and Σ are essentially continuous. If f is only piecewise continuous, for example the blocks data, then jumps in f will cause spikes in the differences but these can be smoothed out by taking a running median. Using the scale model (8.32) the scale can be quantified by one of the several methods discussed above. Having done this the signal can be extracted by one of the methods considered in Chapter 7.

8.7.1 Financial Data

Typically in finance data the volatility will be much larger than the rate of return as in the Standard and Poor's data. This can be seen in Figure 8.1 where the signal is not visible against the background of the noise. The first step is to quantify the noise as in Section 8.2. Once this has been done the signal can in turn be quantified.

In [110] this is done by applying the idea of a piecewise constant function to the noise and to the signal. Interest in [110] centres on 'long term trends' and, although there is no precise definition of 'long term', the approximation region

$$\begin{aligned} &\mathscr{A}_n^*((\mathbf{t},\mathbf{x})_n, \mathscr{I}_n, \sigma, \gamma_n(\alpha)) \\ &= \Big\{g : |w((\mathbf{t},\mathbf{x})_n, g, \sigma, I)| \le \\ &\qquad \sqrt{2\log(n/|I|) + \gamma_n(\alpha)\log(\log(e^e n/|I|))} \\ &\qquad \text{for all } I' \subset I \in \mathscr{I}_n \text{ an interval constancy of } g\Big\} \end{aligned} \tag{8.33}$$

with

$$w((\mathbf{t},\mathbf{x})_n, g, \sigma, I) = \frac{1}{\sqrt{|I|}} \sum_{t_i \in I} \frac{r(t_i) - g(t_i)}{\sigma_n(t_i)}$$

was used. This favours long intervals of constancy.

For a given α, n and $\mathscr{I}_n$ the quantiles $\gamma_n(\alpha)$ of Γ_n of (7.9) may be determined by simulations. Unfortunately the distribution of the limiting Γ_∞ is not known and the convergence would, on the basis of simulations, seem to be slow. With these provisos the approximations

$$\begin{aligned} \gamma_n(0.90) &\approx 8.96 - 13.7/\log n \\ \gamma_n(0.95) &\approx 10.16 - 13.3/\log n \end{aligned} \tag{8.34}$$

will be used for the case that $\mathscr{I}_n$ is the set of all intervals. They are based on 5000 simulations for the sample sizes

$$n = 100, 250, 500, 1000, 2500, 5000, 1000, 20000.$$

It can be decided to treat all intervals of constancy equally and not place emphasis on long intervals. In such a case the approximation region

$$\begin{aligned} \tilde{\mathscr{A}}_n((\mathbf{t},\mathbf{x})_n, \mathscr{I}_n, \sigma, \tau_n(\alpha)) = \{g : |w((\mathbf{t},\mathbf{x})_n, g, \sigma, I')| \le \\ \sqrt{\tau_n(\alpha)\log n} \text{ for all } I' \subset I, I \text{ an interval of constancy}\} \end{aligned} \tag{8.35}$$

can be used. The values of $\tau_n(\alpha)$ can be obtained by simulations. The following approximations will be used for the case that $\mathscr{I}_n$ is the set of all intervals.

$$\begin{aligned} \tau_n(0.90) &\approx 2 + 4.7/\log n, \\ \tau_n(0.95) &\approx 2 + 6.2/\log n. \end{aligned} \tag{8.36}$$

The top panel of Figure 8.17 shows the result of applying the procedure to the Standard and Poor's data using the approximation region (8.33) with $\alpha = 0.9$. There is only one interval. The centre panel shows the result for the approximation region (8.35) with $\alpha = 0.9$. There are 15 intervals of constancy. The bottom panel shows the result of the procedure using the approximation region (8.35) with $\alpha = 0.95$ with 10 intervals of constancy. In all cases the signal has been multiplied by a factor of 20 to make it visible. The zero returns line is shown as a dotted line.

8.7.2 *Diffusion Processes*

The following is based on [223]. A diffusion process X is defined by the stochastic differential equation

$$dX(t) = \beta(X(t))dt + \gamma(X(t))dB(t) \tag{8.37}$$

where β is the drift term, γ the dispersion term and B is standard Brownian motion. If β has a bounded first derivative and γ is continuous then X satisfies (8.32) with the appropriate changes in notation. The difference to the other models considered is that the design points are now given by the process X itself rather than being given times $t_1, \ldots, t_n$. The upper panel of Figure 8.18 shows the first 5000 observations of a diffusion process $X(t)$ with

$$\beta(x) = -5x, \quad \gamma(x) = 0.05(0.2 + \sin(4\pi x)^2).$$

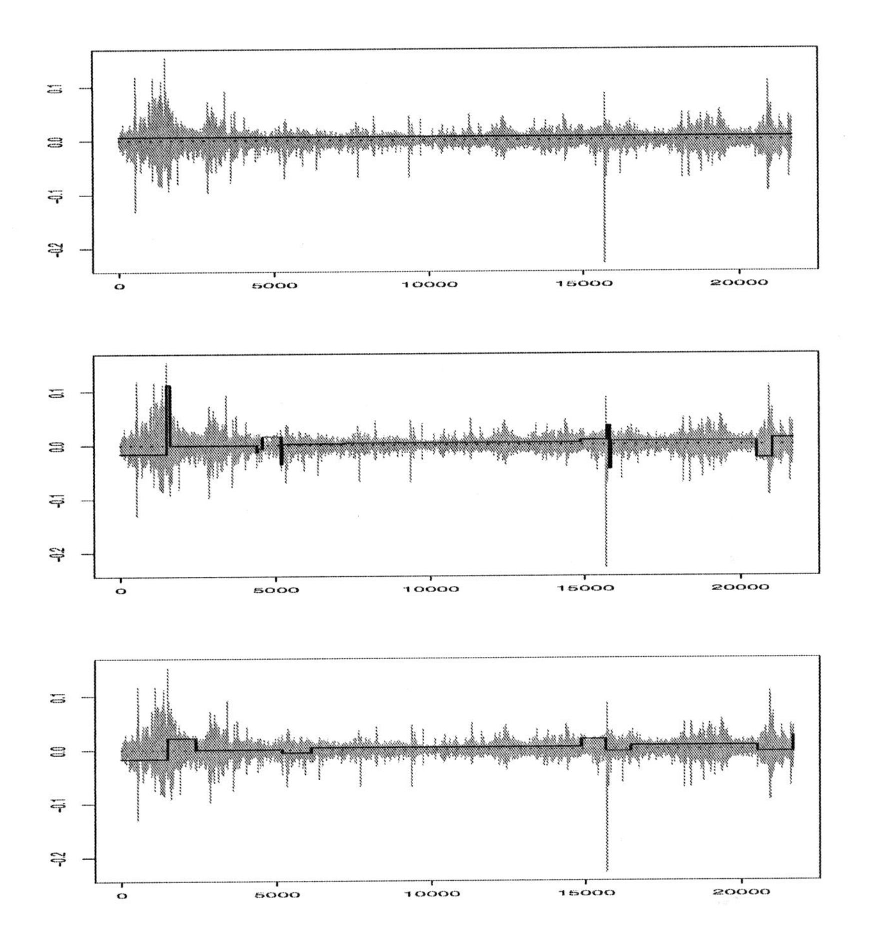

Figure 8.17 *Upper panel: The returns (×20) of the Standard and Poor's data together with the zero returns line for $\alpha = 0.9$ and the approximation region (8.33). Centre panel: the same but for the approximation region (8.35) with $\alpha = 0.90$. Bottom panel: the same but for the approximation region (8.35) with $\alpha = 0.95$.*

The starting point was $X_1 = 0$ and the data were generated by

$$X_{i+1} = X_i - 5x_i/m + 0.05(0.2 + \sin(4\pi X_i)^2)Z_i/\sqrt{m}, i = 1,\ldots,n-1$$

with $n = 50000$, $m = 1000$ and Z_i standard Gaussian white noise. The lower panel of Figure 8.18 shows the differences $X_{i+1} - X_i$ plotted against the X_i. The signal is hardly visible.

The noise can be quantified using any of the methods in this chapter. As however the support does not form a lattice the piecewise constant and taut string procedures

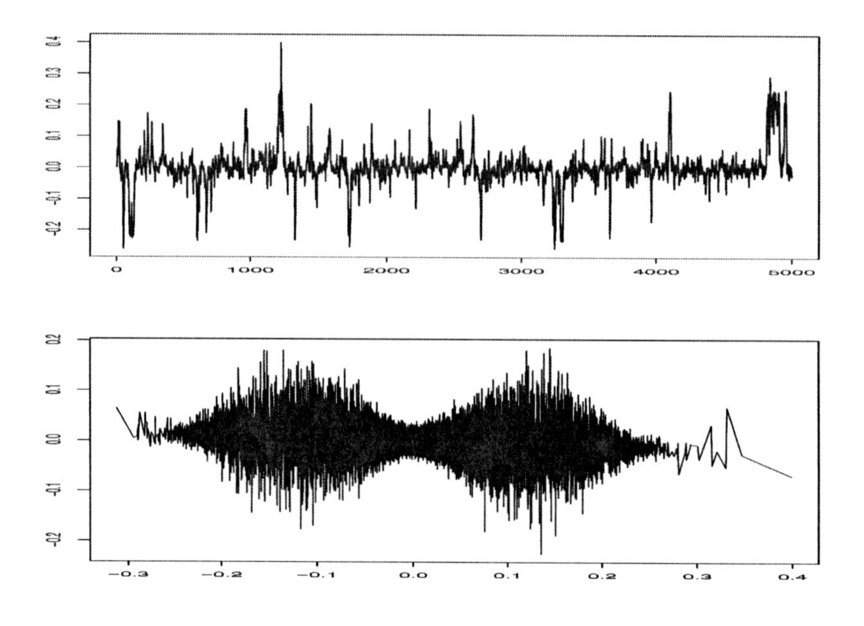

Figure 8.18 *The first 5000 values of the diffusion process.*

are more stable. The same applies to the quantification of the signal. The upper panel of Figure 8.19 shows the quantification of the noise using the piecewise constant method. The lower panel of shows the true and estimated signal again using the piecewise constant method.

Theoretical considerations may be found in [223].

8.7.3 *Differencing*

If the signal $f(t)$ is not negligible compared to the scale $\Sigma(t)$ then the scale can be quantified using (8.32) and then the signal quantified given the noise level. One example will suffice. Figure 8.20 shows the doppler function contaminated with block noise (see [62]) and the reconstruction using the using the piecewise constant approach for the scale and the signal.

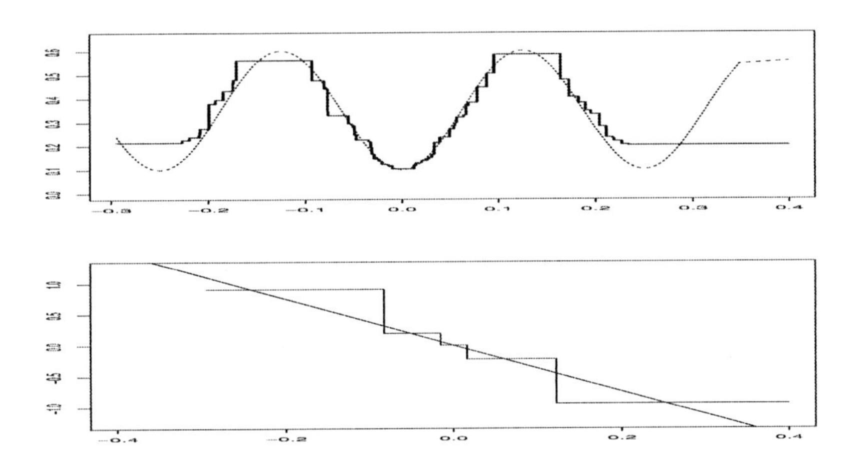

Figure 8.19 *Upper panel: estimated and true diffusion terms for the diffusion process of Figure 8.18. Lower panel: estimated and true drift terms for the diffusion process of Figure 8.18.*

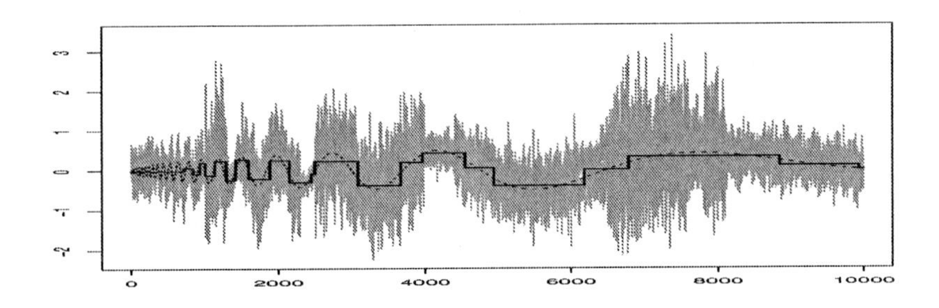

Figure 8.20 *The doppler function (dashed) contaminated with block noise together with the piecewise constant reconstruction.*

Chapter 9

Image Analysis

9.1 Two and Higher Dimensions

Image analysis is the two dimensional version of the one dimensional non-parametric regression discussed in the last two chapters. The extension to two dimensions is conceptually straight forward. Just as in one dimension an approximation region is defined followed by some form of regularization. The extra dimension does however cause difficulties both with respect to the definition of the approximation region and to the form of regularization. In addition there is a considerable increase in algorithmic complexity.

In the one-dimensional case the approximation region is based on intervals. This could be extended to say the unions of disjoint intervals but this does not alter the fact that intervals are the natural and the only practical choice in one dimension. The situation is different in two dimensions. The natural extension of intervals is rectangles, but it is also possible to use ellipsoids and intersections of ellipsoids. The only theoretical restriction on the subsets to be used in the definition of the approximation region is that they form a Vapnik-Cervonenkis class or a class with polynomial discrimination (see [172]). Practical considerations are of utmost importance and these reduce the possibilities more than the restriction to V-C classes does. Another property of the real line is that it is well ordered. Many algorithms of make use of this as do the two fast algorithms requiring little memory space developed by Kovac for minimizing the total variation of the function $TV(g)$ and its first derivative $TV(g^{(1)})$.

Shape constraints are a useful and intuitive form of regularization in one dimension but it is not easy to extend the idea to two dimensions. The simplest case is that of a monotone increasing function. In one dimension the meaning is unambiguous but in two dimensions it is necessary to specify the directions in which the function is monotone. Some work has been done on shape constraints in higher dimensions but only in the context of densities rather than non-parametric regression (see [190] and [168]).

Difficulties also arise in the definition of total variation or other measures of smoothness. For discrete data in one dimension the definition of the total variation $TV(g)$ of a function g is

$$TV(g) = \sum_{i=1}^{n-1} |g(t_{i+1}) - g(t_i)|.$$

In two dimensions on a regular lattice there are several possibilities such as

$$TV(g) = \sum_{\mathbf{i}_n} \left(\sum_{\mathbf{j}_n, |\mathbf{i}_n - \mathbf{j}_n| = 1/n} |g(\mathbf{i}_n) - g(\mathbf{j}_n)| \right)$$

or

$$TV(g) = \sum_{\mathbf{i}_n} \left(\sum_{\mathbf{j}_n, |\mathbf{i}_n - \mathbf{j}_n| = 1/n} (g(\mathbf{i}_n) - g(\mathbf{j}_n))^2 \right)^{1/2}.$$

The third and main problem in image analysis is the size of the data sets. In one dimension there are many data sets of size $n \le 1000$; a sample of size $n = 50000$ is considered large. In image analysis a sample of size $n = 1024$ converts to a square image of 32×32 pixels, that is a coarse image. An image of 256 pixels corresponds to a sample size of $n = 65536$, a 1024×1024 image corresponds to $n = 1048576$.

9.2 The Approximation Region

The basic model is

$$X(\mathbf{t}) = f(\mathbf{t}) + \sigma Z(\mathbf{t}), \mathbf{t} \in [0, 1]^2 \tag{9.1}$$

where $Z(\mathbf{t}), \mathbf{t} \in [0, 1]^2$ is standardized Gaussian white noise. Given data set $\mathbf{x}_n = \{x(\mathbf{t}_i) : 1 \le i \le n\}$ and a Vapnik-Cervonenkis family $\mathscr{C}$ of subsets of $[0, 1]^2$ ([172, 214]) an α-approximation region is given by

$$\mathscr{A}(\mathbf{x}_n, \mathscr{C}, \alpha, \sigma) = \Big\{ g : g : [0, 1]^2 \to \mathbb{R}, \max_{C \in \mathscr{C}} \left(\frac{1}{\sqrt{|C|}} \left| \sum_{\mathbf{t}_i \in C} (x(\mathbf{t}_i) - g(\mathbf{t}_i)) \right| \right) \le \sigma \sqrt{\tau_n(\alpha) \log n} \Big\} \tag{9.2}$$

where $\tau_n(\alpha)$ is such that

$$\mathbf{P}\left(\max_{C \in \mathscr{C}} \left(\frac{1}{\sqrt{|C|}} \left| \sum_{\mathbf{t}_i \in C} Z(\mathbf{t}_i) \right| \right) \le \sqrt{\tau_n(\alpha) \log n} \right) = \alpha. \tag{9.3}$$

The value of $\tau_n(\alpha)$ can be determined by simulations.

Typically a value for σ is required and must be obtained from the data. In general the $\mathbf{t}_i$ form a regular lattice in which case one possibility is

$$\begin{aligned}\hat{\sigma}_n &= \text{median}\{|x((i+1)/m,(j+1)/m)+x(i/m,j/m)-x((i+1)/m,j/m) \\ &\qquad -x(i/m,(j+1)/m)| : 1 \le i,j \le m-1\}. \qquad (9.4)\end{aligned}$$

Just as in one dimension if the median in (9.4) is replaced with an appropriate higher quantile (see (7.13) and [48]) then the approximation region becomes honest instead of exact irrespective of $\mathbf{f}$. This will be ignored and the $\hat{\sigma}_n$ of (9.4) will be used.

For data generated under the model (9.1) a particular family $\mathscr{C}$ can only work well if the function f can be well approximated by $\mathscr{C}$, that is if

$$f \approx \sum_i f_i C_i. \qquad (9.5)$$

In principle any Vapnik-Cervonenkis family of subsets $\mathscr{C}$ can be used in (9.2) but in practice the choice is limited by the ability to calculate $\sum_{\mathbf{t}_i \in C}(x(\mathbf{t}_i) - g(\mathbf{t}_i))$ for all $C \in \mathscr{C}$. In one dimension where $\mathscr{C}$ consists only of intervals the calculation of $\sum_{t_i \in I}(x(t_i) - g(t_i))$ is done most efficiently by calculating the partial sums $S_0 = 0, S_j = \sum_{i=1}^{j}(x(t_i) - g(t_i)), j = 1,\ldots,n$ and then taking differences

$$\sum_{t_i \in I}(x(t_i) - g(t_i)) = S_j - S_{i-1}, \text{ for } I = [i/n, j/n].$$

This is a version of Green's formula. In two dimensions a similar simplification is available for rectangles. However in general it is prohibitively expensive to consider all rectangles: there are approximately 10^9 rectangles for a 256×256 image. One possibility is to consider only squares rather than rectangles. For a 256×256 image this reduces the number to approximately $10^{4.5}$ but at the cost of flexibility. A better approach is to use a multiresolution scheme as in (7.4). As an example consider a $2^{m_1} \times 2^{n_1}$ image and let $\mathscr{C}$ be the set of rectangles with indices of the form

$$[(i_1-1)2^{k_1}+1, i_1 2^{k_1}] \times [(j_1-1)2^{\ell_1}+1, j_1 2^{\ell_1}]$$

with $1 \le i_1 \le 2^{m_1-k_1}, 0 \le k_1 \le m_1$ and $1 \le j_1 \le 2^{n_1-\ell_1}, 0 \le \ell_1 \le n_1$. This results in $4 \cdot 2^{m_1+n_1}$ rectangles. This corresponds to the choice $\lambda = 2$ in (7.4). A finer scheme results if the 2^{k_1} and 2^{ℓ_1} are replaced by λ^{k_1} and λ^{ℓ_1} respectively with $1 < \lambda < 2$ with the obvious modifications. The number of rectangles remains $O(2^{m_1+n_1})$.

The restriction to rectangles means that a good reconstruction is only possible for function f which can be well approximated as in (9.5) with the C_i rectangles. One way of improving the degree of approximation without a large increase in computational cost is to use so called wedgelets (see [64], [90], [91]). Given a rectangle wedgelets in the sense of this section are the two polygons which are created when the rectangle is bisected by a line as in Figure 9.1. For any given slope α the sums of $x((t_{i_1/m}, t_{j_1/n})) - g((t_{i_1/m}, t_{j_1/n}))$ over the wedgelets obtained by bisecting the rectangles by lines of slope α can be efficiently calculated ([90], [91]). The total computational cost is proportional to the number of slopes considered.

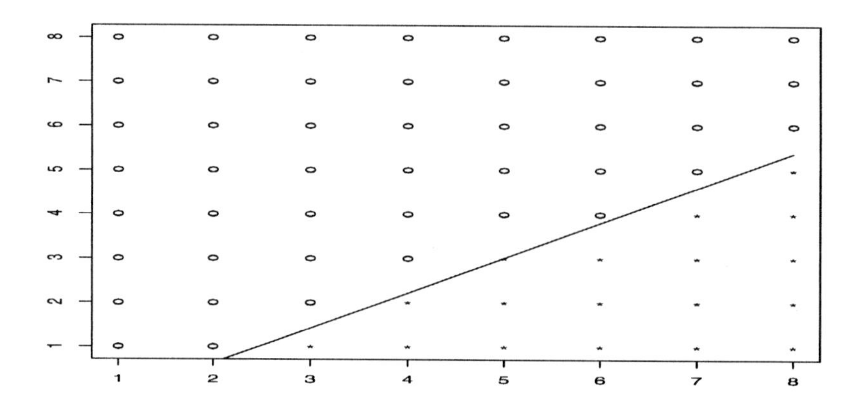

Figure 9.1 *Two wedgelets obtained from bisecting a rectangle with a line.*

9.3 Linear Programming and Related Methods

The basic optimization problem (7.15) of the one dimensional case can be carried over without change to the two (or higher) dimensional case namely,

$$\text{minimize } J(g) \text{ subject to } g \in \mathscr{A}_n$$

where now $g : [0,1]^2 \to \mathbb{R}$, J is a convex functional and $\mathscr{A}_n$ is a family of subsets of $[0,1]^2$ (see (5.2) in [25]). In [192] the functional $J(g) = \|Tg\|_{\ell_1}$ is considered where T is a linear transformation obtained by combining the linear transformations $(T_j)_1^\nu$, in particular wavelets and curvelets, and $\|\cdot\|_{\ell_1}$ is the ℓ_1 norm of the coefficients. An exact solution of the optimization problem is difficult to calculate but an algorithm for an approximate solution is given. In [192] total variation minimization is mentioned as a further possibility. It is considered in [25] and [144] where the approximation regions are defined in terms of curvelets and cubic spline wavelets respectively. As mentioned in [25] for a 512×512 image the optimization problem is 'daunting' and in both papers algorithms for approximate solutions are given.

A detailed analysis of the optimization problem is given in [88] together with a convergent algorithm and application to two images one of which is given in Figure 9.2. It is a 512×512 black and white image on the range 0-1 (top left panel) contaminated with $\mathfrak{N}(0, 0.1^2)$ noise (top right panel). The approximation region is given by squares with side lengths from 1 to 25. The reconstruction is the bottom panel of Figure 9.2.

A further example given in [88] comes from confocal microscopy. The original image is blurred by a Gaussian kernel and the data are modelled as Poisson counts. Figure 9.3 shows a recording of a PtK2 cell taken from a kidney. The top image shows the original data, the bottom image the deconvolved and denoised image. A detailed description of the method is given in [88].

Figure 9.2 *Top left: original. Top right: noisy image. Bottom: de-noised.*

9.4 Choosing Smoothing Parameters

The methods of the previous section lead to high quality denoising of images. However they lead to difficult optimization problems which can be very time consuming: in [88] a time of two hours is reported for each image. Faster but lower quality denoising can be achieved by adopting the ideas of Section 7.4 to images.

9.4.1 *Diffusion*

The following is the two dimensional version of the diffusion method of Section 7.4.5. It is based on [195] and the more comprehensive treatment in [112].

The notation in this section will be different. The data will be denoted by y_{ij}, the spatial coordinates by $\mathbf{x} = (x_1, x_2)$ and time by t. The inhomogeneous two-dimensional heat equation is

$$\frac{\partial u(\mathbf{x},t)}{\partial t} = a(\mathbf{x})\frac{\partial^2 u(\mathbf{x},t)}{\partial x_1 \partial x_2} \tag{9.6}$$

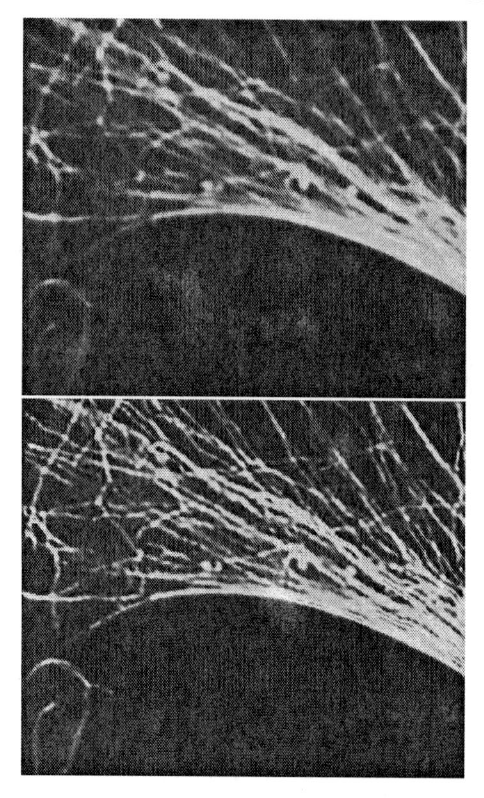

Figure 9.3 *Top: original. Bottom: deconvolved and denoised.*

where $a \geq 0$ is the diffusivity. The method is as follows. For a given diffusivity function a the differential equation (9.6) is solved for $t = 1$ with

$$u(\mathbf{x}_{ij}, 0) = y_{ij}$$

and the Neumann boundary conditions

$$\frac{\partial u(\mathbf{x}, t)}{\partial \mathbf{n}} = 0, \, \mathbf{x} \in \partial\Omega$$

where $\mathbf{n}$ denotes the normal on the boundary $\partial\Omega$ of the domain of definition which is here $\Omega = [0, 1]^2$. The denoised version of the data is then $\hat{f}(\mathbf{x}_{ij}) = u(\mathbf{x}_{ij}, 1)$.

The diffusivity function a is data dependent and chosen as follows. Initially a has a large value which causes the heat to diffuse quickly. The result is that $u(\mathbf{x}_{ij}, 1)$ is almost constant and equal to the mean of the data. The multiresolution inequalities which define the approximation region are checked for the function $u(\mathbf{x}_{ij}, 1)$. On points $\mathbf{x}_{ij} \in C \in \mathscr{C}$ where the conditions are violated the diffusivity $a(\mathbf{x}_{ij})$ at that point is reduced. This is continued until finally $u(\mathbf{x}, 1) \in \mathscr{A}(\mathbf{y}_n, \mathscr{C}, \alpha, \sigma)$.

Figure 9.4 shows four images of Rahel Stichtenoth. From top to bottom they are the original image, the noisy image, the denoised picture using local diffusivity and the denoised image using global diffusivity. Figure 9.5 taken from [112] shows the corresponding images for an artificial data set. In both cases the denoised image with constant diffusivity was obtained by altering the diffusivity until all the multiresolution conditions were fulfilled. There are many different regimes doing this and different ones were used in Figure 9.4 and Figure 9.5. All regimes however have strengths and weaknesses. The local adaption works well and a high degree of denoising is achieved but edge preservation is not as good as it should be. The human brain can denoise many images better than algorithms and if the noise level is not too high it can even prefer the noisy image to the denoised ones.

9.4.2 *Total Variation Penalty*

The following is due to Arne Kovac and Andrew Smith and is a form of taut string procedure in two dimensions. They consider the penalized least squares problem

$$\begin{aligned}\text{argmin}\Big(&\sum_{i,j}(x(i,j)-f(i,j))^2+\sum_{i,j}\lambda^1(i,j)|f(i+1,j)-f(i,j)|\\&+\sum_{i,j}\lambda^2(i,j)|f(i,j+1)-f(i,j)|+\sum\lambda^3(i.j)|f(i+1,j+1)-f(i,j)|+\\&\sum_{i,j}\lambda^4(i,j)|f(i+1,j)-f(i,j+1)|\Big)\end{aligned}$$

The weights $\lambda^k, k=1,2,3,4$ are determined iteratively and the solution calculated at each iteration. This may be seen as a two-dimensional version of the taut string with local squeezing.

From top to bottom the first three images of Figure 9.6 are the original image (see [173]), the noisy version and the version denoised with total variation penalized least squares.

The same image was considered in [132] where the image was treated as a graph. Again a penalized least squares method was used with an L_1 penalty on each edge of the graph. The only information used were the neighbouring points of each point on the graph: the distance between points played no role. The bottom image of Figure 9.6 shows a denoised version based on this graphical procedure.

Figure 9.4 *Four images of Rahel Stichtenoth.*

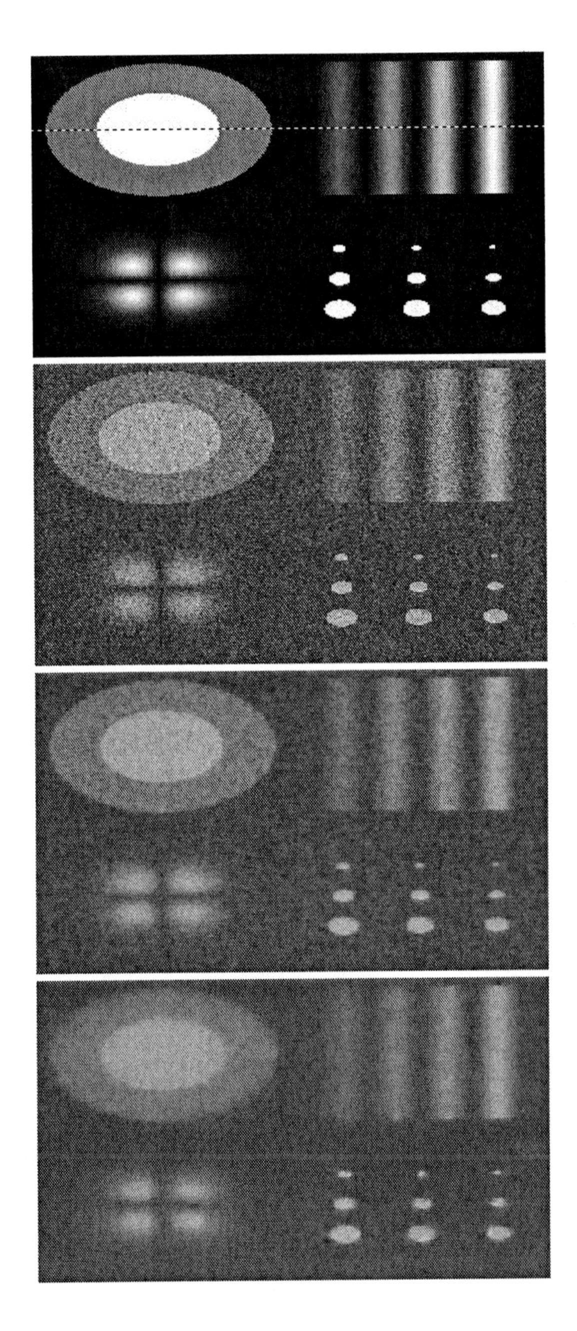

Figure 9.5 *From top to bottom: original, noisy image, local diffusivity, constant diffusivity.*

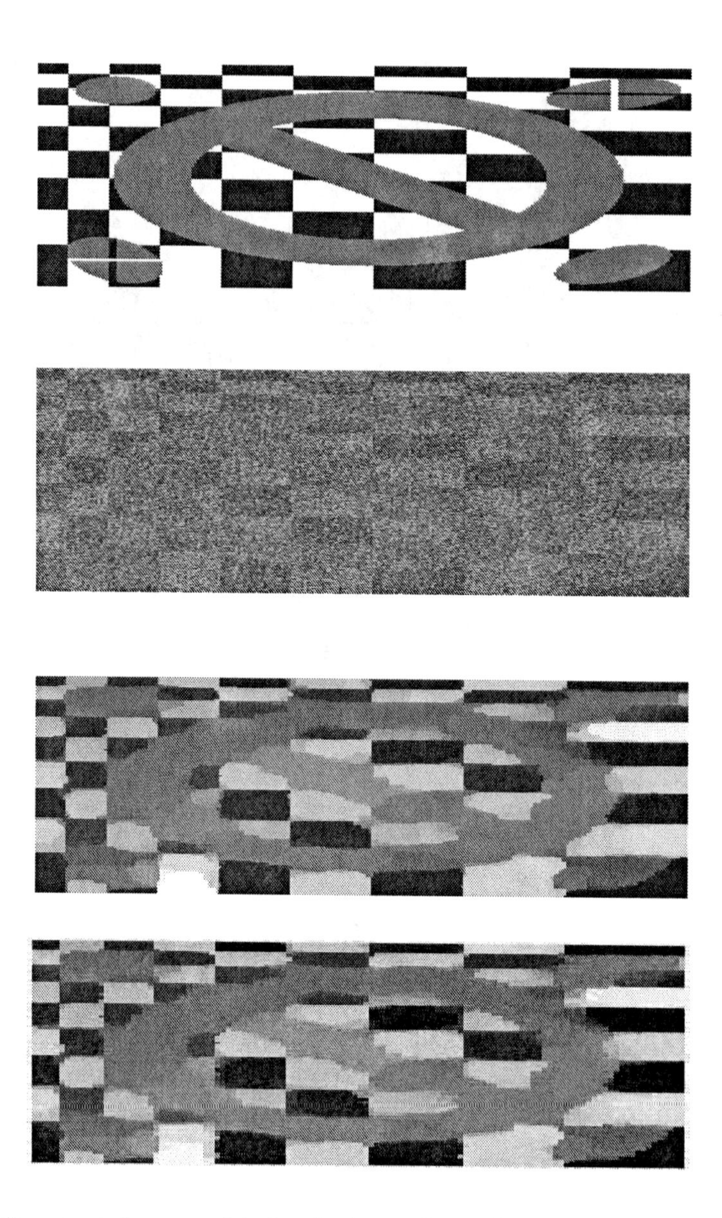

Figure 9.6 *From top to bottom: original, noisy image, denoised image using total variation penalized least squares, denoised image based on penalizing edges in graph with the* L_1 *norm.*

Chapter 10

Non-parametric Densities

The following is mainly based on [47] and [52].

10.1 Introduction

The problem of providing an adequate non-parametric density for a data set is much more difficult than that of providing an adequate non-parametric regression function. The claims made for any particular method are correspondingly modest and this applies to the method described below. There are several reasons for this. Firstly, it is more of an ill-posed problem so to speak than non-parametric regression as it involves differentiation. Secondly, a density is a measure of closeness of points whereas non-parametric regression has an additive error. Thirdly, the additive error is often modelled by some parametric family which makes matters easier as the approximation region can be based on this. There is no equivalent for the density problem. Fourthly, although distribution free metrics such as the Kuiper metrics can be used to measure the closeness of the density as expressed by its distribution function to the empirical distribution function, the resulting bounds are typically too weak to pick up relevant features of the density. The situation is not satisfactory and although the taut string procedure to be described works well it is not entirely clear why this is so.

10.2 Approximation Regions and Regularization

A useful test density is the claw density of [149]. It is a mixture of six normal densities and is defined by

$$f_{\text{cw}}(x) = \sum_{i=1}^{6} w_i \varphi((x-\mu_i)/\sigma_i)/\sigma_i \tag{10.1}$$

where φ is the density of the standard normal distribution,

$$w_1 = 0.5, w_2 = \ldots = w_6 = 0.1, \quad \sigma_1 = 1, \sigma_2 = \ldots = \sigma_6 = 0.1$$

and

$$\mu_2 = -\mu_6 = 1, \mu_3 = -\mu_5 = 0.5, \mu_1 = \mu_4 = 0.$$

The density has five peaks and is shown in the upper panel of Figure 10.1. The lower panel shows a histogram of a sample of size $n = 1000$.

The in a sense simplest 0.95-approximation region is that based on the Kolmogorov metric

$$\mathscr{A}(\mathbf{x}_n, 0.95, d_{\text{ko}}) = \{P : d_{\text{ko}}(P, \mathbb{P}_n) \leq 1.36/\sqrt{n}\} \tag{10.2}$$

and the simplest form of regularization is to minimize the number of peaks

$$\min J(P) \quad \text{subject to} \quad P \in \mathscr{A}(\mathbf{x}_n, 0.95, d_{\text{ko}}) \tag{10.3}$$

where $J(P)$ denotes the number of peaks of the density f of P. The problem can be solved using the taut string. The Kolmogorov ball $\{P : d_{\text{ko}}(P, \mathbb{P}_n) < \varepsilon\}$ is simply the tube of diameter 2ε centred at the empirical distribution function $\mathbb{F}_n$ of $\mathbb{P}_n$. A distribution which miminizes the number of peaks of the density in this ball is the taut string (see [45] and [141]).

There are two surprising results: firstly, how poor the method is, and secondly, how close it is possible to get to the empirical distribution without the number of peaks being over estimated.

The first surprise: In a simulation of 1000 samples of size $n = 5000$ the solution of (10.3) had one peak in all cases. For samples of size $n = 8000$ the solutions of (10.3) had one peak in about 52.2% of the cases. It required samples of size $n = 13000$ before the correct number of peaks was detected in 90% of the cases.

The second surprise: If the factor 1.36 in (10.2) is replaced by 0.18 then for samples of size $n = 500$ the correct number of peaks is found in about 94% of the cases. In other words although the taut string is extremely close to the empirical distribution function there are no artifacts. A possible explanation of this is a result given in [128] where it was shown that for data generated with a nonincreasing density function

$$\min_{F, F \text{ concave}} d_{\text{ko}}(F, \mathbb{F}_n) = o(n^{-1/2}).$$

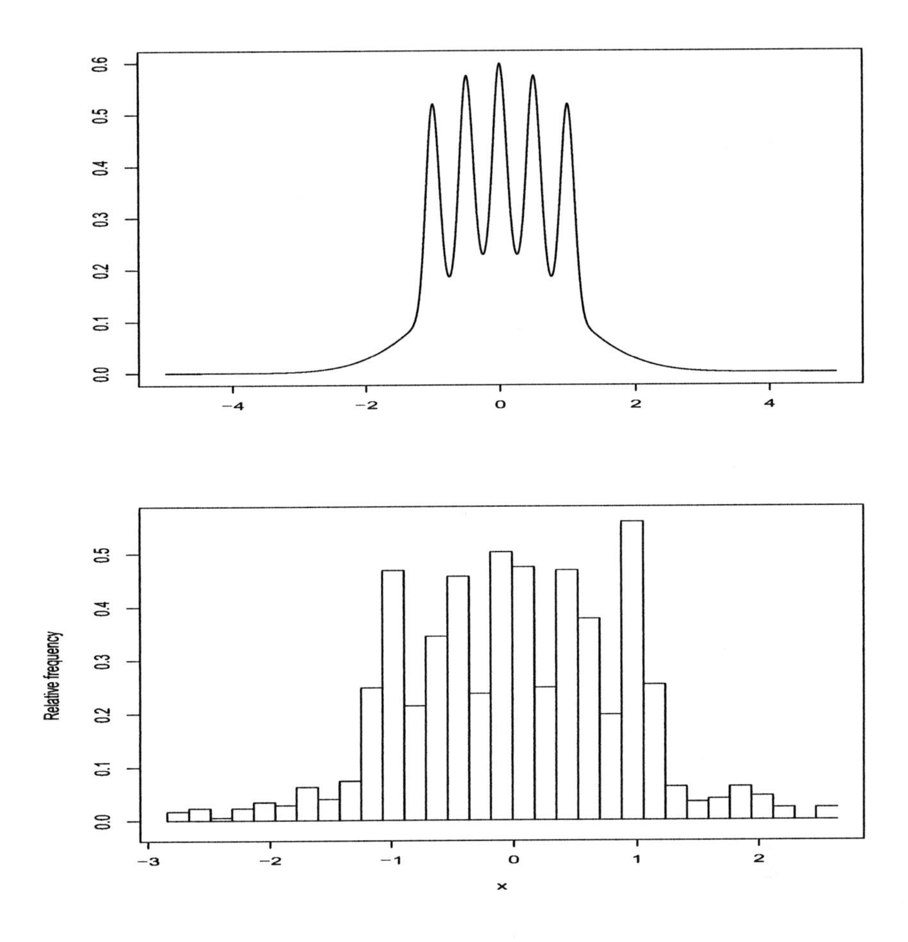

Figure 10.1 *Upper panel: the claw density f_{cl} of (10.1). Lower panel: histogram of a sample of size $n = 1000$.*

The Kolmogorov metric in (10.2) can be replaced as in [47] by a Kuiper metric d_{ku}^k of some order k (2.25) to give a Kuiper approximation region

$$\mathscr{A}(\mathbf{x}_n, 0.95, d_{\mathrm{ku}}^k) = \left\{ P : d_{\mathrm{ku}}^k(P, \mathbb{P}_n) \leq \mathrm{quku}(\alpha, k) \right\} \tag{10.4}$$

where $\mathrm{quku}(\alpha, k)$ is the α-quantile of d_{ku}^k. For $k = 1$ a simple expression for the asymptotic quantiles is known (2.24). For other values of k the asymptotic quantiles can be simulated using a Brownian Bridge. The quantiles for small samples can be obtained by simulations using uniform random variables. In [47] the default value $k = 19$ was used.

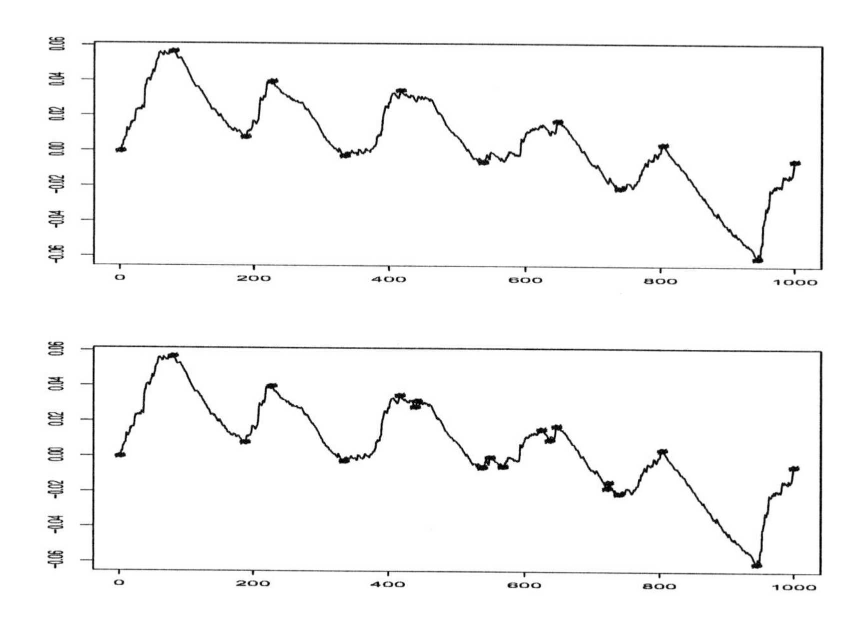

Figure 10.2 *Upper panel: a plot of* $\Phi - \mathbb{F}_n$ *for a sample of size* $n = 1000$ *generated under the claw distribution together with the intervals picked out by the* d_{ku}^{11} *metric. Lower panel: the same for the* d_{ku}^{19} *metric.*

The usefulness of a Kuiper metric for detecting peaks is seen in Figure 10.2. The upper panel shows the plot of the difference of the distribution function Φ of the standard normal distribution and the empirical distribution function $\mathbb{F}_n$ of $n = 1000$ claw random variables together with the 11 intervals which give the value of $d_{\mathrm{ku}}^{11}(\Phi, \mathbb{F}_n) = 0.464$. The five peaks are apparent and it is seen that the Kuiper metric picks out then and the troughs between them. In this case the number 11 corresponds to the five peaks $5 = (11-1)/2$. The lower panel shows the same for the d_{ku}^{19} metric. It also picks out the major peaks and troughs but also some small intervals which however do not contribute very much to $d_{\mathrm{ku}}^{19}(\Phi, \mathbb{F}_n) = 0.500$. The default value of $k = 19$ means that the metric is geared to distributions with at most $(19-1)/2 = 9$ peaks. The total variation of $TV(\Phi - \mathbb{F}_n)$ for the data of Figure 10.2 is 0.814 more than half of which is due to the eleven intervals picked out by the d_{ku}^{11} metric.

In the case of non-parametric regression it was pointed out in Chapter 7.2 that minimizing the number of peaks is strongly related to minimizing the total variation. Applying this to the density problem suggest minimizing $TV(f)$ subject to $P_f \in \mathscr{A}(\mathbf{x}_n, 0.95, d_{\mathrm{ku}}^k)$ where P_f has density f. This can be reduced to a linear programming problem but in practice it is not solvable due to the large number of constraints. Given this one plausible strategy is the one adopted for the taut string in the case of non-parametric regression, namely to decrease the width of the tube until the taut string lies in the approximation region. This will be treated in more detail in the next section.

The approximation regions (10.2) and (10.4) are not multiscale but multiscale approximation regions can be defined using the results of [78] and [183]. They are based on ratios of order statistics $x_{(19} < x_{(2)} < \ldots < x_{(n)}$

$$T_{jk}(\mathbf{x}_n) = \sqrt{\frac{3}{k-j-1}} \sum_{i=j+1}^{k-1} \left(2\frac{x_{(i)} - x_{(j)}}{x_{(k)} - x_{(j)}} - 1 \right). \qquad (10.5)$$

The critical values of $T_{jk}(\mathbf{X}_n)$ for any i.i.d.sample $\mathbf{X}_n$ are majorized by those of $T_{jk}(\mathbf{U}_n)$ where the U_i are i.i.d. and uniformly distributed on $[0,1]$. Upper bounds for the critical values can therefore be obtained by simulating under the uniform distribution. Thus for any α critical values $c_{jk}(\alpha)$ can be determined so that

$$\mathbf{P}(T_{jk}(\mathbf{X}_n) \le c_{jk}(\alpha) \text{ for all } 1 \le j < k-1 \le n-1) \ge \alpha. \qquad (10.6)$$

The method does not give a density but specifies intervals on which any adequate density must be either decreasing or increasing. The results for the claw density are much better than those based on the Kolmogorov metric in that they detect the correct number of intervals of decrease or increase, namely 10, for much smaller samples sizes. For samples of size $n = 1000$ the correct number is found in about 53% of the simulations, for samples of size $n = 2000$ this increases to 95%. Although the results are better they are still disappointing: the histogram of the lower panel of Figure 10.1 'clearly' shows that there are five peaks, and such data are typical.

The results given in [78] and [183] show that the method is asymptotically optimal but in the sense of Chapter 11.5 the asymptotics are strictly non-applicable, that is whatever the sample size there is no guarantee that the asymptotics are sufficiently accurate for them to be applied.

Although the method does not specify a density it leads naturally to the multiscale approximation region

$$\mathscr{A}(\mathbf{x}_n, 0.95, T) = \left\{F : T_{jk}(F(\mathbf{x}_n)) \le c_{jk}(\alpha), 1 \le j < k-1 \le n-1\right\}. \qquad (10.7)$$

Again there must be some form of regularization and one that is in principle tractable is to minimize the total variation $TV(f)$ of the density f of P_f with $P_f \in \mathscr{A}(\mathbf{x}_n, 0.95, T)$ which is a standard linear programming problem. The 'in principle tractable' refers to the large number of inequalities: in practice it can only be done for the smallest of samples.

10.3 The Taut String Strategy for Densities

Minimizing the number of peaks of a density or its total variation subject to the corresponding distribution lying in either of the approximation regions (10.4) or (10.7) is numerically impossible except for such small sample sizes that make nonsense of a non-parametric approach to specifying a density. However the taut string methodology of Chapter 10.7 whereby the width of the tube is decreased until the distribution lies in the approximation region can also be applied to densities. The approximation region (10.7) even permits the use of local squeezing by squeezing the tube only on those intervals $[x_{(j)}, x_{(k)}]$ for which the inequality

$$T_{jk}(F(\mathbf{x}_n)) \leq c_{jk}(\alpha)$$

does not hold. As there is nothing new in this only the results are given for $\alpha = 0.95$.

For $n = 500$ the correct number of peaks of the claw density is returned in 8.66% of the cases for the approximation region (10.4) and in 12.3% of the cases for the approximation region (10.7). The raw mode hunting algorithm of [78] and [183], that is without the taut string and based solely on the number of interval of increase/decrease, returns the correct number of peaks in 4.38% of the cases. The corresponding percentages for $n = 1000$ are 76.5%, 65.6% and 55.9%.

The results for the approximation region (10.4) and $n = 500$ are particularly disappointing. During the proof reading stage of [47] Arne Kovac suggested using the increments of the Kuiper metrics

$$d_{\text{ku}}^1, d_{\text{ku}}^2 - d_{\text{ku}}^1, d_{\text{ku}}^3 - d_{\text{ku}}^3, \ldots$$

rather than the d_{ku}^k directly. The corresponding approximation region is

$$\mathscr{A}^*(\mathbf{x}_n, 0.95, d_{\text{ku}}^k) = \{P : d_{\text{ku}}^i(P, \mathbb{P}_n) - d_{\text{ku}}^{i-1}(P, \mathbb{P}_n) \leq \text{quku}(\alpha, i, k), i = 1, \ldots, k\} \tag{10.8}$$

with $d_{\text{ku}}^0 = 0$. The $\text{quku}(\alpha, i, k)$ are chosen to give an α-approximation region. Figure 10.3 shows the Kuiper distances (upper panel) and their differences (lower panel) for a sample of size $n = 500$ after global squeezing. The method results in a considerable improvement as for samples of size $n = 500$ the correct modality is found in 78.9% of the cases, rising to 92.1% for $n = 1000$. Why this works so well remains a mystery.

10.4 Smoothing the Taut String Approximation

Figure 10.4 shows the claw density and an estimate using the taut string methodology for a sample of size $n = 500$. A smooth density can be calculated by adapting the methods of Chapter 7.2.2.

Denote the ordered data set by $x_{(1)}, \ldots, x_{(n)}$, empirical distribution by $\mathbb{F}_n$, the taut string solution by $F_{n,ts}$ with density $f_{n,ts}$ and the smoothed version by $F_{n,s}$ with density $f_{n,s}$. To prevent the solution moving too far away from the data the requirement

$$d_{ko}(F_{n,s}, \mathbb{F}_n) \leq \frac{1.36}{\sqrt{n}} \tag{10.9}$$

will be imposed together with $F_{n,s}(x_{(1)}) = 0, F_{n,s}(x_{(n)}) = 1$. The number, positions and the values of the peaks and troughs are made to agree with those of the taut string density $f_{n,ts}$. Furthermore the density must be nonnegative and integrate to one. All restrictions are linear inequalities of which there are at least $3n$ and at most

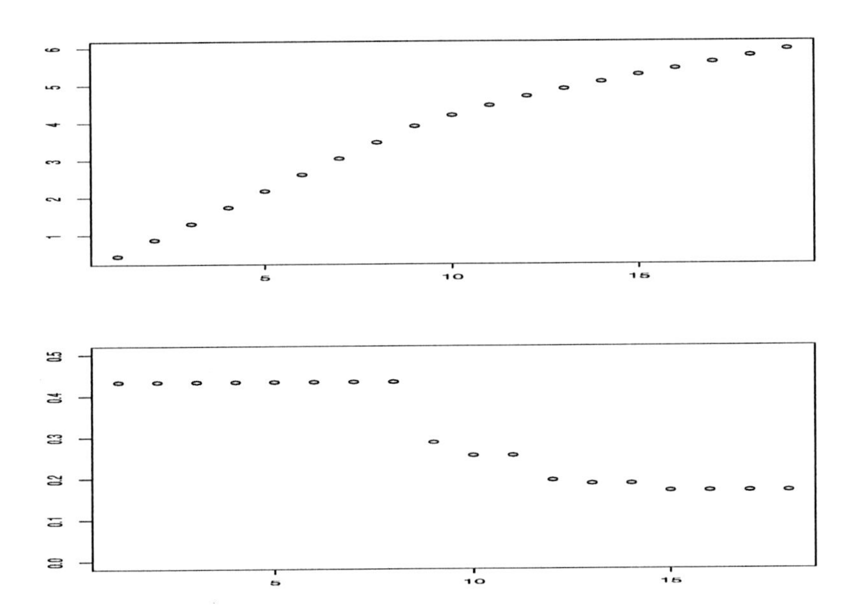

Figure 10.3 *Upper panel: plot of* $d^i_{ku}, i = 1,\ldots,199$ *for a sample size* $n = 500$ *after global squeezing. Lower panel: the differencess* $d^1_{ku}(P,\mathbb{P}_n)$ *and* $d^i_{ku}(P,\mathbb{P}_n) - d^{i-1}_{ku}(P,\mathbb{P}_n), i = 2,\ldots,19.$

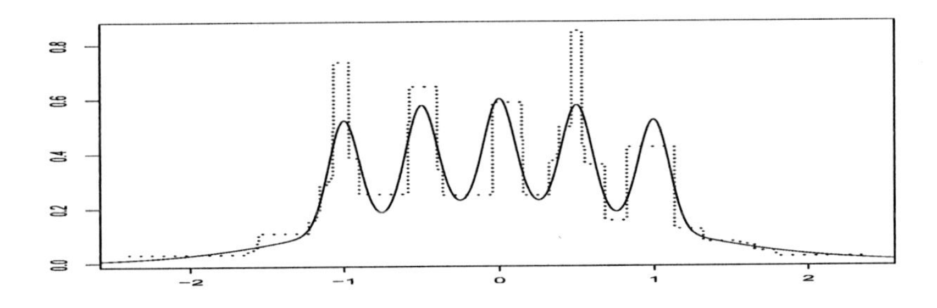

Figure 10.4 *The claw density (solid) and a non-parametric estimate (dashed) using the taut string methodology for a sample of size* $n = 500$.

$4n$. Given these inequalities the total variation of $TV(f_{n,s}^{(k)})$ of the kth derivative $f_{n,s}^{(k)}$ of $f_{n,s}$ is miminized. For a smooth density $k \geq 2$ is required but values up to $k = 4$ are in general possible. To reduce the numerical complexity there is no attempt to check any of the conditions of the approximation region used to calculate the taut string density.

For large sample sizes it is not possible to solve the resulting linear programme. The problem can be overcome by either using a subsample of the sample which includes the extreme points and the locations of the local extremes values or by replacing the sample by a grid . The problem is solved for this subsample or grid and, as the solution is a spline, it can be interpolated to the points of the whole sample.

Figure 10.5 shows the taut string density, the smoothed taut string density and the true density based on samples of size $n = 500$ for, from top to bottom the normal, exponential and claw distributions.

The raw taut string density can be viewed as an adaptive histogram with variable bin width (see [141]). The algorithm can be modified to give equal bin widths although in this case there will in general be more peaks. Figure 10.6 shows the result for the claw data of Figure 10.5.

The Hidalgo stamp data

The Hidalgo stamp data (see [121] and [13]) is available from the R-package 'bootstrap' [82]. The description given there is the following:

> Thickness in millimetres of 485 postal stamps, printed in 1872. The stamp issue of that year was thought to be a "philatelic mixture", that is, printed on more than one type of paper. It is of historical interest to determine how many different types of paper were used.

In contrast to data simulated from models with a density the stamp data are highly discrete: there are 42 observations with the value 0.079. This is in general not a problem for mixture models and kernel estimators but both the taut string of [47] and the multiscale approach of [78] and [183] cannot cope with this degree of discretization. To overcome this the data were jittered by adding a $0.001U$ random variable with U uniform on $[0, 1]$ to each data point. The taut string density, the smooth density and the equal bin width histogram are given in Figure 10.7.

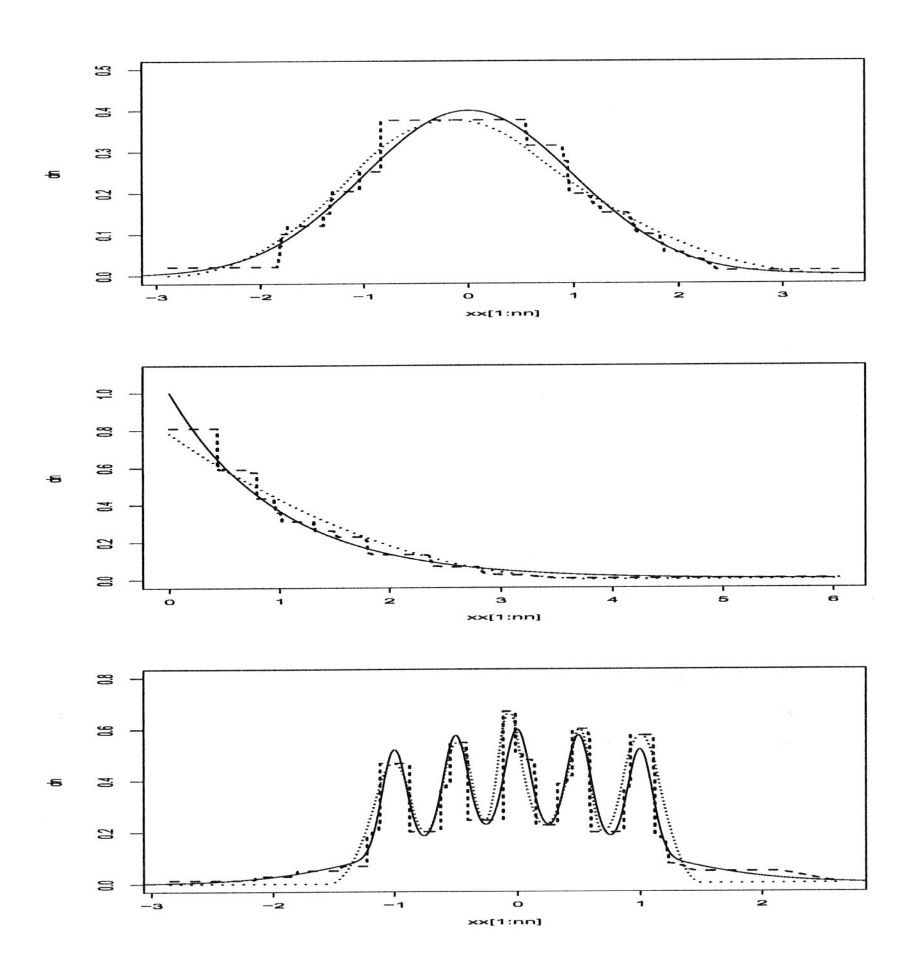

Figure 10.5 *The taut string (dashed), the smoothed taut string (points) and the true density (solid) based on samples of size* $n = 500$ *and the minimization of* $TV(f_{n,s}^{(2)})$ *for, from top to bottom, the normal, the exponential and the claw distributions*

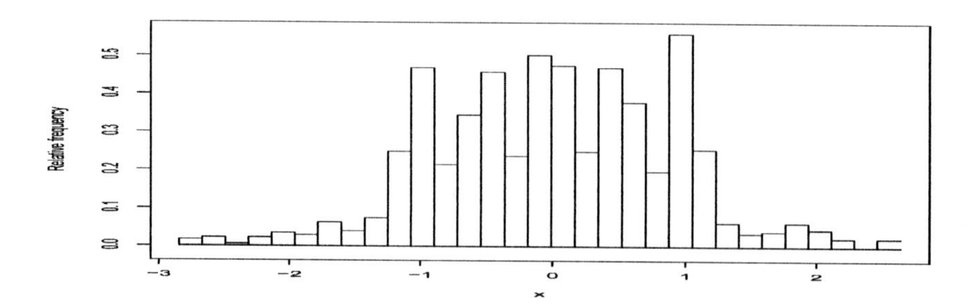

Figure 10.6 *Automatic histogram of the claw data of Figure 10.5 with equal bin widths.*

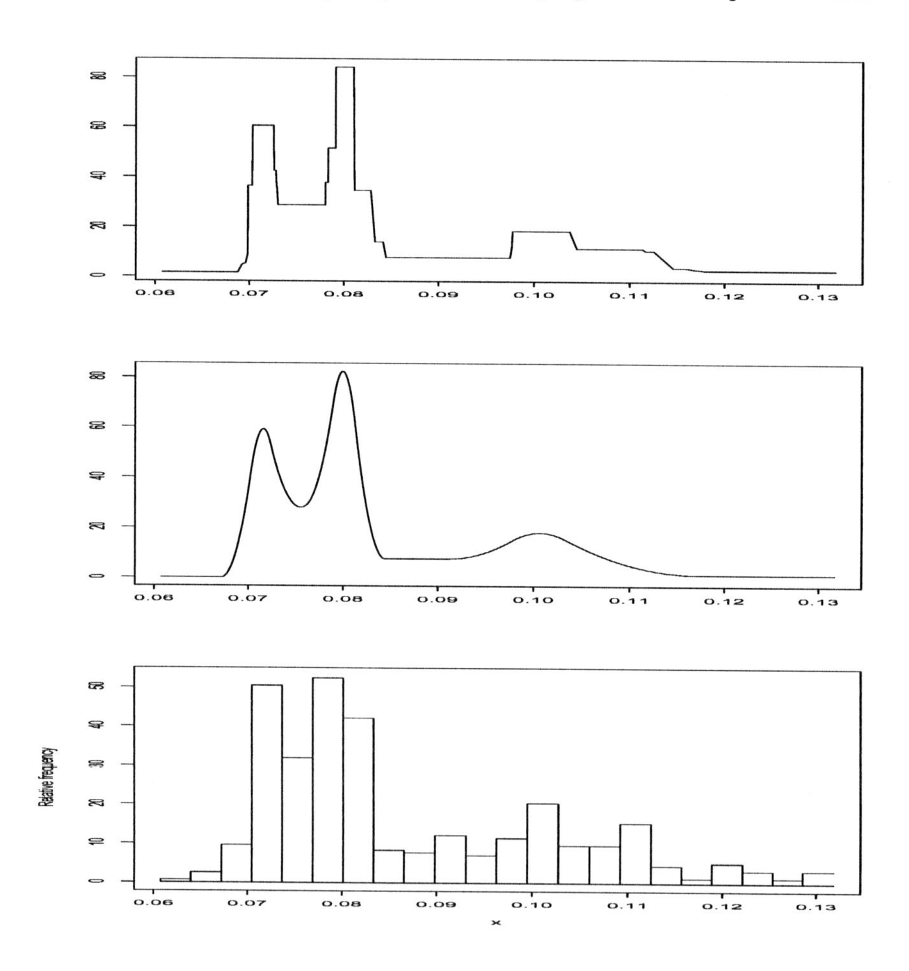

Figure 10.7 *The jittered Hidalgo stamp data of [121]: from top to bottom, the taut string, the smoothed taut string and the histogram with equal bin widths.*

Chapter 11

A Critique of Statistics

This chapter is largely critical as the title suggests. The negative criticisms expressed in it were the spur to all previous chapters. Without a feeling of dissatisfaction, there is no reason to try and do things differently. Although each of the sections below covers a particular topic, they are of necessity interrelated.

11.1 Likelihood

The following is taken from [219] and was already cited in Chapter 1.3.6:

The **Likelihood Function**

$$\mathrm{lhd}(\theta;\mathbf{y}) = \mathrm{lhd}(\theta;y_1,\ldots,y_n) = f(y_1|\theta)\ldots f(y_n|\theta)$$

is the second most important thing in (both Frequentist and Bayesian) Statistics, the first being common sense,

For this reason the first section is on likelihood. One of the main theses of this book is that it is possible to discard likelihood to a large extent. Its use must be restricted to very special situations where the model has been regularized and is consistent with the data. An optimal likelihood based procedure is then a border post delimiting what is the best possible under the most unfavourable circumstances. Common sense will always be needed. It is common sense which prevents likelihood from going astray.

Given a data set $\mathbf{x}_n$ and a model P_θ with density (with respect to either counting or Lebesgue measure) $f(\cdot|\theta)$ the likelihood function is defined by

$$L(\theta,\mathbf{x}_n) = f(\mathbf{x}_n|\theta) \tag{11.1}$$

and the log-likelihood by

$$\ell(\theta,\mathbf{x}_n) = \log f(\mathbf{x}_n|\theta). \tag{11.2}$$

It is to be noted: in this definition there is no such thing as the 'wrong' likelihood. If the data $\mathbf{x}_n$ are in fact i.i.d. $\mathfrak{b}(1,p)$ but the model used is the $\mathfrak{N}(\mu,\sigma^2)$ then the likelihood is

$$L((\mu,\sigma),\mathbf{x}_n) = \frac{1}{\left(\sqrt{2\pi}\sigma\right)^n}\exp\left(-\sum_{i=1}^{n}\frac{(x_i-\mu)^2}{2\sigma^2}\right)$$

and not

$$L(p,\mathbf{x}_n) = \prod_{i=1}^{n} p^{x_i}(1-p)^{1-x_i}.$$

Without this definition it would only be possible to write down the likelihood for simulated data where the true distribution is known. Strictly speaking even this would not be possible as simulations are (i) of finite precision and, (ii) it is not known how to generate random numbers.

11.1.1 Likelihood as a Measure of Fit

Given discrete data $\mathbf{x}_n$ and a family of probability models P_θ with densities p_θ the likelihood

$$L(\theta,\mathbf{x}_n) = p_\theta(\mathbf{x}_n)$$

is the probability of the data observed under the model P_θ. If the data are transformed using a 1-1-transformation $\mathbf{y}_n = h(\mathbf{x}_n)$ and the models P_θ are transformed accordingly to P_θ^h given by

$$P_\theta^h(B) = P_\theta(h^{-1}(B))$$

it follows that

$$L(\theta,\mathbf{x}_n) = p_\theta(\mathbf{x}_n) = p_\theta^h(\mathbf{y}_n) = L(\theta,\mathbf{y}_n)$$

and both give the probability of the data under the respective model. This is as it should be. With this interpretation of likelihood it can be considered as a measure of fit of the model to the data: the higher the probability of the data under the model, the better the model fits the data.

For discrete models it is possible to use $L(\theta,\mathbf{x}_n)$ to check the adequacy of the model by simulation. For any given θ and samples $\mathbf{X}_n(\theta)$ are generated under the model the values of $L(\theta,\mathbf{X}_n(\theta))$ can be compared with $L(\theta,\mathbf{x}_n)$. This is not however a good idea. Consider a Poisson model with $\lambda = 5.356352$ and suppose that the data take on only the values 0 and 6 with probabilities 0.1072746 and 0.8927254 respectively. These are so chosen that $\mathbf{E}(\mathbf{X})$ and $\mathbf{E}(\log(\mathbf{X}!))$ agree with the Poisson model. In this situation likelihood will not reject the Poisson model although the data take on only the values 0 and 6. The total variation metric does not suffer from this deficiency: if the model is not rejected then the model probabilities and the empirical relative frequencies agree to the order of $1/\sqrt{n}$.

In the continuous case the interpretation as a probability is lost, as is the invariance of the likelihood with respect to 1-1-transformations of the data. There is no objective scale any more and at best likelihood is now a relative measure of fit: given two parameter values θ_1 and θ_2 and a $1-1$ transformation h then

$$\frac{f(\mathbf{x}_n,\theta_1)}{f(\mathbf{x}_n,\theta_2)} = \frac{g(\mathbf{y}_n,\theta_1)}{g(\mathbf{y}_n,\theta_2)}$$

where $\mathbf{y}_n = h(\mathbf{x}_n)$, $f(\cdot,\theta)$ is the density function of the distribution P_θ and $g(\cdot,\theta)$ that of the distribution P_θ^h. One consequence is that in the continuous case likelihood cannot be used to check the adequacy of the model.

11.1.2 Likelihood is (almost) Blind

This follows from the preceding section but is worth emphasizing. The 'almost' in the title refers to the discrete case where it may be possible to detect a poor model on the basis of likelihood alone. Consider data generated as $n = 1000$ i.i.d. $\mathfrak{b}(1,0.5)$ random variables and suppose the model used is the Poisson model $\mathfrak{Po}(\lambda)$. The mean of the data is approximately 0.5 corresponding to a value of $\lambda \approx 1/2$. Under the model, the expected value of the likelihood is about -928 whereas the value observed from the data is about -847. Simulations show that the difference is so large that it will be detected. It is to be noted that in this situation the Bayesian would discard that part of the likelihood $\prod_{i=1}^{1000} x_i!$ which contains the information about the lack of fit but does not depend on the value of the parameter.

If instead of the Poisson model the Gaussian model $\mathfrak{N}(\mu, \sigma^2)$ is used then there is no way in which the lack of fit can be detected. The contour plots are identical to those which would be obtained from Gaussian data with the same mean and variance.

A statistician given a data set $\mathbf{x}_n$, informed only of the sample size n and required to analyse the data using likelihood alone, that is none of the methods listed in Table 1.1, would be in an impossible position.

11.1.3 Data Reduction

The likelihood reduces the measure of fit between a data set $\mathbf{x}_n$ and a statistical model P_θ to a single number irrespective of the complexity of both. If the data is an image it is simply not possible to read off relevant features of the fit such as edge detection and peak identification from this single number.

11.1.4 The Likelihood Principle

The likelihood principle may be formulated as follows [19]. Let an experimental situation be represented by an adequate mathematical statistical model E. When any outcome x of E has been observed, then (E,x) is an instance of statistical evidence. The symbol $\mathrm{E_v}(E,x)$ denotes the evidential meaning of a specified instance (E,x) of statistical evidence. Two likelihood functions, $f(x,\theta)$ and $g(y,\theta)$, are called the same if they are proportional, that is if there exists a positive constant c such that $f(x,\theta) = cg(y,\theta)$ for all θ.

> If E and E' are any two experiments with the same parameter space, represented respectively by density functions $f(x,\theta)$ and $g(y,\theta)$; and if x and y are any respective outcomes determining the same likelihood function; then $\mathrm{E_v}(E,x) = \mathrm{E_v}(E',y)$. That is, the evidential meaning of any outcome x of any experiment E is characterized fully by giving the likelihood function $c(x,y)f(x,\theta)$ (which need be described only up to an arbitrary positive constant factor), without other reference to the structure of E.

It is shown in [19] that the likelihood principle can be derived from the principles of *sufficiency* and *conditionality*. This 'derivation' has been criticized by [80] and [7]. As pointed out in [80] the 'proof' is not a mathematical one, as the concept $\mathrm{E_v}(E,x)$ is not well defined and consequently the 'disproofs' are also not mathematical.

A model in the sense of Birnbaum and in accordance with standard notation in statistics is a family of models indexed by a parameter. It is not a single probability distribution which is the meaning in this book. Birnbaum himself does not assume that the model is true but that it is 'adequate' without specifying what is meant by 'adequate'. The consequence is that a family of models may be judged to be adequate, say the Gaussian family $\{\mathfrak{N}(\mu, \sigma^2) : \mu \in \mathbb{R}, \sigma > 0\}$ but not an individual member of this family, say the $\mathfrak{N}(0,1)$ model. Whatever Birnbaum means by 'adequate' it is difficult to imagine a data set for which both the $\mathfrak{N}(0,1)$ and $\mathfrak{N}(1000,1)$ models are regarded as adequate. Nevertheless both will be admitted. Once an adequate family of models is decided upon all the arguments regarding the likelihood principle take place within the model. This is accepted by all discussants of [19] with the exception of Bross, and is equivalent to behaving as if the model were true, that as if the data were in fact generated by one of the models in the family.

There will in general be many models which are regarded as adequate for the data. If the Gaussian family is adequate so will be some families of t-distributions, the Laplace family may also be adequate, as may be a comb family. Some skewed distributions, such as the gamma or log-normal, may also be adequate. If likelihood were a continuous function of adequacy none of this would be of great importance as the conclusions drawn would be more or less the same for every adequate model. But likelihood is not a continuous function of adequacy, it is pathologically discontinuous deriving from the pathological discontinuity of the differential operator, (1.7) and (2.43). Some form of regularization such as minimizing Fisher information will eliminate some models but not all, and there are no objective rules for deciding what form of regularization to use. The likelihood principle makes no mention of regularization and cannot be rephrased to include it.

Likelihood can be taken as a measure of fit of each individual model to the data as in Chapter 11.1.1, but as shown in Chapter 11.1.2 it is a poor to useless measure of fit. The whole weight of the task of deciding if a model is adequate is thus borne by whatever non-likelihood methods are used to judge the adequacy of the model. The information gained in this process goes beyond anything of which likelihood is capable, and once this task has been accomplished, likelihood has little if anything to add and without due care its use may be malevolent.

One conclusion Birnbaum derives from the likelihood principle is the following:

> One basic consequence is that reports of experimental results in scientific journals should in principle be descriptions of likelihood functions, when adequate mathematical statistical models can be assumed, rather than reports of significance levels or interval estimates.

One consequence of this recommendation is that that which is reported is cut off from the data with little or no information about the way or degree to which the model can be regarded as adequate for the data: the ladder is thrown away (6.54, [221]). There is a gap between the words 'adequate' and 'true' and in the context of the Likelihood principle it cannot be breached by the word 'idealize' [19].

The likelihood principle is an example of what Huber [116] calls the 'Scapegoat Syndrome', confusing the model with the truth. Given a data set there is not one adequate ideal model consistent with the data, there are many and equally many likelihoods. Once this is admitted the intellectual barreness of the concept and the discussions surrounding it become apparent.

11.1.5 Stopping Rules

All discussions about the relevance or otherwise of stopping times for statistical analysis take place within the model. The following example is taken from [15] but may have been used by others (see [7]). Two experiments E_1 and E_2 are performed. In E_1 twelve i.i.d. $\mathfrak{b}(1,\theta)$ random variables are observed of which nine take on the value 0 and three the value 1. The likelihood is

$$L_1(\theta,\mathbf{X}_n) = \binom{12}{3}\theta^3(1-\theta)^9. \tag{11.3}$$

In E_2 the trials are continued until three take on the value 1. Twelve trials are required for this so that the likelihood is

$$L_2(\theta,\mathbf{Y}_n) = \binom{11}{2}\theta^3(1-\theta)^9. \tag{11.4}$$

Berger and Wolpert ([15]) conclude

> As the likelihoods are proportional as functions of θ, the information about θ in experiments E_1 and E_2 is identical.

Consider a third experiment E_3 where the trials are continued until three successive ones followed by nine successive zeros are observed. Again twelve trials are required for this so the likelihood is

$$L_3(\theta,\mathbf{Z}_n) = \theta^3(1-\theta)^9. \tag{11.5}$$

According to the Likelihood Principle the information about θ given by the experiments E_1, E_2 and E_3 is the same. The experiments E_1, E_2 and E_3 are now carried out on a larger scale. In E_1' a sample of size $n = 50$ is taken and results in 25 ones. In E_2' the experiment is continued until 25 ones are observed and for this 50 trials are required. In E_3' the experiment is continued until 25 successive ones followed by 25 successive zeros are observed and for this 50 trials are required. The likelihoods are proportional and so once again the likelihood principle implies that information about θ in experiments E_1', E_2' and E_3' is identical. Following Birnbaum's recommendation and the irrelevance of the stopping rule, in all three cases only the function $\theta^{25}(1-\theta)^{25}$ would be reported. Experimenter 1 would find the result of E_1' slightly amusing, as would Experimenter 2 the result of E_2'. Experimenter 3, Rosencrantz, would be highly delighted, but the result would cause his co-worker Guildenstern to philosophize about the laws of probability ([197]). If the truth of the model is replaced by the adequacy of the model, then the result of experiment E_3' would give grounds for calling the model into question. In the case of E_3' there is no stopping time which would make any i.i.d. $\mathfrak{b}(1,\theta)$ model adequate for the data but this is not always the case.

It may be objected that no statistician would use the Bernoulli i.i.d. model for twenty-five ones followed by twenty-five zeros. There would be no objection for one one follows by a zero, or two ones followed by two zeros, or perhaps even for six ones followed by six zeros. However presumably at some point the appropriateness of the model would be called into question and some explanation for this event would be sought.

Let $\varepsilon = (\varepsilon_1, \varepsilon_2, \ldots)$ denote a sequence of ± 1 with $\varepsilon_n = (\varepsilon_1, \ldots, \varepsilon_n)$ and partial sums $S_n(\varepsilon) = \sum_{i=1}^{n} \varepsilon_i$. For given ε_n the length of the longest sequence of $+1$s or -1s will be denoted by $rn(\varepsilon_n)$. Consider the following stopping times:

$$
\begin{aligned}
\tau_1(\varepsilon) &= \min\{n : rn(\varepsilon_n) \geq 5\}, & (11.6)\\
\tau_2(\varepsilon) &= \min\{n : |S_n(\varepsilon)| = 4\}, & (11.7)\\
\tau_3(\varepsilon) &= \min\left\{n : \left(\sum_{i=1}^{n} \{|S_i(\varepsilon)| > 0\}\right) \geq 42\right\}, & (11.8)\\
\tau_4(\varepsilon) &= \min\left\{n : \left(\sum_{i=1}^{n} \{|S_i(\varepsilon)| \geq 2\}\right) \geq 15\right\}, & (11.9)\\
\tau_5(\varepsilon) &= \min\left\{n+1 : \left(\sum_{i=1}^{n} \{|S_i(\varepsilon)| \geq 3\}\right) \geq 3, \right\}, & (11.10)\\
\tau_6(\varepsilon) &= \min\{\tau_1(\varepsilon), \ldots, \tau_5(\varepsilon)\}, & (11.11)\\
& & (11.12)
\end{aligned}
$$

and the sixty ± 1s $\mathbf{x}_{60}$ whose partial sums are shown in Figure 11.1. For any sequence of ± 1s whose first 60 values coincide with $\mathbf{x}_{60}$ it can be checked that $\tau_1(\mathbf{x}) = \ldots = \tau_6(\mathbf{x}) = 60$. Suppose the data are modelled as i.i.d. data with $\mathbf{P}(X_i = 1) = \theta$, $\mathbf{P}(X_i = -1) = 1 - \theta$ for $0 \leq \theta \leq 1$. As the likelihoods are proportional for all six stopping times a Bayesian would give the same analysis in all six cases. Instead of assuming the there is some true θ-value the problem is to determine which if any θ-values are such that the model $\mathbf{P}(X_i = 1) = \theta$, $\mathbf{P}(X_i = -1) = 1 - \theta$ is an adequate approximation for the data. In general a stopping time can be either too small or too large for the data to be consistent with the model. This leads to an 0.99 approximation region of the form

$$\mathscr{A}(\mathbf{x}_{60}, \tau, \Theta, 0.99) = \{\theta : \mathrm{qu}(0.005, \tau, \theta) \leq \tau(\mathbf{x}) \leq \mathrm{qu}(0.995, \tau, \theta)\}. \qquad (11.13)$$

As in this particular case all the stopping times are too large rather than too small the approximation region will be taken to be

$$\mathscr{A}(\mathbf{x}_{60}, \tau, \Theta, 0.99) = \{\theta : \tau(\mathbf{x}) \leq \mathrm{qu}(0.99, \tau, \theta)\}. \qquad (11.14)$$

Figure 11.2 shows the 0.99-quantiles of τ_1, $\tau 2$ and τ_4 as well as the 0.999-quantiles of τ_4. The quantiles are symmetric about $p = 1/2$. The quantiles of τ_1 were calculated exactly (see for example [87]), those of τ_2 and τ_4 were obtained by simulations. The approximation regions for the data of Figure 11.1 are empty for τ_3, τ_4, τ_5 and τ_6. Those for τ_1 and τ_2 are

$$\mathscr{A}(\mathbf{x}_{60}, \tau_1, \Theta, 0.99) = [0.3, 0.7], \quad \mathscr{A}(\mathbf{x}_{60}, \tau_2, \Theta, 0.99) = [0.46, 0.54].$$

The 0.999- and 0.9999-quantiles of τ_4 give

$$\mathscr{A}(\mathbf{x}_{60}, \tau_4, \Theta, 0.999) = [0.4, 0.6], \quad \mathscr{A}(\mathbf{x}_{60}, \tau_4, \Theta, 0.9999) = [0.33, 0.67].$$

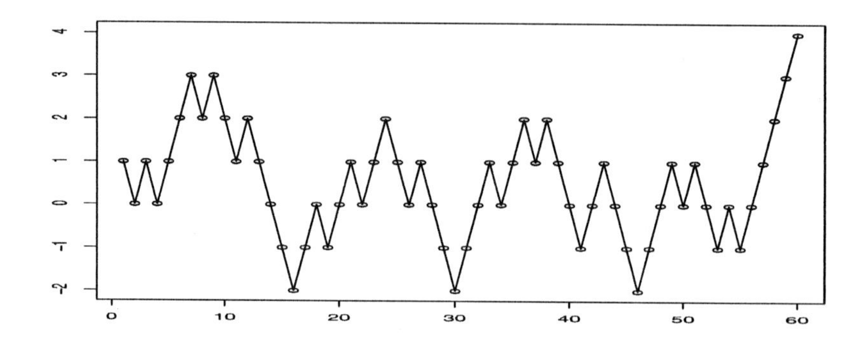

Figure 11.1 *The partial sums of 60 ± 1 random variables.*

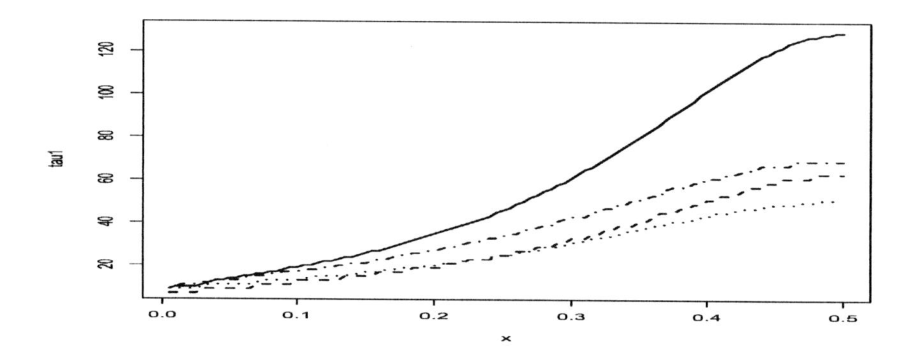

Figure 11.2 *The 0.99-quantiles of $\tau 1$ (solid), $\tau 2$ (dashed), τ_4 (dotted) and the 0.999-quantiles of $\tau 4$ (dot-dash).*

The above shows that the adequacy of a model depends on the stopping time: for some stopping times a given model P_θ may be an adequate approximation for a data set $\mathbf{x}_n$, for others it may not be. Arguments about stopping times which take place within the family of models, that is, the data really are distributed as P_θ for some $\theta \in \Theta$, completely disregard this aspect of a statistical analysis although it is of great importance: if there is no adequate model for a given data set, this will spur the search for reasons and other possible models. One reason could be that the experimenter cheated, that the stopping time was defined after seeing the data but this was not stated. In the above the sceptical reader may suspect that τ_3, τ_4 and τ_5 were defined after seeing the data. This is indeed the case. The data were generated as follows. The 0.99-quantile for τ_2, namely 60, was determine by simulations with

$\theta = 0.5$. The first data set was then taken for which $\tau_2 = 60$. This is the data of (11.1). The other stopping times were then defined so that $\tau_i = 60, i = 1,3,4,5$.

11.1.6 Discontinuity and Plurality

In the case of models with Lebesgue densities likelihood is based on the differential operator and is consequently pathologically discontinuous in any weak topology. As all decisions on the adequacy of a model or a class of models are taken in the weak topology there will be many likelihoods consistent with any data set. Not only that a slight perturbation of the model P_θ to a model P_θ^* can cause the likelihood to vanish.

11.1.7 Maximum Likelihood

One common use of likelihood in frequentist statistics is the maximum likelihood method

$$\hat{\theta}_n = \text{argmax}_\theta\, L(\mathbf{x}_n, \theta).$$

Students are often introduced to the maximum likelihood method by means of an example of discrete models. The interpretation is that the maximum likelihood estimator maximizes the probability of the sample observed. This has a certain intuitive plausibility. It goes back to at least Gauß who proposed the method in 'Theoria motus corporum coelestium' published in Hamburg in 1809:

> Hocce principium, quad in omnibus applicationibus mathesisad philosophicam naturalem usum frequentissimum offert, ubique axiomatis eodem iure valere debet, quo medium arithmeticum inter plures valores observatos eiusdem quantitatis tamquam valor maxime probabilis adoptatur.

About thirty years later he retracted in a letter to Bessel:

> Ich muss es nämlich in alle Wege für weniger wichtig halten, denjenigen Werth einer unbekannten Größe auszumitteln, dessen Wahrscheinlichkeit die größte ist, die ja doch immer nur unendlich klein bleibt, als vielmehr denjenigen, an welchen sich haltend man das am wenigsten nachtheilige Spiel hat.

Roughly speaking he thought it more important to minimize one's losses rather than to maximize a quantity which is infinitesimally small. Minimizing the loss is a decision-theoretic approach to statistics.

The interpretation of maximizing the probability of the data obtained is lost in the case of continuous models. For such, and also for discrete models, maximum likelihood can be given a decision-theoretic justification, namely that under the model maximum likelihood estimates are optimal, or at least asymptotically optimal, in that they minimize the expected quadratic loss. Gauß was not aware of this connection.

Likelihood gives no indication of the appropriateness of the model. Consequently the use of maximum likelihood in any given situation cannot be justified alone by its asymptotic optimality under the model. Given a data set there is in general an interest in making the conclusion to be drawn from the data as precise as possible. As argued in Chapter 1.3.6 precision depends not only on the data but also on the model.

If the model has certain peculiarities these will be used by any optimal estimation procedure, and thus also by maximum likelihood, to increase the precision. Unless the peculiarities can be justified on the basis of substantive knowledge about the data, the increase in precision due to the model alone is a free lunch. To avoid free lunches the model must not exhibit peculiarities, in Tukey's sense it must be bland.

This suggests the following procedure: given the data find an adequate bland model and then use maximum likelihood. However precision is not the only consideration. Another is stability: small changes in the data should not produce large changes in the conclusions. As already mentioned in Chapter 5.3 the maximum likelihood functionals for the normal and Huber distributions ((5.6)) have a finite sample breakdown point of $1/n$. To guard against this the maximum likelihood estimates can be replaced by robustified versions. This is the subject matter of much of robust statistics as expounded in [114], [103], [118] and Chapter 5. There is a cost in doing so but this can be quantified by comparing the robust functionals with the maximum likelihood functionals at the regularized models. Thus maximum likelihood combined with regularization may be justified if the functional is sufficiently stable with respect to the data. If it is not sufficiently stable it can be used as a measure of the loss of efficiency at unfavourable models, that is as a border post.

11.1.8 Likelihood and Features

Given data and a model the likelihood associated with the model will emphasize certain features of the data and neglect others. Two different likelihoods will emphasize different features of the data and it can be the case that the model with the highest likelihood emphasizes features of the data that are of no concern to the analyst. The easiest such example is to take data which are counts but may reasonably described by a Gaussian model. The second model is any Poisson model with a non-zero mean. As the data are counts the Poisson model will always be better, in fact infinitely better, in terms of likelihood although the fact that the data are counts is of little importance to the analyst.

Another example is data with long tails which make the Gaussian model inappropriate. In such a situation the Cauchy will often be chosen on the grounds that it assigns a higher likelihood to large observations. This is one feature of the Cauchy likelihood. Nevertheless it is not the only one. The Cauchy density is highly peaked at the origin and the likelihood will consequently favour observations which are very close to the centre of symmetry. The data analyst may not even be aware of this. Figure 11.3 shows the densities of the standard Cauchy distribution (solid) and those of a slash (X/U with X a standard Gaussian distribution and U uniformly distributed over $[0,1]$) (dashed). The scales are chosen so that the densities are the same for large x. The peakedness of the Cauchy distribution is apparent. In Tukey's terms the slash distribution is blander than the Cauchy and is to be preferred for this reason (see [211]).

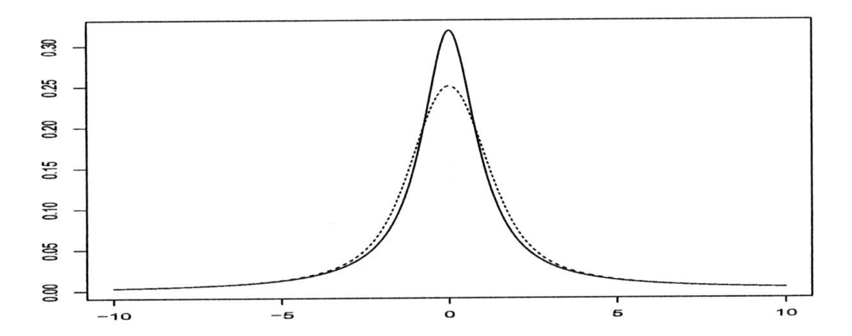

Figure 11.3 *The Cauchy (solid) and slash (dashed) densities chosen to have the same asymptotic behaviour.*

11.2 Bayesian Statistics

11.2.1 The Bayesian Kernel

There are several versions of Bayesian statistics but they overlap sufficiently to allow consideration to be restricted to its simplest form. Given data $\mathbf{x}_n$ a parametric model $p(\cdot,\theta)$ and a prior distribution π over the parameter space $\Theta \subset \mathbb{R}^k$ the posterior distribution of θ given the data is

$$p(\theta|\mathbf{x}_n) = \frac{p(\mathbf{x}_n,\theta)\pi(\theta)}{\int p(\mathbf{x}_n,\gamma)\pi(\gamma)\,d\gamma}. \tag{11.15}$$

In a sense that is the end of the story. All questions concerning the parameter θ can be answered using $p(\theta|\mathbf{x}_n)$. Moreover, as $p(\theta|\mathbf{x}_n)$ is a probability density over Θ, the answers to the questions will be consistent, that is, there is no Dutch book. This is known as coherence.

An excellent critique from a Bayesian point of view of this simplest form but which applies to (almost) all variations may be found in [95]. In essence the critique points out the necessity of model checks and states that 'the usual philosophical story of Bayes as inductive inference is faulty'. Bayesian methodology requires that for at least one θ the model $p(\cdot,\theta)$ is adequate for the data in terms of the analysis. As already pointed out in Section 11.1.2 a model can be hopelessly wrong, an i.i.d. $\mathfrak{N}(\mu,\sigma^2)$ model for i.i.d. $\mathfrak{b}(1,1/2)$ data, and the posterior distribution $p(\theta|\mathbf{x}_n)$ would not indicate this. It is also easy to construct hopelessly bad models which have a higher posterior probability than very good models. All that is needed is for the likelihood to focus on some irrelevant feature of the model.

11.2.2 The (Non-) Additivity of Bayesian Priors

Few if any statisticians regard models as being true. In spite of this the coherence of the Bayesian approach depends on some value of $\theta \in \Theta$ being true. If the fiction of truth is replaced by some concept of adequacy the prior cannot be expressed by a probability distribution. In Bayesian statistics the questions concerning the parameter θ are of the form, 'What are your odds that the true value θ_0 which generated the data lies in the subset Θ' of Θ?'. If the Bayesian answer is $\int_{\Theta'} p(\theta|\mathbf{x}_n)\,d\theta$ then the answers will be coherent, no Dutch book. If there is no true θ_0 which will always be the case, then the only possible answer is that the probability in question is zero. If how ever the questions are of the form 'What are your odds that the model θ_0 is an adequate approximation for the data?' then this question can be answered, the odds quoted and the bet realized. All that is necessary is that it be made precise what 'adequate approximation' means. The previous chapters of this book show that this can be done. Further, if say odds of 1-1 are quoted for the $\mathfrak{N}(0,1)$ model being adequate almost the same odds will be offered for the $\mathfrak{N}(10^{-100},1)$ model being adequate and so on. The quoted probabilities will in general sum to either 0 or ∞, but in any case will not be describable by a probability measure. The conclusion is that if sensible questions are asked and sensible realizable bets are on offer, the prior will not be additive.

In [71] the authors make the same point when they explicitly state that the *p*-values they use for classification are non-additive in contrast to the Bayesian posterior probabilities.

11.2.3 p-values

One Bayesian argument against *p*-values and which is often cited is the following due to Jeffreys [122] page 385:

> .... gives the probability of departures, measured in a particular way, equal to or greater than the observed set, and the contribution from the actual value is nearly always negligible. What the use of *P* implies, therefore, is that a hypothesis that may be true may be rejected because it has not predicted observable results that have not occurred. This seems to be a remarkable procedure. On the face of it, the evidence might more reasonably be taken as evidence for the hypothesis, not against it.

The first problem is to understand the argument which is not made any easier by the use of a double negative. It is not clear what is meant by 'observable results' but in the context it seems plausible that the individual measurements are intended. The next problem is to decide what is meant by 'not predicted observable results'. Presumably the 'not predicted observable results' is the complement of the 'predicted observable results'. Given a model P one possibility is to take the set of predicted results $Pre(P)$ to be the support $supp(P)$ of P. This choice is however essentially vacuous: if $P = \mathfrak{N}(0,1)$ then $Pre(P) = supp(P) = \mathbb{R}$ which is not a great deal of help. To avoid this one must specify α such that $P(X(P) \in Pre(P)) \geq \alpha$. Once this is done $Pre(P)$ is simply another notation for the sets $E_n(\alpha,P)$ of Chapter 2.13. The arguments given there for the different choices of $E_n(\alpha,P)$ apply to the idea of prediction. As an example put $P = \mathfrak{N}(0,1)$ and predict one value Z with $\alpha = 0.95$. If there is an interest in making the prediction as precise as possible given the restriction imposed by $\alpha = 0.95$ the result is $Pre(\mathfrak{N}(0,1)) = E_1(0.95, \mathfrak{N}(0,1)) = (-1.96, 1.96)$.

The following example is a slightly modified version of Example 1 of [217] with values $0-5$ instead of $1-6$. In its essentials the example is not new (see the references given in [217]). Two discrete models P_1 and P_2 have densities as given in Table 11.1. The value $X = 4$ is observed. In [217] it is pointed out that although the observed value $X = 4$ has the same probability under the both models the p-values are different being 0.04 for P_1 and 0.06 for P_2. Thus if a statistician tested both models at the 0.05 level, P_1 would be rejected but P_2 accepted. It is claimed this is an example of a not predicted and not observed value, namely $X = 5$, influencing the statistical analysis. Suppose that the numbers are the number of incorrectly answered questions

	0	1	2	3	4	5
p_1	0.04	0.30	0.31	0.31	0.03	0.01
p_2	0.04	0.30	0.30	0.30	0.03	0.03

Table 11.1 *Two discrete densities taken from [217].*

of five questions and that the two models refer to two different training methods. Of interest is the number of incorrectly answered questions and not whether a particular question was answered correctly or incorrectly. This would require more information and a different analysis. The sets of predicted values will therefore be taken to be of the form $Pre(P_1) = \{0,1,2,\dots,k_1\}$ and $Pre(P_2) = \{0,1,\dots,k_2\}$ where k_1 and k_2 depend of the value of α. On setting $\alpha = 0.95$ it is seen that $k_1 = 3$ and $k_2 = 4$. Thus under model P_1 a not predicted value namely 4 has occurred. Under model P_2 a predicted value, namely 4, has occurred. Thus in contrast to the claim of a not predicted value not occurring it is an example of a not predicted value occurring. This interpretation also has the merit of removing the double negative.

11.3 Sufficient Statistics

Given a parametric family $\mathscr{P}_\Theta$ of probability measures $P_\theta, \theta \in \Theta$, a statistic $s(\mathbf{X}_n(\theta))$ is called a sufficient statistic if the conditional distribution of $\mathbf{X}_n(\theta)$ given $s(\mathbf{X}_n(\theta))$

$$\mathbf{P}(\mathbf{X_n}(\theta) \in \cdot \,|\, \mathbf{s}(\mathbf{X_n}(\theta)))$$

is independent of θ. A statistic $s(\mathbf{X}_n(\theta))$ is a minimal sufficient statistic if it can be expressed as a function of any other sufficient statistic $\tilde{s}(\mathbf{X}_n(\theta))$, that is if $s(\mathbf{X}_n(\theta)) = h(\tilde{s}(\mathbf{X}_n(\theta)))$ for some function h. The interpretation of a sufficient statistic is that it contains all the information about the parameter θ which is contained in the sample. A minimal sufficient statistic represents the greatest reduction of the data which still contains all the relevant information about the parameter. The Rao-Blackwell inequality shows that an estimate of θ not based on a sufficient statistic can be improved, or at least not made worse, by conditioning with respect to one. The new statistic is then a function of the sufficient statistic.

If s is a sufficient statistic the likelihood can be written in the form

$$\text{lhd}(\theta;\mathbf{y}_n) = \text{lhd}(\theta;y_1,\ldots,y_n) = h(\mathbf{y}_n)g(s(\mathbf{y}_n),\theta)$$

so that Bayesian methods and maximum likelihood estimates are based on sufficient statistics. Birnbaum's derivation of the likelihood principle [19] is based on the principles of sufficiency and conditionality both of which are taken as being more plausible, at least at first glance, than the likelihood principle itself. The sufficiency principle states simply that the statistical analysis should be based on the sufficient statistics without specifying how this is to be done.

The objections to the prescribed use of the sufficient statistics are the same as those for likelihood. No model is ever known to be true and is at best a good approximation. A small perturbation of the model can lead to completely different sufficient statistics, for example from the mean and standard deviation for the normal model to the ordered sample for a small perturbation. The definition of a sufficient statistic is a mathematical one and therefore often accompanied by an interpretation intended to help the student to understand the (supposed) practical implications of the concept. Here are some such interpretations to be found on the web:

> In statistics, a sufficient statistic for a statistical model is a statistic that captures all information relevant to statistical inference within the context of the model.

> We say T is a sufficient statistic if the statistician who knows the value of T can do just as good a job of estimating the unknown parameter μ as the statistician who knows the entire random sample.

> The knowledge of the sufficient statistic $\mathbf{x}$ yields exhaustive material for statistical inferences about the parameter θ, since no complementary statistical data can add anything to the information about the parameter contained in the distribution of $\mathbf{X}$.

> In other words, all the information about the unknown parameter is captured in the sufficient statistic T. If, say, we are interested in finding out the percentage of defective light bulbs in a shipment of new ones, it is enough, or sufficient, to count the number of defective ones (sum of the X_i's), rather than worrying about which individual light bulbs are the defective ones (the vector $(X_1,\ldots,X_n)$).

The first quotation even seems to accept that the model is a model rather than the truth but nevertheless claims that all relevant information is captured by the sufficient statistics. Whether the model fits or not is eminently relevant information but it is not captured by the sufficient statistics. Similarly if the advice taken in the last comment (which has to be interpreted correctly if it is not meant as a tautology) were to be acted upon there would be no possibility of discovering that defective light bulbs may occur in batches rather than at random throughout the sample.

11.4 Efficiency

The role and importance of the concept of efficiency has already been discussed in Chapters 1.3.6, 2.10 and 5 . Suppose a data set $\mathbf{x}_n$ is modelled by a parametric class $\mathscr{P}_\Theta$ of models P_θ where θ is identified with a quantity of interest such as the actual concentration of copper in a sample of drinking water. With this identification precision in estimating θ within the model translates into precision for the quantity of interest when the procedure is applied to the actual data. The final precision claimed for the quantity of interest depends not only on the data but also on the model as 'precision' can be imported through the model. To avoid this free lunch the model used should be bland, for example a Gaussian model. This is another form of trying to be honest. An optimal procedure for such a bland model then delimits the amount of precision obtainable using models which import little or no precision into the analysis. An optimal procedure is provided by the maximum likelihood estimator, at least asymptotically. This requires some smoothness conditions to be placed on the distributions P_θ but typically these will hold for bland models. From this point of view the use of the Gaussian model and the mean can be seen as a minimax strategy. Even when this programme has been followed there is no obligation to use the optimal procedure for the bland model. There are nearly always other considerations such as stability under perturbations of the data.

11.5 Asymptotics

Asymptotic results can be divided into several classes. There are pointwise asymptotics, uniform and non-uniform asymptotics. There are asymptotics which are known to hold with sufficient accuracy for samples larger than some specified sample size n_0. There are asymptotics which are known to hold with sufficient accuracy for sample sizes larger than some n_0 but n_0 is not known. Finally there are some asymptotics for which there is no such n_0. The concept of approximation developed in this book requires an approximation for the given sample $\mathbf{x}_n$. It follows that asymptotics are only applicable if there is some known n_0 and $n \geq n_0$. In principle this can always be determined but in practice the calculations may be too difficult. An example is the use of M-functionals in the location-scale problem. The set of adequate models is contained in a Kolmogorov ball of known radius centred at the empirical distribution function. To use the normal approximation it should be known that the approximation is sufficiently accurate for samples of size n for all the distributions in the ball. Theoretically this is possible using the Berry-Esséen bound but in practice even if it were possible, the bounds would be so large to be of no help (see [109] on the Berry-Esséen bound).

One advantage of the approximation approach is that there are no unknown parameters. In the standard formulation of frequentist statistics there are unknown parameters and this complicates the use of asymptotics. An example is the determination of an optimal bandwidth for a kernel estimation in the non-parametric regression model

$$X(t) = f(t) + \sigma Z(t).$$

An asymptotic expression can be given for the optimal bandwidth h_n but as this involves the unknown regression function f there is no n_0 for which it is known that the asymptotic result is sufficiently accurate. In other words no matter how large the sample size n is, it is not known whether the asymptotics are applicable.

The following remarks of Huber in [116] compliment the above:

- Remainder terms? With asymptotic expansions, the first neglected term gives an indication of the size of the error. In statistics, asymptotic results hardly ever are complemented by remainder terms, however crude. That is, whatever the actual sample size is, we never know whether an asymptotic result is applicable.
- What kind of asymptotics is appropriate? In regression, for example, we have n observations and p parameters. Should the asymptotics be for p fixed, $n \to \infty$, or for $p/n \to 0$, or for what?
- Order of quantifiers and limits? Usually, one settles on an order that makes proofs easy.

11.6 Model Choice

11.6.1 Likelihood Based Methods

The AIC model choice criterion is perhaps he best known likelihood based method. It was introduced by Akaike ([5],[6]). In its simplest form there are k models $P_{\Theta_i}, i = 1,\ldots,k$ with nested parameter spaces $\Theta_1 \subset \Theta_2 \cdots \subset \Theta_k \subset \mathbb{R}^k$. Each parameter space Θ_i has a natural embedding in an open subset of $\mathbb{R}^i$, that is, there are i free parameters. Given data $\mathbf{x}_n$ the AIC criterion chooses that model $P_{\Theta_{i_0}}$ which minimizes

$$-2\log(p(\mathbf{x}_n, \hat{\theta}_i)) + 2i \tag{11.16}$$

where $\hat{\theta}_i$ is the maximum likelihood estimator of $\theta_i \in \Theta_i$. The derivation and justification of (11.16) assumes that the 'truth' Q is one of the models under consideration, $Q = P_{\theta_i}$ for some $\theta_i \in \Theta_i$. It is also derived from the asymptotics and so assumes that the sample size n is large.

There are various versions of AIC which include finite sample corrections ([201], [120]), focusing on some parameter $\mu(\theta_i)$ rather than on θ_i itself (FIC, [26]) and the replacement of $2i$ in (11.16) by $\log(n)i$ (BIC [188]). The assumptions can be weakened somewhat. In [106] the 'true' model Q with density f_{true} need not be one of the models under consideration. The 'true' parameters $(\theta_{\text{true}}, \gamma_{\text{true}})$ are replaced by the 'least false' parameters which minimize the Kullback-Leibler discrepancy

$$\int f_{\text{true}} \log(f_{\text{true}}/f_{\theta,\gamma})\, dy.$$

The analysis now goes through with the 'least false' parameters replacing the 'true' parameters.

11.6.1.1 The Student and Copper Data

The choice of model for simple one-dimensional data sets is a useful test bed for illustrating some aspects of this approach. The simplest case will be considered where all models under consideration have two parameters so that the $2i$ in (11.16) can be ignored. The data sets are the logarithm of the of the student data of Figure1.6 with $n = 258$ and the copper data of Exercise 1.3.3 with $n = 27$. The six models under consideration are

(i) the Gauss model,
(ii) the Laplace model,
(iii) the t_3 model,
(iv) the lognormal model,
(v) the gamma model,
(vi) the comb model of (1.8).

For both data sets and each of the six models the maximum likelihood estimators of the parameters were calculated. In the interest of comparability the means and standard deviations of the six models based on the maximum likelihood parameters are stated rather than the parameter values. The results are given in Table 11.2. The left values are for the student data, the right ones in italics are for the copper data. If

Model	μ		σ		log-like		Kuiper	
Gauss	2.61	*2.02*	0.23	*0.11*	18.3	*20.3*	1.52	*1.06*
t_3	2.59	*2.03*	0.30	*0.15*	11.7	*19.7*	1.37	*0.99*
Laplace	2.58	*2.03*	0.25	*0.12*	11.4	*20.1*	1.78	*1.04*
log-normal	2.61	*2.03*	0.23	*0.12*	17.8	*19.8*	1.27	*1.09*
gamma	2.61	*2.02*	0.23	*0.12*	18.3	*20.0*	1.33	*1.08*
comb	2.67	*2.03*	0.34	*0.14*	36.5	*31.4*	4.62	*1.29*

Table 11.2 *Results for the logarithm of the student data of Figure1.6 (left) and for the copper data of Exercise 1.3.3 (right in italics).*

AIC, or more generally likelihood, is taken seriously then for both data sets the best model by far is the comb model.

In Birnbaum's formulation of the Likelihood Principle given above there is a restriction, namely, the Likelihood Principle applies 'when adequate statistical models can be assumed'. The restriction to adequate models is rarely if ever made in discussions of the likelihood principle and not always made when discussing likelihood based procedure of model choice such as AIC. The analysis almost always take place with 'adequate' replaced by 'true'. But even in those rare cases where adequacy is mentioned there is no attempt to to define it. A general definition of adequacy is not possible but at least one could expect a definition in simple cases, such as the student and copper data considered in this section, if only for pedagogical reasons. The refusal to do so constitutes an intellectual failure.

It is not easy to combine adequacy with likelihood. The standard use of the world 'model' in statistics refers to a whole family of individual models in the sense of this book. Thus the 'normal model' includes all normal distributions over the parameter set. With this usage it is not clear what an adequate model is. Given a parametric family $\mathscr{P}_\Theta = \{P_\theta : \theta \in \Theta\}$ it will be rarely the case that all $P_\theta, \theta \in \Theta$ are adequate models for the data. In general there will be a strict subset $\Theta' \subset \Theta$, $\Theta' \neq \Theta$ of adequate models which may indeed by empty. Strange though it may seem applying the words 'adequate model' to $\mathscr{P}_\Theta$ may simply mean that $\Theta' \neq \emptyset$. The chi-squared goodness-of-fit test is often based on this: the test statistic is calculated only for one value θ_0 of the parameter θ, for example the mean of the data in the case of the Poisson model. It is possible for P_{θ_0} not to be an adequate approximation although for there are other values of θ for which P_θ is an adequate approximation. Suppose the definition of adequacy for the models of Table 11.2 is taken to be a normalized Kuiper distance of at most 2.303 this number being the 0.999 quantile for continuous models. If the goodness-of-fit is based on the maximum likelihood values then the comb model with $\theta_0 = (2.67, 0.34)$ is not adequate. However the value of the Kuiper metric for the comb model with $\theta_0 = (2.61, 0.23)$ is 1.67 and so this particular model is adequate. One possibility would be to restrict the parameter space to the those values which yield adequate models. This would however not solve the problem. There will still be a conflict between the adequacy of the model and the likelihood. Thus although the comb model with $\theta_0 = (2.61, 0.23)$ is adequate the value of the log-likelihood is -187.5.

11.6.1.2 Low Birth Weight Data

One of the models considered in [106] and [26] is the logistic regression model

$$\mathbf{P}(Y_i = 1) = p(\mathbf{x}_i^t \beta) = \frac{\exp(\beta_0 + \mathbf{x}_i^t \beta)}{1 + \exp(\beta_0 + \mathbf{x}_i^t \beta)}, \quad i = 1, \ldots, n \tag{11.17}$$

where $\mathbf{x}_i = (x_{i1}, \ldots, x_{ik})^t$ are the regressors. The data used are a subset of the data on low birth weight given in [111]. The data consists of 189 cases with the description given in Table 11.3.

Hjort and Claeskens considered the variables 2, 3, 4, 10 and 11 but numbered them differently. The variable 3 was always included in their analysis. The AIC method for model choice chose the model with only the variables 3 and 10. The BIC method chose the model with only variable 3. If the model with no variable is included, that is, all the β_i are zero apart from β_0 then there are 8 models in all. The no variable model is the 8th in the AIC case but the 1st if BIC is used. For all eight models the values of AIC range from 231.0749 to 237.0187 and those of BIC from 234.6720 to 249.0945.

Likelihood provides no evidence about how well a model fits the data. The value of the log-likelihood for the birth weight data where the model only includes β_0

Variable	Description	Codes/Values	Name
1	Birth Weight	Grams	BWT
2	Low Birth Weight	0 = BWT>2500g 1 = BWT≤2500g	LOW
3	Age of Mother	Years	AGE
4	Weight of Mother at Last Menstrual Period	Pounds	LWT
5	Smoking Status During Pregnancy	0 = No, 1 = Yes	SMOKE
6	History of Premature Labor	0 = None, 1 = One, 2 = Two, etc.	PTL
7	History of Hypertension	0 = No, 1 = Yes	HT
8	Presence of Uterine Irritability	0 = No, 1 = Yes	UI
9	Number of Physician Visits During the First Trimester	0 = None, 1 = One, 2 = Two, etc.	FTV
10	Race	1 = Black, 0=otherwise	RACE
11	Race	1 = White, 0=otherwise	RACE

Table 11.3 *Low birth weight data.*

and β_4 (weight of mothers) is 115.51. As a measure of goodness of fit this number conveys nothing as the following example shows.

Put

$$x_{1i} = i/189, \quad i = 1, \ldots, 189,$$

and generate binomial data Y_i by

$$\mathbf{P}(Y_i = 1) = p(x_i) = \begin{cases} 1, & 1 \le x_i < 29/189, \\ 0, & 30/189 \le x_i \le 114/189, \\ 1, & 115/189 \le x_i \le 1. \end{cases} \tag{11.18}$$

and consider the two logistic regression models, one with just β_0 and the full model which includes x_1. If AIC is used to choose between the two models the full model is chosen, the AIC values being 235.30 for the full model and 262.10 for the offset model. The chosen model is

$$\mathbf{P}(Y_i = 1) = \tilde{p}(x_i) = \frac{\exp(-1.235 + 2.927x_{1i})}{1 + \exp(-1.235 + 2.927x_{1i})}, \quad i = 1, \ldots, 189. \tag{11.19}$$

The value of the likelihood is 115.65 and one can ask whether this model is consistent with the data. Here is one method of doing this. Firstly the data are ordered according to the values of the $\tilde{p}(x_i)$. Denote this data and the corresponding $\tilde{p}(x_i)$ by Y_i^* and $p^*(x_i)$ respectively and define the standardized data by

$$W_i^* = \frac{Y_i^* - p^*(x_i)}{\sqrt{p^*(x_i)(1 - p^*(x_i))}}.$$

The definition of approximation is taken from the multiscale definition of approximation in the context of non-parametric regression (see [73]). The general model is

$$\mathbf{P}(Y_i = 1) = p_i, \quad i = 1, \ldots, n, \tag{11.20}$$

with approximation region

$$\mathscr{A}_n = \left\{ \mathbf{p} : \max_I \frac{1}{\sqrt{|I|}} \left| \sum_{i \in I} w_i^* \right| \leq \sqrt{\tau_n(\mathbf{p}) \log n} \right\}$$

where $\mathbf{p} = (p_1, \ldots, p_n)$. The value of $\tau_n(\mathbf{p})$ can be determined by simulation. If the sorted p_i are constant on some interval I then this interval is collapsed to a single point j with

$$w_j^* = \frac{1}{\sqrt{|I|}} \sum_{i \in I} w_i^* .$$

Using this it can be checked if the model (11.19) is adequate for the data generated under (11.18). based on 100000 simulations the 0.999 quantile of $\tau_n(\mathbf{p})$ under (11.19) is approximately 5.02. This gives an upper bound of

$$\sqrt{\tau_n(\mathbf{p}) \log n} = \sqrt{5.02 \log 189} = 5.13.$$

The value for the data was 8.91 so that the model (11.19) with $p_i = \tilde{p}(x_i)$ is not an adequate approximation for the data generated under (11.18). The i.i.d. model that contains only the offset is an adequate approximation for the data as the maximum likelihood estimate gives probabilities $p_i = \bar{y}_i$, the mean of the y_i. The data set collapses to a single point as described above and the value of w_1^* is zero.

If one tests all the maximum likelihood models considered by Hjort and Claeskens and also the offset model using the multiscale statistic it turns out that they are all adequate approximations for $\alpha = 0.9$. If all 512 models for the complete data set are tested for adequacy then 486 are adequate for $\alpha = 0.9$, 508 are adequate for $\alpha = 0.95$ and for $\alpha = 0.99$ all the models are adequate.

The problem with the multiscale definition of adequacy is that the adequacy of a model depends only on the regressors contained in that model and does not depend on those omitted. The problem cannot be solved by choosing the simplest adequate model as this will always be the model of i.i.d. binomial random variables. Another definition of adequacy is required. If a model is chosen then implicitly it contains those regressors thought to be important and omits those thought not to be so. This should be respected in the concept of approximation. Here is one definition of approximation which attempts to reflect this idea.

Let $\mathscr{S}$ be a subset of the regressors $x_i, i = 0, \ldots, k$ where x_0 is the offset and denote the set of all regressors by $\mathscr{S}_{\text{all}}$. Given a subset $\mathscr{S}$ denote by $\beta(\mathscr{S})$ a logistic model for which $\beta_j(\mathscr{S}) = 0, j \notin \mathscr{S}$ and write

$$s^2(\beta(\mathscr{S})) = \sum_{i=1}^{n} (y_i - p_i(\mathbf{x}_i^t \beta))^2 . \tag{11.21}$$

The least squares parameter $\beta_{\mathrm{lsq}}(\mathscr{S})$ f is defined by

$$\beta_{\mathrm{lsq}}(\mathscr{S}) = \mathrm{argmin}_{\beta(\mathscr{S})}\ s^2(\beta(\mathscr{S}))\,. \tag{11.22}$$

It is clear that if $\mathscr{S} \neq \mathscr{S}_{\mathrm{all}}$ then $s^2(\beta(\mathscr{S})) > s^2(\beta_{\mathrm{lsq}}(\mathscr{S}_{\mathrm{all}}))$ and the difference

$$s^2(\beta(\mathscr{S})) - s^2(\beta_{\mathrm{lsq}}(\mathscr{S}_{\mathrm{all}}))$$

is the maximal reduction in the sum of squared deviations which can be obtained by including the regressors not in $\mathscr{S}$. This is in turn is minimized by putting $\beta(\mathscr{S}) = \beta_{\mathrm{lsq}}(\mathscr{S})$ in which case the reduction of the sum of squares obtained by including all variables is

$$s^2(\beta_{\mathrm{lsq}}(\mathscr{S})) - s^2(\beta_{\mathrm{lsq}}(\mathscr{S}_{\mathrm{all}}))\,. \tag{11.23}$$

If the responses y_i are influenced by a regressor in $\mathscr{S}_{\mathrm{all}} \setminus \mathscr{S}$ then one expects a larger decrease in the sum of squares than in the case where the responses are independent of the regressors in $\mathscr{S}_{\mathrm{all}} \setminus \mathscr{S}$. The latter case can be modelled by modelling the regressors in $\mathscr{S}_{\mathrm{all}} \setminus \mathscr{S}$ as random variables independent of the y_i. The variables 3 and 4 can be satisfactory modelled as Gaussian random variables, the variable 5 as a binomial random variable, the variable 6 as a Poisson random variable and so on. The reduction in the sum of squares can be simulated and some quantile say the 0.95 quantile of the reduction can be chosen as a cut-off point. If the reduction (11.23) exceeds this for the real data the conclusion is that $\mathscr{S}_{\mathrm{all}} \setminus \mathscr{S}$ contains some regressor which should be included in the model. In other words no model based on $\mathscr{S}$ can be an adequate model. It turns out that in general it is not necessary to perform simulations. If $\mathscr{S}_{\mathrm{all}} \setminus \mathscr{S}$ contains ν regressors which are independent of the y_i then asymptotically under the model $\beta(\mathscr{S})$

$$s^2(\beta_{\mathrm{lsq}}(\mathscr{S})) - s^2(\beta_{\mathrm{lsq}}(\mathscr{S}_{\mathrm{all}})) \sim \left(\frac{\sum_{i=1}^n p_i^3(1-p_i)^3}{\sum_{i=1}^n p_i^2(1-p_i)^2}\right)\chi_\nu^2 \tag{11.24}$$

where p_i are the probabilities calculated under the model. Simulations show that the approximation is sufficiently good for the purpose. A model $\beta(\mathscr{S})$ will therefore be judged to be adequate if

$$s^2(\beta(\mathscr{S})) - s^2(\beta_{\mathrm{lsq}}(\mathscr{S}_{\mathrm{all}})) \le \left(\frac{\sum_{i=1}^n p_i^3(1-p_i)^3}{\sum_{i=1}^n p_i^2(1-p_i)^2}\right)\chi_\nu^2(\alpha) \tag{11.25}$$

for some α for example $\alpha = 0.95$. If

$$s^2(\beta_{\mathrm{lsq}}(\mathscr{S})) - s^2(\beta_{\mathrm{lsq}}(\mathscr{S}_{\mathrm{all}})) > \left(\frac{\sum_{i=1}^n p_i^3(1-p_i)^3}{\sum_{i=1}^n p_i^2(1-p_i)^2}\right)\chi_\nu^2(\alpha) \tag{11.26}$$

then no model based on $\mathscr{S}$ can be adequate. The full model cannot be checked in this fashion and must be checked by other means such as the multiscale procedure described above.

The interpretation of the results at one level is clear. Given real data, a subset $\mathscr{S}$ and a model $\beta(\mathscr{S})$ put, with a slight abuse of notation,

$$p = 1 - \text{pchisq}\left(\frac{\left(s^2(\beta(\mathscr{S})) - s^2(\beta_{\text{lsq}}(\mathscr{S}_{\text{all}}))\right)\sum_{i=1}^n p_i^2(1-p_i)^2}{\sum_{i=1}^n p_i^3(1-p_i)^3}, \nu\right).$$

If the regressors in $\mathscr{S}_{\text{all}} \setminus \mathscr{S}$ are replaced by independent copies then in approximately $100(1-p)\%$ of the cases there will be a larger reduction in the sum of squares than that actually observed. On a higher level it may not be the case that a model is satisfactory even if the observed p-value is high: the set $\mathscr{S}$ may include a regressor or regressors whose contribution to the decrease in the sum of squares (11.21) is no more than would be expected from independent copies of that regressor or regressors. For the reduced data set considered by Hjort and Claeskens the model with $\mathscr{S} = \{0,4\}$ and $\beta(\mathscr{S})$ the maximum likelihood parameter value has an observed p-value of 0.152 so that the model may be considered adequate. The model $\mathscr{S} = \{0\}$ and again maximum likelihood has a p-value of 0.023. This model is not adequate and in fact there does not exist a model with $\mathscr{S} = \{0\}$ which is adequate. The conclusion that the variable 4, the weight of the mother, has an effect on the birth weight of the child. All models judged to be adequate with $\alpha = 0.95$, include variable 4. In a Bayesian context it has been suggested that odds of less than 3:1 should be regarded as weak evidence. This corresponds to an observed p-value of 0.25. With this interpretation there is weak evidence that some other variable apart from variable 4 should be included. There are two maximum likelihood models with p-values of more than 0.25, namely $\mathscr{S} = \{0,4,10\}$ with a p-value of 0.316 and $\mathscr{S} = \{0,4,10,11\}$ with a p-value of 0.526. The conclusion is that variable 4 has a significant effect on the weight of the child and weak evidence that 'race=BLACK' also has an effect. All other variables 'explain' no more than would be expected from including x-variables which are independent of the y_i.

The results for the full data set are as follows. In all cases the model is the least squares model based on the regressor subset. Of the 512 models (the full model is given an attained p-value of 1) 89 are adequate with $\alpha = 0.95$. One model has three variables, namely the model $\mathscr{S} = \{0,4,6,7\}$ with an attained p-value of 0.073. All the other models have at least four variables. Of the models with four variables the best with $p = 0.23$ is the model $\mathscr{S} = \{0,4,5,7,11\}$, second best with $p = 0.11$ was the model $\mathscr{S} = \{0,4,6,7,11\}$ and third best with $p = 0.10$ was the model $\mathscr{S} = \{0,4,6,7,10\}$. If one regards a p-value of $0.23 < 0.25$ as very weak evidence that some variable is missing one considers those models with $p \geq 0.25$. There are 19 such models. All contain the variables 4, 5 and 7. The variable 11 is contained in 18 of the 19 models. There are two models which contain just five variables namely the models $\{0,4,5,6,7,11\}$ with $p = 0.53$ and $\{0,4,5,7,8,11\}$ with $p = 0.45$. The occurrence of the variable 8 here is somewhat unexpected but the discussion has now reached a level of speculation without the benefit of expert knowledge. On this basis the two preferred models are $\mathscr{S} = \{0,4,5,7,11\}$ and $\mathscr{S} = \{0,4,5,6,7,11\}$.

It is interesting to compare the above analysis with the results of model choice based on AIC and BIC. The upper panel of Figure 11.4 shows the plot of the attained p-values against the AIC values, the lower panel the same for BIC. There

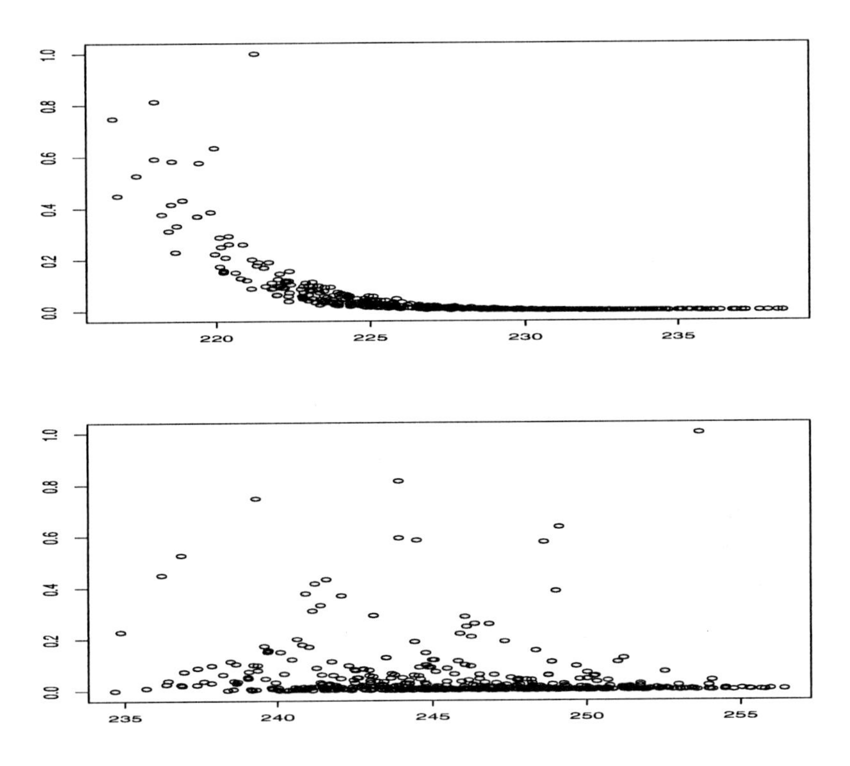

Figure 11.4 *Upper panel: attained p-values against AIC values. Lower panel: attained p-values against BIC values.*

is a clear correlation between the AIC values and the p-values. The model $\mathscr{S} = \{0,4,5,7,11\}$ is 10th on the list of AIC models with a value of 218.66. The model $\mathscr{S} = \{0,4,5,6,7,11\}$ is 3rd on the list of AIC models with an value of 217.39. The correlation between the p-values and the BIC values is much weaker. The best BIC model is the offset $\mathscr{S} = \{0\}$ with a BIC value of 234.67 and a p-value of 0.00023. The second best BIC model with a BIC value of 234.87 is the model $\mathscr{S} = \{0,4,5,7,11\}$. The model $\mathscr{S} = \{0,4,5,6,7,11\}$ is 8th in the list of BIC models with a value of 236.84.

Measures of closeness other than (11.21) are possible. One such is

$$s_{\mathscr{S}}^2 = \min_{\beta_j, j\in\mathscr{S}} \sum_{i=1}^{n} \frac{(y_i - p_i(\mathbf{x}_i^t\beta))^2}{p_i(\mathbf{x}_i^t\beta)(1-p_i(\mathbf{x}_i^t\beta))}. \tag{11.27}$$

Another, if one is addicted, is the maximum value of the likelihood. The differences are only minor for the low birth weight data set.

Given a subset $\mathscr{S}$ and a α-value the approximation region $\mathscr{A}(\alpha)$ for the parameter $\beta(\mathscr{S})$ is given by

$$\mathscr{A}(\alpha) = \left\{ \beta(\mathscr{S}) : s^2(\beta(\mathscr{S})) - s^2(\beta_{\text{lsq}}(\mathscr{S}_{\text{all}})) \leq \left(\frac{\sum_{i=1}^n p_i^3(1-p_i)^3}{\sum_{i=1}^n p_i^2(1-p_i)^2} \right) \chi_\nu^2(\alpha) \right\} \tag{11.28}$$

where the p_i are calculated under the model $\beta(\mathscr{S})$. The only asymptotics involved in this definition is the use of the chi-squared quantile χ_ν^2, but this can be checked using simulations and, if necessary, replaced by simulated values. There are no 'true but unknown' parameters, everything can in principle be calculated exactly. In practice the 'in principle' is tempered by algorithmic constraints but as long as the number of parameters is small so that the approximation region can be calculated on a sufficiently fine lattice. This will give a set of points in $\mathbb{R}^{|\mathscr{S}|}$ where $|\mathscr{S}|$ is the number of elements of $\mathscr{S}$. The set of points can very often be well approximated by the ellipsoid in $\mathbb{R}^{|\mathscr{S}|}$ defined by the mean and the covariance matrix of the lattice points. Figure 11.5 shows part of the results for the regressor subset $\mathscr{S} = \{0,4,5,7,11\}$ with $\alpha = 0.95$. The upper panel shows the plot of the parameter values for the variables 4 (weight of mother) and 7 (hypertension). The lower panel shows the same for the variables 5 (Smoking) and 11 (race=white). The upper panel of Figure 11.6 shows the plot of the largest attained p-values against the parameter values for variable 4. The lower panel shows the same for the variable 7.

If desired the whole approximation region can be described by an ellipsoid. The mean values of the parameters are

$$\mu = (1.935, -0.023, 1.034, 2.178 - 1.076)$$

with covariance matric

$$\Sigma = \begin{pmatrix} 0.7579 & -0.0062 & -0.0239 & 0.3461 & -0.0402 \\ -0.0062 & 0.00005 & -0.00002 & -0.0031 & 0.0002 \\ -0.0239 & -0.00002 & 0.1000 & -0.0037 & -0.0574 \\ 0.3461 & -0.0031 & -0.0037 & 0.4398 & -0.0186 \\ -0.0402 & 0.0002 & -0.0574 & -0.0186 & 0.1109 \end{pmatrix}$$

and inverse

$$\Sigma^{-1} = \begin{pmatrix} 55.52 & 6617 & 25.85 & 3.99 & 21.25 \\ 6617 & 820502 & 3081 & 698.9 & 2507 \\ 25.85 & 3081 & 25.65 & 2.28 & 17.00 \\ 3.99 & 698.9 & 2.28 & 4.15 & 1.96 \\ 21.25 & 2507 & 17.00 & 1.96 & 20.94 \end{pmatrix}.$$

The approximation region is then given by

$$\{\beta = (\beta_0, \ldots, \beta_4) : (\beta - \mu)^t \Sigma^{-1} (\beta - \mu) \leq 8.64\}.$$

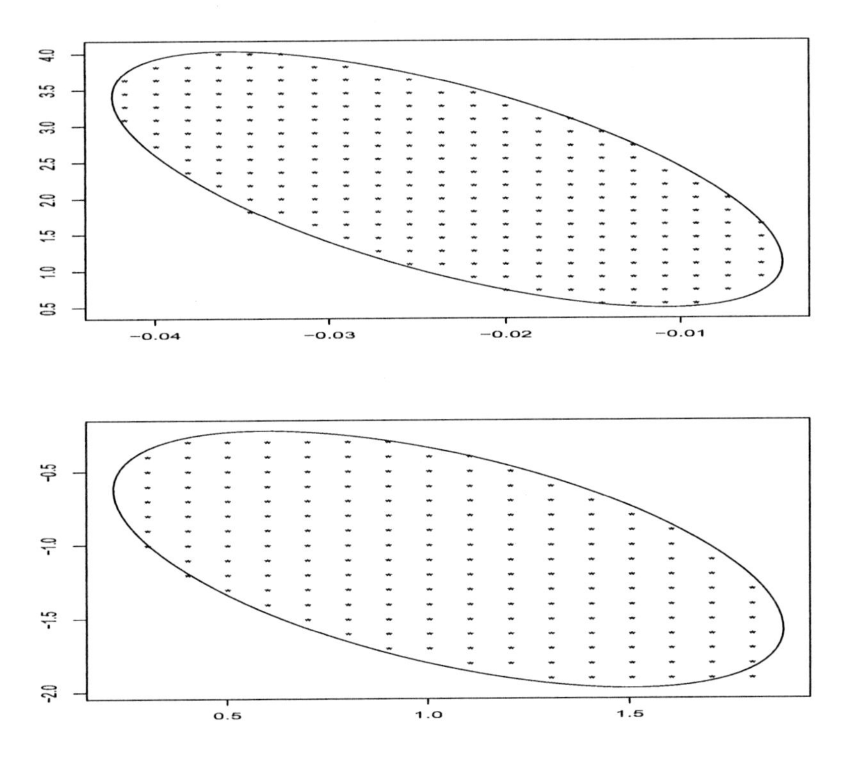

Figure 11.5 *Upper panel: the lattice points for the variables 4 (x-axis, weight of mother) and 7 (y-axis hypertension) together with the approximating ellipse. Lower panel: the same for the variables 5 (smoking status) and 11 (race=white).*

11.7 What can actually be estimated?

This short section is best read together with Chapter 8.2 of [102] which is entitled 'What can actually be estimated?'.

Suppose a statistician is interested in some aspect of the real world that can be expressed as a numerical quantity or quantities. Examples already considered are the amount of copper in a water sample, the speed of light and the low birth weight data. If such data are modelled by a probability model $P = P_\theta$ consistent with the data and belonging to a parametric family $\mathscr{P}_\Theta$ then the aspect of the world is identified with some function $h = h(\theta)$ of $\theta \in \Theta$. There will always be more than one parametric family consistent with the data leading to the problem of consistency of interpretation across different models, say $\mathscr{P}_\Theta$ and $\mathscr{P}_\Gamma$:

> Such a parameter, say μ, must have a definition making it meaningful across competing models.

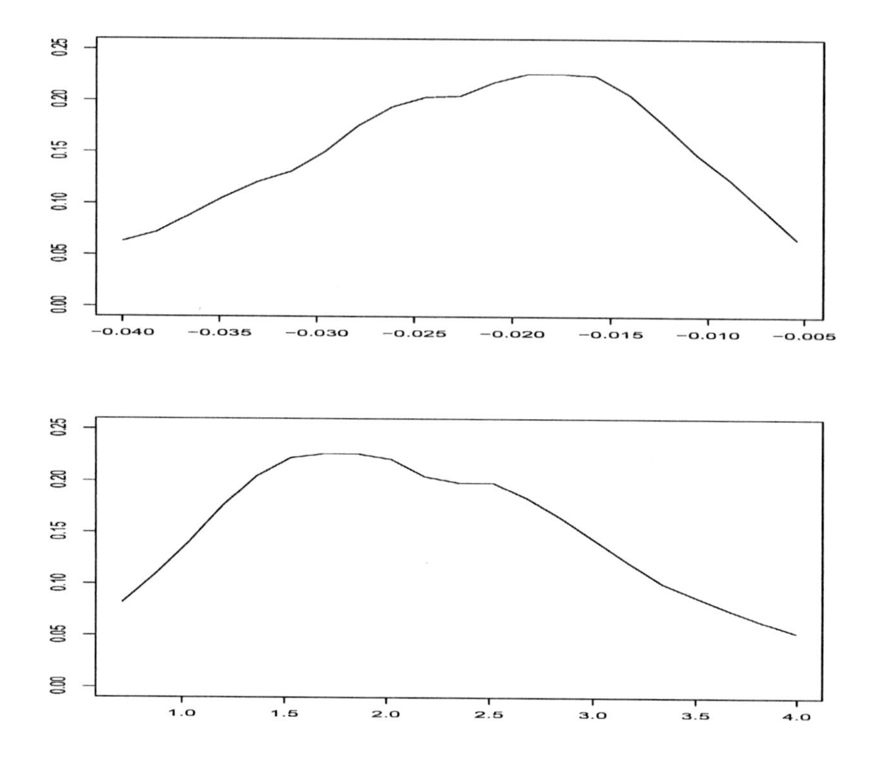

Figure 11.6 *Upper panel: attained p-values plotted against the parameter values of variable 4 (weight of mother). Upper panel: the same for variable 7 (hypertension).*

(page 901 of [26]). In the discussion of [26] Raferty and Zheng make the same point on page 932:

> AS HC point out, for this to be valid, Q must have the same interpretation under all the models. What precisiely this means has not been spelled out, as far as we know. We suggest one meaning: that Q be interpretable as a quantity that could be calculated from future data, at least asymptotically.

The concentration on parametric models tunnels vision and limits imagination. Once the eyes have been opened sufficiently to consider all models P consistent with the data the obvious solution is to identify the aspect of the world with the value $T(P)$ of some functional T. If this is done then $T(P)$, $T(P_\theta)$ and $T(P_\gamma)$ all have the same interpretation. If the individual data values are costs then the mean is the functional of choice, perhaps modified to take extreme outliers into account. In other cases there is no single functional of preferred choice and several different ones can be considered and compared (see the eleven examples in [196] and Chapter 5.5). The speculative nature of the identification may be seen from the examples in [196]: any identification is provisional and is always subject to revision.

The parameters $\beta = (\beta_0, \dots, \beta_k)$ of the low birth weight data can be seen as the values of a functional. Given a model $P = P(y, x_1 \dots, x_k)$ with $y \in \{0, 1\}$ the functional can be defined by

$$T(P) = \text{argmin}_\beta \int \left(y - \frac{\exp(\beta_0 + \beta_1 x_1 + \cdots + \beta_k x_k)}{1 + \exp(\beta_0 + \beta_1 x_1 + \cdots + \beta_k x_k)} \right)^2 dP(y, x_1 \dots, x_k). \tag{11.29}$$

The functional T is well defined and interpretable for models P. Other definitions are possible leading to different functionals Fisher consistent at particular models. If the model P_0 holds

$$P_0(Y = 1 | x_1, \dots, x_n) = \frac{\exp(\beta_0 + \beta_1 x_1 + \cdots + \beta_k x_k)}{1 + \exp(\beta_0 + \beta_1 x_1 + \cdots + \beta_k x_k)}$$

where the Y_i are conditionally independently distributed given $\mathbf{x}$, then $T(P_0) = (\beta_0, \dots, \beta_k)$.

Section 4 of [26] gives four examples all based on the regression model

$$Y = a + bx + \sigma\varepsilon \tag{11.30}$$

where ε has the skewed distribution with density

$$\lambda \Phi(u)^{\lambda - 1} \varphi(u).$$

This defines a parametric model with parameter $\theta = (a, b, \sigma, \lambda)$. It is so specific that it is not clear whether any sense can be made of λ outside of the model. One possibility would be to replace λ by some general measure of skewness $\kappa(P)$ of the distribution P of ε. In this case $\theta = (a, b, \sigma, \kappa)$ could be treated as the value of a functional along the lines of (11.29) for the birth weight model.

In certain situations the parametric model may not be ad hoc but have an interpretation in its own right. The Poisson distribution is one example of this where it is used to check the appropriateness of randomness. The actual value of the parameter λ of the Poisson distribution is, in such a situation, of minor concern. Of greater relevance is the metric or measure of dicrepancy which will to some extent depend on whether attraction or repelance is the alternative to randomness. In such cases it does not make sense to consider all models consistent with the data.

Finally the model could be used as a convenient and approximate description of the shape of the data without its being used to answer any substantial question. Here again some form of discrepancy or metric can be used to decide whether the approximation is sufficiently good for the purpose. It may be desired to approximate the shape of the student data of Figure 1.6 by a simple parametric model. For the normal model with parameters $(\mu, \sigma) = (13.12, 2.864)$, the log-normal model with parameters $(\mu, \sigma) = (2.58, 0.222)$ and the gamma model with parameters $(\gamma, \lambda) = (18.01, 0,74)$ the Kuiper distances to the empirical distribution function are 0.082, 0.074 and 0.094 respectively. If the log-normal model is chosen then it could be stated that the data may be roughly described by a log-normal distribution with parameters $(\mu, \sigma) = (2.58, 0.222)$ the maximal error over any interval being 0.074.

Bibliography

[1] DIN 38402-45:2003-09 German standard methods for the examination of water, waste water and sludge - General information (group A) - Part 45: Interlaboratory comparisons for proficiency testing of laboratories (A 45). Deutsches Institut für Normierung, 2003.

[2] D. Adams. *The Hitchhiker's Guide to the Galaxy*. Pan Books, 1979.

[3] M. Aitkin. Statistical Inference: An Integrated Bayesian/Likelihood Approach. Technical report, University of Melbourne, Melbourne, N.S.W., Australia, 2009.

[4] H. Akaike. Information theory and an extension of the maximum likelihood principle. In *Proceedings of the Second International Symposium on Information Theory Budapest*, pages 267–281. Akademiai Kiado, 1973.

[5] H. Akaike. Information theory and an extension of the maximum likelihood principle. In B.N. Petrov and F. Csaki, editors, *Second international symposium on information theory*, pages 267–281, Budapest, 1973.

[6] H. Akaike. A new look at the statistical model identification. *IEEE Transactions on Automatic Control*, 19:716–723, 1974.

[7] H. Akaike. On the fallacy of the likelihood principle. *Statistics & probability letters*, 1:75–78, 1982.

[8] D. F. Andrews, P. J. Bickel, F. R. Hampel, W. H. Rogers, and J. W. Tukey. *Robust Estimates of Location: Survey and Advances*. Princeton University Press, Princeton, N.J., 1972.

[9] R. R. Bahadur and J. S. Savage. The nonexistence of certain statistical procedures in nonparametric problems. *Annals of Mathematical Statistics*, 27:1115–1122, 1955.

[10] S. Banasiak. Ein nichtparametrisches Verfahren zur Bestimmung von Intervallen einer stückweise konstanten Volatilitätsfunktion. Diplomarbeit, Fakultät Mathematik, Universität Duisburg-Essen, Essen, Germany, 2011.

[11] R. E. Barlow, D. J. Bartholomew, J. M. Bremner, and H. D. Brunk. *Statistical Inference under Order Restrictions*. Wiley, New York, 1972.

[12] V. Barnett and T. Lewis. *Outliers in Statistical Data*. Wiley, Chichester, third edition, 1994.

[13] K. E. Basford, G. J. Mclachlan, and M. G York. Modelling the distribution of stamp paper thickness via finite normal mixtures: The 1872 Hidalgo stamp issue of Mexico revisited. *Journal of Applied Statistics*, 1997.

[14] R. Beran and P. W. Millar. Confidence sets for a multivariate distribution. *Annals of Statistics*, 14(2):431–443, 1986.

[15] J. O. Berger and R. L. Wolpert. *The Likelihood Principle*, volume 6 of *Lecture Notes, Monograph Series*. Institue of Mathematical Statistics, Hayward, California, 1988.

[16] R. L. Berger and D. D. Boos. p values maximized over a confidence set for the nuisance parameter. *Journal of the American Statistical Association*, 89:1012–1016, 1994.

[17] Thorsten Bernholt and Thomas Hofmeister. An algorithm for a generalized maximum subsequence problem. In *LATIN 2006: Theoretical informatics*, volume 3887 of *Lecture Notes in Computer Science*, pages 178–189. Springer, Berlin, 2006.

[18] A. C. Berry. The accuracy of the Gaussian approximation to the sum of independent variates. *Transactions of the American Mathematical Society*, 49(1):122–136, 1941.

[19] A. Birnbaum. On the foundations of statistical inference. *Journal of the American Statistical Association*, 57:269–326, 1962.

[20] T. Bollerslev. Generalized autoregressive conditional heteroscedasticity. *Journal of Econometrics*, 31(3):307–327, 1986.

[21] A. Boysen, L.and Kempe, L. Liebscher, A. Munk, and O. Wittich. Consistencies and rates of convergence of jump-penalized least squares estimators. *Annals of Statistics*, 37(1):157–183, 2009.

[22] D. R. Brillinger. The 2005 Neyman Lecture: Dynamic Indeterminism In Science. *Statistical Science*, 2008. to appear.

[23] A. Buja, D. Cook, H. Hofmann, M. Lawrence, E-K. Lee, D.F. Swayne, and H. Wickham. Statistical inference for exploratory data analysis and model diagnostics. *Philosophical Transactions of the Royal Society A*, 367:4361–4383, 2009.

[24] P. J. Burt and A. E. Adelson. The Laplacian pyramid as a compact image code. *IEEE Tansactions on Commenications*, 31:523–540, 1983.

[25] E. J. Candès and F. Guo. New multiscale transforms, minimum total variation synthesis: Applications to edge-preserving image reconstruction. *Signal Processing*, 82:1519–1543, 2002.

[26] G. Claeskens and N. L. Hjort. Focused information criterion. *Journal of the American Staistical Association*, 98:900–916, 2003.

[27] R. D. Clarke. An application of the Poisson distribution. *Journal of the Institute of Actuaries*, 72:481, 1946.

[28] R. R. Coifman and A. Sowa. Combining the calculus of variations and

wavelets for image enhancement. *Applied and Computational Harmonic Analysis*, 9:1–18, 2000.

[29] J. B. Copas. On the unimodality of the likelihood for the Cauchy distribution. *Biometrika*, 62:701–704, 1975.

[30] D. R. Cox. *Principles of Statistical Inference*. Cambridge University Press, Cambridge, 2006.

[31] C. Daniel. Patterns in residuals in the two-way-layout. *Technometrics*, 20:385–395, 1978.

[32] L. Davies and U. Gather. The identification of multiple outliers (with discussion). *Journal of the American Statistical Association*, 88:782–801, 1993.

[33] P. L. Davies. Asymptotic behaviour of S-estimates of multivariate location parameters and dispersion matrices. *Annals of Statistics*, 15:1269–1292, 1987.

[34] P. L. Davies. Statistical evaluation of interlaboratory tests. *Fresenius Zeitschrift für analytische Chemie*, 331:513–519, 1988.

[35] P. L. Davies. Aspects of robust linear regression. *Annals of Statistics*, 21:1843–1899, 1993.

[36] P. L. Davies. Data features. *Statistica Neerlandica*, 49:185–245, 1995.

[37] P. L. Davies. On locally uniformly linearizable high breakdown location and scale functionals. *Annals of Statistics*, 26:1103–1125, 1998.

[38] P. L. Davies. The one-way table: In honour of John Tukey 1915-2000. *Journal of Statistical Planning and Inference*, 122:3–13, 2004.

[39] P. L. Davies. Approximating data (with discussion). *Journal of the Korean Statistical Society*, 37:191–240, 2008.

[40] P. L. Davies. Interactions in the analysis of variance. *Journal of the American Statistical Association*, 107:1502–1509, 2012.

[41] P. L. Davies, R. Fried, and U. Gather. Robust signal extraction for on-line monitoring data. *Journal of Statistical Inference and Planning*, 122:65–78, 2004.

[42] P. L. Davies and U. Gather. Breakdown and groups (with discussion). *Annals of Statistics*, 33:977–1035, 2005.

[43] P. L. Davies, U. Gather, M. Meise, D. Mergel, and T. Mildenberger. Residual based localization and quantification of peaks in x-ray diffractograms. *Annals of Applied Statistics*, 2(3):861–886, 2008.

[44] P. L. Davies, C. Höhenrieder, and W. Krämer. Recursive estimation of piecewise constant volatilities. *Computational Statistics and Data Analysis*, 56(11):3623–3631, 2012.

[45] P. L. Davies and A. Kovac. Local extremes, runs, strings and multiresolution (with discussion). *Annals of Statistics*, 29(1):1–65, 2001.

[46] P. L. Davies and A. Kovac. Densities, spectral densities and modality. Technical Report 1, SFB 475, University of Dortmund, Dortmund, Germany, 2003.

[47] P. L. Davies and A. Kovac. Densities, spectral densities and modality. *Annals of Statistics*, 32(3):1093–1136, 2004.

[48] P. L. Davies and A. Kovac. Quantifying the cost of simultaneous non-parametric approximation of several samples. *Electronic Journal of Statistics*, 3:747–780, 2009.

[49] P. L. Davies and A. Kovac. ftnonpar: Features and strings for nonparametric regression. http://CRAN.R-project.org/package=ftnonpar, 2010. R package version 0.1-84.

[50] P. L. Davies, A. Kovac, and M. Meise. Nonparametric regression, confidence regions and regularization. *Annals of Statistics*, 37(5B):2597–2625, 2009.

[51] P. L. Davies and M. Meise. Approximating data with weighted smoothing splines. *Journal of Nonparametric Statistics*, 20(3):207–208, 2008.

[52] P. L. Davies and M. Meise. Smoothing densities under shape constraints. Universität Duisburg-Essen, Essen, Germany, 2008.

[53] M. A. Delgado. Testing the equality of nonparametric regression curves. *Statistics and Probability Letters*, 17:199–204, 1992.

[54] H. Dette and N. Neumeyer. Nonparametric analysis of covariance. *Annals of Statistics*, 29:1361–1400, 2001.

[55] L. Devroye. Bounds for the uniform deviation of empirical measures. *Journal of Multivariate Analsysis*, 12:72–79, 1982.

[56] P. Diaconis, S. Holmes, and R. Montgomery. Dynamical bias in the coin toss. *SIAM Review*, 49(2):211–235, 2007.

[57] D. L. Donoho. *Breakdown properties of mulitvariate location estimators*. PhD thesis, Department of Statistics, Harvard University, Harvard, Mass., 1982.

[58] D. L. Donoho. One-sided inferences about functionals of a desnity. *Annals of Statistics*, 1988.

[59] D. L Donoho. De-noising by soft-thresholding. *IEEE Transactions on Information Theory*, 1995.

[60] D. L Donoho. On the marriage of multiscale analysis and statistics. *Statistica Sinica*, 19:1361–1365, 2009.

[61] D. L. Donoho and P. J. Huber. The notion of breakdown point. In P. J. Bickel, K. A. Doksum, and Jr. J. L. Hodges, editors, *A Festschrift for Erich L. Lehmann*, pages 157–184, Belmont, California, 1983. Wadsworth.

[62] D. L. Donoho and I. M. Johnstone. Ideal spatial adaptation by wavelet shrinkage. *Biometrika*, 81:425–455, 1994.

[63] D. L. Donoho and I. M. Johnstone. Adapting to unknown smoothness via wavelet shrinkage. *Journal of the American Statistical Association*, 90(432):1200–1224, 1995.

[64] D.L. Donoho. Wedgelets: nearly minimax estimation of edges. *Annals of Statistics*, 27(3):859–897, 1999.

[65] R. M. Dudley. *Real Analysis and Probability*. Wadsworth and Brooks/Cole, California, 1989.

[66] R. M. Dudley, Sergiy Sidenko, and Zuoqin Wang. Differentiability of t-functionals of location and scatter. *Annals of Statistics*, 37(2):939–960, 2009.

[67] L. Dümbgen. On Tyler's M-functional of scatter in high dimensions. *Annals of the Institue of Statistical Mathematics*, 50:471–491, 1997.

[68] L. Dümbgen. New goodness-of-fit tests and their application to nonparametric confidence sets. *Annals of Statistics*, 26:288–314, 1998.

[69] L. Dümbgen. Optimal confidence bands for shape-restricted curves. *Bernoulli*, 9(3):423–449, 2003.

[70] L. Dümbgen. Confidence bands for convex median curves using sign-tests. In E. Cator, G. Jongbloed, C. Kraaikamp, R. Lopuhaä, and J.A. Wellner, editors, *Asymptotics: Particles, Processes and Inverse Problems*, volume 55 of *IMS Lecture Notes - Monograph Series 55*, pages 85–100. IMS, Hayward, USA, 2007.

[71] L. Dümbgen, B-W. Igl, and A. Munk. p-values for classification. *Electronic Journal of Statistics*, 2:468–493, 2008.

[72] L. Dümbgen and R. B. Johns. Confidence bands for isotonic median curves using sign-tests. *J. Comput. Graph. Statist.*, 13(2):519–533, 2004.

[73] L. Dümbgen and A. Kovac. Extensions of smoothing via taut strings. *Electronic Journal of Statistics*, 3:41–75, 2009.

[74] L. Dümbgen, K. Nordhausen, and H. Schuhmacher. New algorithms for m-estimation of multivariate location and scatter. Technical report, Institue of Mathematical Statistics and Actuarial Science, University of Bern, Switzerland, 2014.

[75] L. Dümbgen, M. Pauly, and T. Schweizer. A survey of m-functionals of multivariate location and scatter. Technical Report 77, Institue of Mathematical Statistics and Actuarial Science, University of Bern, Switzerland, 2014.

[76] L. Dümbgen and V. G. Spokoiny. Multiscale testing of qualitative hypotheses. *Annals of Statistics*, 29(1):124–152, 2001.

[77] L. Dümbgen and D. E. Tyler. On the breakdown properties of some multivariate M-functionals. *Scandinavian Journal of Statistics*, 32:247–264, 2005.

[78] L. Dümbgen and G. Walther. Multiscale inference about a density. *Annals of Statistics*, 36:1758–1785, 2008.

[79] S. Durand and J. Froment. Artifact free signal denoising with wavelets. In *Proceedings IEEE International Conference on Acoustics, Speech and Signal Processing*, pages 3685 – 3688, 2001.

[80] J. Durbin. On Birnbaum's theorem on the relation between sufficiency, conditionality anf likelihood. *Journal of American Statistical Association*, 65(329):395–398, 1970.

[81] D. Dürr and S. Teufel. *Bohmian Mechanics - The Physics and Mathematics of Quantum Theory*. Springer, 2009.

[82] B. Efron and R. Tibshirani. *An Introduction to the Bootstrap*. Chapman and Hall, 1993.

[83] R. F. Engle. Autoregressive conditional heteroscedasticity with estimates of the United Kingdom inflation. *Econometrica*, 50(4):987–1007, 1982.

[84] C-G Esseen. On the Liapunoff limit of error in the theory of probability. *Arkiv för matematik, astronomi och fysik*, A28:1–19, 1942.

[85] D. Esteban and C. Galand. Application of quadrature mirror filter to split-band voice coding schemes. In *Proceedings IEEE International Conference on Acoustics, Speech and Signal Processing*, pages 191–195, 1977.

[86] J. Fan and S. K. Lin. Test of significance when data are curves. *Journal of American Statistical Association*, 93:1007–1021, 1998.

[87] I. Fazekas, Z. Karácsony, and Z. Libor. Longest runs in coin tossing. Comparison of recursive formulae, asymptotic theorems, computer simulations. *Acta Universitatis Sapientiae, Mathematica*, 2(2):215–228, 2010.

[88] K. Frick, P. Marnitz, and A. Munk. Statistical multiresolution estimation in imaging: fundamental concepts and algorithmic framework. arXiv:1101.4373v2 [stat.AP], June 2011.

[89] K. Frick, A. Munk, and H. Sieling. Multiscale change point inference. *Journal of the Royal Statistical Society: Series B*, 76(2):1–46, 2014.

[90] F. Friedrich. *Complexity Penalized Segmentations in 2D*. PhD thesis, Department of Mathematics, Technical University Munich, Munich, Germany, 2005.

[91] F. Friedrich, L. Demaret, H. Führ, and K. Wicker. Efficient moment computation over polygonal domains with an application to rapid wedgelet approximation. *Siam J. Sci. Comput.*, 29(2):842–863, 2007.

[92] F. Friedrich, A. Kempe, V. Liebscher, and G. Winkler. Complexity penalized M-estimation: Fast computations. *Journal of Computational and Graphical Statistics*, 17:201–224, 2008.

[93] W. A. Gardner. *Statistical Spectral Analysis: A Nonprobabilistic Theory*. Prentice Hall, New Jersey, 1988.

[94] A. Gelman. A Bayesian forulation of exploratory data analysis and goodness-of-fit testing. *International Statistical Review*, 71(2):369–382, 2003.

[95] A. Gelman and C. R. Shalizi. Philosophy and the practice of Bayesian statistics. *British Journal of Mathematical and Statistical Psychology*, 66(1):8–38, 2013.

[96] C. Granger and C. Stărică. Nonstationarities in stock returns. *The Review of Economics and Statistics*, 87(3):523–538, 2005.

[97] J. Grossman and A. Morlet. Decomposition of Hardy functions into square integrable wavelets of constant shape. *SIAM Journal of Mathematical Analysis*, 15:723–726, 1984.

[98] P. Hall and D. H. Hart. Bootstrap test for difference between means in nonparametric regression. *Journal of the American Statistical Association*, 85(412):1039–1049, 1990.

[99] F. R. Hampel. *Contributions to the theory of robust estimation.* PhD thesis, University of California, Berkeley, 1968.

[100] F. R. Hampel. Beyond location parameters: robust concepts and methods (with discussion). In *Proceedings of the 40th Session of the ISI*, volume 46, Book 1, pages 375–391, 1975.

[101] F. R. Hampel. The breakdown points of the mean combined with some rejection rules. *Technometrics*, 27:95–107, 1985.

[102] F. R. Hampel, E. M. Ronchetti, P. J. Rousseeuw, and W. A. Stahel. *Robust statistics.* Wiley Series in Probability and Mathematical Statistics: Probability and Mathematical Statistics. John Wiley & Sons Inc., New York, 1986. The approach based on influence functions.

[103] F. R. Hampel, E. M. Ronchetti, P. J. Rousseeuw, and W. A. Stahel. *Robust Statistics: The Approach Based on Influence Functions.* Wiley, New York, 1986.

[104] W. Härdle and J. S. Marron. Semiparametric comparison of regression curves. *Annals of Statistics*, 18(1):63–89, 1990.

[105] J. A. Hartigan and P. M. Hartigan. The dip test of unimodality. *Annals of Statistics*, 13:70–84, 1985.

[106] N.L. Hjort and G. Claeskens. Frequentist model average estimates. *Journal of the American Staistical Association*, 98:879–899, 2003.

[107] D. C. Hoaglin, F. Mosteller, and J. W. Tukey. *Understanding Robust and Exploratory Data Analysis*. Wiley, New York, 1983.

[108] D. C. Hoaglin, F. Mosteller, and J. W Tukey. *Understanding Robust and Exploratory Data Analysis*, chapter analysis of two-way tables by medians. Wiley, New York, 1983.

[109] J. Hoffmann-Jørgensen. *Probability with a View Toward Statistics, Volumes I and II.* Chapman and Hall Probability Series. Chapman and Hall, New York, 1994.

[110] C. Höhenrieder. *Nichtparametrische Volatilitäts- und Trendapproximation von Finanzdaten.* PhD thesis, Department of Mathematics, University Duisburg-Essen, Germany, 2008.

[111] D. W. Hosmer and S. Lemeshow. *Applied Logistic Regression.* Wiley, New York, 1989.

[112] T. Hotz, P. Marnitz, R. Stichtenoth, L. Davies, Z. Kabluchko, and A. Munk. Locally adaptive image denoising by a statistical multiresolution criterion. *Computational Statistics and Data Analysis*, 56(3):543–558, 2012.

[113] P. J. Huber. Robust estimation of a location parameter. *Annals of Mathematical Statistics*, 35:73–101, 1964.

[114] P. J. Huber. *Robust Statistics*. Wiley, New York, 1981.

[115] P. J. Huber. Finite sample breakdown of M- and P-estimators. *Annals of Statistics*, 12(1):119–126, 1984.

[116] P. J. Huber. On the non-optimality of optimal procedures. In J. Rojo, editor, *Optimality: The Third Erich L. Lehmann Symposium*, volume 57 of *IMS Lecture NotesMonograph Series*, pages 31–46, Beachwood, Ohio, USA, 2009. Institute of Mathematical Statistics.

[117] P. J. Huber. *Data Analysis*. Wiley, New Jersey, 2011.

[118] P. J. Huber and E. M. Ronchetti. *Robust Statistics*. Wiley, New Jersey, second edition, 2009.

[119] M. Hubert, P. J. Rousseeuw, , and S. Van Aelst. High-breakdown robust multivariate methods. *Statistical Science*, 23(1):92–119, 2008.

[120] C. M. Hurvich and C. L. Tsai. Regression and time series model selection in small samples. *Biometrika*, 76:297–307, 1989.

[121] A. J. Izenman and C. J. Sommer. Philatelic mixtures and multimodal densities. *Journal of the American Statistical Association*, 83:941–953, 1988.

[122] H. Jeffreys. *Theory of Probability*. Oxford Classic Texts in the Physical Sciences. Oxford University press, third edition, 1961.

[123] Z. Kabluchko. Extreme-value analysis of standardized Gaussian increments. arXiv:0706.1849, 2007.

[124] J. T. Kent and D. E. Tyler. Redescending M-estimates of multivariate location and scatter. *Annals of Statistics*, 19:2102–2119, 1991.

[125] J. T. Kent and D. E. Tyler. Constrained M-estimation for multivariate location and scatter. *Annals of Statistics*, 24:1346–1370, 1996.

[126] J. T. Kent and D. E. Tyler. Regularity and uniqueness for constrained M-estimates and redescending M-estimates. *Annals of Statistics*, 29(1):252–265, 2001.

[127] J. Kiefer. The foundations of statistics - Are there any? *Synthese*, 36:161–176, 1977.

[128] J. Kiefer and J. Wolfowitz. Asymptotically minimax estimation of concave and convex distribution functions. *Zeitschrift fr Wahrscheinlichkeitstheorie und Verwandte Gebiete*, 34(1):73–85, 1976.

[129] E. King, J. D. Hart, and T. E. Wehrly. Testing the equality of two regression curves using linear smoothers. *Statistics and Probability Letters*, 12:239–247, 1990.

[130] R. Koenker and I. Mizera. The alter egos of the regularized maximum likelihood density estimators: deregularized maximum-entropy, Shannon, Renyi, Simpson, Gini, and stretched strings. In M. Hušková and M. Janžura, editors, *Prague Stochastics 2006: Proceedings of the joint session of 7th Prague Symposium on Asymptotic Statistics and 15th Prague conference on Information Theory, Statistical Decision Functions and Random Processes*. Matfyzpress, 2006.

[131] A. Kovac. *ftnonpar*. The R Project for Statistical Computing, Contributed Packages, 2007.

[132] A. Kovac and A.D.A.C. Smith. Nonparametric regression on a graph. *Journal of Computational Graphics and Statistics*, 20:432–447, 2011.

[133] N. H. Kuiper. On a metric in the space of random variables. *Statistica Neerlandica*, 16(3):231–235, 1962.

[134] K. B. Kulasekera. Comparison of regression curves using quasi-residuals. *Journal of the American Statistical Association*, 90(431):1085–1093, 1995.

[135] K. B. Kulasekera and J. Wang. Smoothing parameter selection for power optimality in testing of regression curves. *Journal of the American Statistical Association*, 92(438):500–511, 1997.

[136] D. Landers and L. Rogge. Natural choices for L1 approximations. *Journal of Approximation Theory*, 33:268–280, 1981.

[137] P. Lavergne. An equality test across nonparametric regressions. *Journal of Econometrics*, 103:307–344, 2001.

[138] P. Lévy. *Théorie de l'addition des variables aléatoires*. Gauthier-Villars, 1937.

[139] K-C. Li. Honest confidence regions for nonparametric regression. *Annals of Statistics*, 17:1001–1008, 1989.

[140] M. Li and P. Vitányi. *An Introduction to Kolmogorov Complexity and Its Applications*. Springer-Verlag, New York, 3rd edition, 2008.

[141] V Liebscher. Constrained optimization and adaptive histograms. Faculty of Mathematics, University of Greifswald, 2010.

[142] B. Lindsay and J. Liu. Model assessment tools for a model false world. *Statistical Science*, 24(3):303–318, 2009.

[143] F. Malgouyres. Mathematical analysis of a model which combines total variation and wavelet for image resoration. *Journal of Information Processes*, 2:1–10, 2002.

[144] F. Malgouyres. Minimizing the total variation under a general convex constraint for image resoration. *IEEE Transactions on Image Processing*, 11(12):1450–1456, 2002.

[145] E. Mammen and S. van de Geer. Locally adaptive regression splines. *Annals of Statistics*, 25:387–413, 1997.

[146] T. Märkeläinen, K. Schmidt, and G. P. H. Styan. On the existence and uniqueness of the maximum likelihood estimate of a vector-valued parameter in fixed sample sizes. *Annals of Statistics*, 9:758–767, 1981.

[147] R. A. Maronna. Robust M-estimators of multivariate location and scatter. *Annals of Statistics*, 4:51–67, 1976.

[148] R. A. Maronna, R. D. Martin, and V. J. Yohai. *Robust Statistics: Theory and Methods*. Wiley Series in Probability and Statistics. Wiley, 2006.

[149] J. S. Marron and M. P. Wand. Exact mean integrated squared error. *Annals of Statistics*, 20:712–736, 1992.

[150] P. Martin-Löf. The denition of random sequences. *Information and Control*, 9:602–619, 1966.

[151] P. Massart. The tight constant in the Dvoretzky-Kiefer-Wolfowitz inequality. *Annals of Probability*, 18(3):1269–1283, 1990.

[152] J. A. McCullagh, P.a nd Nelder. *Generalized Linear Models*. Chapman and Hall, 1983.

[153] M. Meise. Lokale Bandbreitenwahl in der nichtparametrischen Regression. Master's thesis, University Duisburg-Essen, 2001.

[154] M. Meise. *Residual Based Selection of Smoothing Parameters*. PhD thesis, Department of Mathematics, University Duisburg-Essen, Essen, Germany, 2004.

[155] D. Mercurio and V. Spokoiny. Statistical inference for time-inhomogeneous volatility models. *Annals of Statistics*, 32:577–602, 2004.

[156] T. Mikosch and C. Stărică. Change of structure in financial time series, long range dependence and the Garch model. Technical report, Laboratory of Actuarial Mathematics,University of Copenhagen, Denmark, 1999.

[157] T. Mikosch and C. Stărică. Stock market risk-return inference: an unconditional non-parametric approach. http://dx.doi.org/10.2139/ssrn.882820, 2003.

[158] T. Mikosch and C. Stărică. Nonstationarities in financial time series, the long-range dependence, and the IGARCH effects. *The Review of Economics and Statistics*, 86:378390, 2004.

[159] R. G. Miller. *Beyond ANOVA, Basics of Applied Statistics*. Wiley, New York, 1986.

[160] F. Mosteller and J. W. Tukey. *Data Analysis and Regression*. Wiley, Reading, Massachusetts, 1977.

[161] A. Munk and H. Dette. Nonparametric comparison of several regression functions: Exact and asymptotic theory. *Annals of Statistics*, 26(6):2339–2368, 1998.

[162] T. Murier and V. Rousson. On the randomness of the decimals of π. *Student*, 2(3):237–246, 1998.

[163] A. Nemirovski. On nonparametric estimation of smooth regression functions. *Soviet Journal of Computer and Systems Sciences*, 23(6), 1985.

[164] A. Nemirovski. *Lectures on Probability Theory and Statistics - Ecole d'Ete de Probabilites de Saint-Flour XXVIII*. Springer, 1998.

[165] N. Neumeyer and H. Dette. Nonparametric comparison of regression curves - an empirical process approach. *Annals of Statistics*, 31:880–920, 2003.

[166] J. Neyman, E. L. Scott, and C. D. Shane. On the spatial distribution of galaxies a specific model. *Astrophysical Journal*, 117:92–133, 1953.

[167] J. Neyman, E. L. Scott, and C. D. Shane. The index of clumpiness of the distribution of images of galaxies. *Astrophysical Journal Supplement*, 8:269–294, 1954.

[168] M.G. Pavlides and J.A. Wellner. Nonparametric estimation of multivariate scale mixtures of uniform densities. *Journal of Multivariate Analysis*, 107:71–89, 2012.

[169] L. I. Pettit. Bayes methods for outliers in exponential samples. *Journal of the Royal Statistical Society: Series B*, 50:371–380, 1988.

[170] J. Pinkse. A consistent nonparametric test for serial independence. *Journal of Econometrics*, 84:205–231, 1998.

[171] A. Pintore, P. Speckman, and C. C. Holmes. Spatially adaptive smoothing splines. *Biometrika*, 93(1):113–125, 2006.

[172] D. Pollard. *Convergence of Stochastic Processes*. Springer-Verlag, New York, 1984.

[173] J. Polzehl and V. Spokoiny. Adaptive weights smoothing with applications to image resoration. *Journal of the Royal Statistical Society: Series B*, 62:335–354, 2000.

[174] Yu. V Prokhorov. Convergence of random processes and limit theorems in probability theory. *Theory of Probability and Its Applications*, 1:157–214, 1956.

[175] F. Pukelsheim. Robustness of statistical gossip and the Antarctic ozone hole. *Institute of Mathematical Statistics Bulletin*, 19:540–542, 1990.

[176] J. A. Rice. *Mathematical Statistics and Data Analysis*. Wadsworth and Brookes/Cole, Belmont, California, 1988.

[177] P. J. Rousseeuw. Least median of squares regression. *Journal of the American Statistical Association*, 79:871–880, 1984.

[178] P. J. Rousseeuw. Unconventional features of positive-breakdown estimates. *Statistics and Probability Letters*, 19:417–431, 1994.

[179] P. J. Rousseeuw and C. Croux. Explicit scale estimators with high breakdown point. In Y.Dodge, editor, *L_1-Statistical Analysis and Related Methods*, pages 77–92, Amsterdam, 1992. North Holland.

[180] P. J. Rousseeuw and K. van Driessen. A fast algorithm for the minimum covariance determinant estimator. *Technometrics*, 41:212–223, 1999.

[181] Peter Rousseeuw, Christophe Croux, Valentin Todorov, Andreas Ruckstuhl, Matias Salibian-Barrera, Tobias Verbeke, Manuel Koller, and Martin Maechler. *Robustbase: Basic Robust Statistics*, 2012. R package version 0.9-4.

[182] J. P. Royston. Some techniques for assessing multivariate normality based on the Shapiro-Wilk W. *Applied Statistics*, 32:121–133, 1983.

[183] K. Rufibach and G. Walther. The block criterion for multiscale inference about a density, with applications to other multiscale problems. *Journal of Computational and Graphical Statistics*, 19(1):175190, 2010.

[184] H. Scheffé. A method for judging all contrasts in the analysis of variance. *Biometrika*, 40:87–104, 1953.

[185] H. Scheffé. *The Analysis of Variance*. Wiley, New York, 1959.

[186] F. W. Scholz. *Comparison of optimal location estimators*. PhD thesis, Department of Statistics, University of California,Berkley, 1971.

[187] V. Schultze and J. Pawlitschko. The identification of outliers in exponential samples. *Statistica Neerlandica*, 56(1):41–57, 2002.

[188] G. Schwartz. Estimating the dimension of a model. *Annals of Statistics*, 6:461–464, 1978.

[189] A. H. Seheult and J. W. Tukey. Towards robust analysis of variance. In A. K. Md. E. Saleh, editor, *Data Analysis from Statistical Foundations: A Festschrift in Honor of the 75th Birthday of D.A.S. Fraser*, pages 217–244. Nova Science, New York, 2001.

[190] A. Seregin and J.A. Wellner. Nonparametric estimation of multivariate convex-transformed densities. *Annals of Statistics*, 38:3751–3781, 2010.

[191] W. A. Stahel. Breakdown of covariance estimators. Research Report 31, Fachgruppe für Statistik, ETH, Zurich, 1981.

[192] J. L. Starck, E. J. Candès, and D. L. Donoho. Very high quality image restoration by combining wavelets and curvelets. In A. Aldroubi, A. F. Laine, and M. A. Unser, editors, *Wavelet Applications in Signal and Image Processing*, number 4478 in SPIE Conference Proceedings. SPIE, 2001.

[193] R. G. Staudte and S. J. Sheather. *Robust Estimation and Testing*. John Wiley and Sons, Hoboken, NJ, USA, 1990.

[194] J. Steinebach. On a conjecture of Révész and its analogue for renewal processes. In S. Szyszkowicz, editor, *Asymptotic Methods in Probability and Statistics: A Volume in Honour of Mikló Csörgő*, pages 311–322. Elsevier, 1998.

[195] R. Stichtenoth. *Signal and Image Denoising using Inhomogeneous Diffusion*. PhD thesis, Department of Mathematics, University of Duisburg-Essen, Essen, 2007.

[196] S. M. Stigler. Do robust estimators work with real data? (with discussion). *Annals of Statistics*, 5(6):1055–1098, 1977.

[197] T. Stoppard. *Rosencrantz and Guildenstern are Dead*. Faber and Faber, London, 1967.

[198] A. J. Stromberg. Computing the exact least median of squares estimate and stability diagnostics in multiple linear regression. *SIAM Journal of Scientific Computing*, 14(6):1289–1299, 1993.

[199] J. Strzałko, J. Grabski, A. Stefański, P. Perlikowski, and T. Kapitaniak. Dynamics of coin tossing is predictable. *Physics Reports*, 469, 2008.

[200] C Stărică. Is GARCH(1,1) as good a model as the Nobel prize accolades would imply? Technical report, Department of Mathematical statistics, Chalmers University of Technology, Gothenburg, Sweden, 2003.

[201] N. Sugiura. Further analysis of the data by Akaike's information criterion and the finite corrections. *Communications in Statistics - Theory and Methods*, 7(1):13–26, 1978.

[202] K. S. Tatsuoka and D. E. Tyler. On the uniqueness of s-functionals and m-functionals under non-elliptic distributions. *Annals of Statistics*, 28(4):1219–1243, 2000.

[203] G. Tennenbaum. *Introduction to analytic and probabilistic number theory*, volume 46 of *Cambridge Studies in Advanced Mathematics*. Cambridge University Press, 1995.

[204] W. Terbeck. Interaktionen im Mehrfaktoren-Fall. Unpublished manuscript, Department of Mathematics, University of Essen, 1996.

[205] W. Terbeck. *Interaktionen in der Zwei-Faktoren-Varianzanalyse*. PhD thesis, University of Essen, Germany, 1996.

[206] W. Terbeck and P. L. Davies. Interactions and outliers in the two-way analysis of variance. *Annals of Statistics*, 26:1279–1305, 1998.

[207] D. Thurm. Sparsity, l_1, l_2 and the analysis of variance. Diplomarbeit, Fakultät Mathematik, Universität Duisburg-Essen, Essen, Germany, 2009.

[208] S. Tu and E. Fischbach. A study on the randomness of the digits of π. *International Journal of Modern Physics C*, 16(2):281–294, 2005.

[209] J. W. Tukey. The problem of multiple comparisons. Unpublished manuscript, 1952.

[210] J. W. Tukey. *Exploratory Data Analysis*. Addison-Wesley, Massachusetts, 1977.

[211] J. W. Tukey. Issues relevant to an honest account of data-based inference, partially in the light of Laurie Davies's paper. Princeton University, Princeton, 1993.

[212] D. E. Tyler. A distribution-free M -estimator of multivariate scatter. *Annals of Statistics*, 15:234–251, 1987.

[213] E. Uhrmann-Klingen. Minimal Fisher information distributions with compact-supports. *Sankhya Series A*, 57:360–374, 1995.

[214] V. N. Vapnik. *Statistical Learning Theory*. Wiley, New York, 1998.

[215] R. von Mises. Sur les fonctions statistiques. In *Conférence de la Réunion Internationale des Mathématiques*, pages 1–8, Paris, 1939. Gauthier-Villars.

[216] R. von Mises. On the asymptotic distribution of differentiable statistical functions. *Annals of Mathematical Statistics*, 18:309–348, 1947.

[217] E-J. Wagenmakers. A practical solution to the pervasive problem of p values. *Psychonomic Bulletin and Review*, 14(5):779–804, 2007.

[218] B. L. Welch. The significance of the difference between two means when the population variances are unequal. *Biometrika*, 29:350–362, 1938.

[219] D. Williams. *Weighing the Odds: A Course in Probability and Statistics*. Cambridge University Press, Cambridge, 2001.

[220] G. Winkler, O. Wittich, V. Liebscher, and A. Kempe. Don't shed tears over breaks. *Jahresbericht der DMV*, 107(2):57–87, 2005.

[221] L. Wittgenstein. *Tractatus logico-philosophicus*. Surhkamp, 9th edition, 1973.

[222] C. F. J. Wu and M. S. Hamada. *Experiments, Planning, Analysis and Optimization*. Wiley Series in Probability and Mathematical Statistics. Wiley, second edition, 2009.

[223] E. Zoldin. *Nonparametric modelling of interest rates*. PhD thesis, Department of Mathematics, University of Duisburg-Essen, Essen, 2007.

Index